Peer review in an Era of Evaluation

Eva Forsberg • Lars Geschwind
Sara Levander • Wieland Wermke
Editors

Peer review in an Era of Evaluation

Understanding the Practice of Gatekeeping in Academia

palgrave
macmillan

Editors
Eva Forsberg
Department of Education
Uppsala University
Uppsala, Sweden

Sara Levander
Department of Education
Uppsala University
Uppsala, Sweden

Lars Geschwind
Department Learning in Engineering
Sciences
KTH Royal Institute of Technology
Stockholm, Sweden

Wieland Wermke
Department of Special Education
Stockholm University
Stockholm, Sweden

ISBN 978-3-030-75262-0   ISBN 978-3-030-75263-7   (eBook)
https://doi.org/10.1007/978-3-030-75263-7

This Palgrave Macmillan imprint is published by the registered company Springer Nature Switzerland AG.
The registered company address is: Gewerbestrasse 11, 6330 Cham, Switzerland

We dedicate this book to our beloved colleague and friend, professor Rita Foss Lindblad, who was involved in the project but passed away in 2018.

Contents

Part I   Peer Review: Introduction and Its Context   1

1   **Peer Review in Academia**   3
*Eva Forsberg, Lars Geschwind, Sara Levander, and
Wieland Wermke*

2   **Hierarchies and Universal Inclusion in Scientific
Communities**   37
Rudolf Stichweh

3   **"'Disciplining' Educational Research in the Twentieth
Century"**   53
Raf Vanderstraeten

**Part II   Peer Review and the Higher Education Evaluation
Machinery**   77

4   **Gatekeepers on Campus: Peer Review in Quality
Assurance of Higher Education Institutions**   79
Don F. Westerheijden

5   The Many Faces of Peer Review   107
     Hanne Foss Hansen

6   Your Brother's Gatekeeper: How Effects of Evaluation
     Machineries in Research Are Sometimes Enhanced   127
     Peter Dahler-Larsen

7   Peer Review in Public Administration: The Case of the
     Swedish Higher Education Authority   147
     Agnes Ers and Kristina Tegler Jerselius

8   Performance-Based Evaluation Metrics: Influence
     at the Macro, Meso, and Micro Level   173
     Gustaf Nelhans

9   Peer Advocacy: Expressions of Loyalty in Peer Review   203
     Lars Geschwind and Kristina Edström

10   Is Peer Review Fit for Purpose?   223
     Malcolm Tight

Part III   Specificities of Different Peer-Review Practices   243

11   Peer Review in Academic Promotion of Excellent
     Teachers   245
     Eva Forsberg, Sara Levander, and Maja Elmgren

12   Peers in Systematic Review: Gate Keeping
     Understandings of Research in the Field   275
     Tine S. Prøitz

13   The Decision-Making Constraints and Processes
     of Grant Peer Review, and Their Effects on the Review
     Outcome   297
     Liv Langfeldt

14 Typecasting in the Recruitment of Full Professors   327
*Sara Levander, Eva Forsberg, Sverker Lindblad, and Gustaf
J. Bjurhammer*

**15 Assessing Academic Careers: The Peer Review of
Professorial Candidates**   347
Björn Hammarfelt

**16 Bureaucratic, Professional and Managerial Power in
University Tenure Track Recruitment**   371
*Tea Vellamo, Jonna Kosonen, Taru Siekkinen,
and Elias Pekkola*

**Correction to: "Disciplining" Educational Research in the
Twentieth Century**   C1
Raf Vanderstraeten

Notes on Contributors

Gustaf J. Bjurhammer is a PhD student in Education at Uppsala University. He is a member of the research group Studies in Educational Policy and Educational Philosophy (STEP). In his research he addresses the issue of knowledge production and science policy within the field of Educational Sciences.

Peter Dahler-Larsen is a professor at the Department of Political Science, University of Copenhagen, where he is the leader of CREME, Center for Research on Evaluation, Measurement and Effects. He is author of *The Evaluation Society* (2013) and *Quality: From Plato to Performance* (Palgrave 2019).

Kristina Edström is Associate Professor of Engineering Education Development at KTH Royal Institute of Technology, and Editor-in-Chief of the *European Journal of Engineering Education*. She is active in educational development at KTH, in Sweden and internationally. Her research takes a critical perspective on the why, what and how of engineering education development.

Maja Elmgren is Associate Professor of Physical Chemistry and an academic developer at Uppsala University. Her research interests are leadership in higher education, educational proficiency in academia, and physics and chemistry education research. She is the director of the

Council for Educational Development at the Faculty of Science and Technology, Uppsala University.

Agnes Ers holds a PhD in Ethnology from Stockholm University. She has a long experience of working in the area of higher education and research, as well as in the area of cultural policy, in public governmental administration in Sweden. She is currently a strategist at the Director General's Office at the Swedish Higher Education Authority.

Eva Forsberg is Professor of Education, Uppsala University, and the scientific leader of STEP. Her research focuses on the interface between education governance, practice and research. Evaluation policy and assessment cultures have been the object of several of her research projects and publications. Lately, her interest concerns higher education, academic work and faculty evaluation specifically.

Lars Geschwind is Professor of Engineering Education Policy and Management and a coordinator of the research group HEOS (Higher Education Organization Studies) at KTH Royal Institute of Technology, Stockholm. His main research interests are higher education policy, institutional governance, academic leadership and academic work.

Björn Hammarfelt is an associate professor at the Swedish School of Library and Information Science (SSLIS), University of Borås, Sweden. His research is situated at the intersection of information studies and sociology of science, with a focus on the organization, communication and evaluation of research.

Hanne Foss Hansen is Professor of Public Administration and Organization at the Department of Political Science, University of Copenhagen. Her main interests are public organization and management, evaluation, evidence-based policy and practice as well as higher education and research policy. She is co-editor of *Reforms, Organizational Change and Performance in Higher Education* (Palgrave 2019).

Kristina Tegler Jerselius holds a PhD in History from Uppsala University. She has extensive experience in working with quality in higher education and research, and is interested in governance and evaluative practices and autonomy in public administration. She works as a senior analyst at the Swedish Higher Education Authority.

Jonna Kosonen is a lecturer at the Law School, University of Eastern Finland (UEF). In addition, she works as a legal expert at UEF's Office of Legal Affairs. She is writing her doctoral thesis on university autonomy. Her latest publication discusses equality in the student selection process from a legal point of view.

Liv Langfeldt is a research professor at the Nordic Institute for Studies in Innovation, Research and Education (NIFU) in Oslo, and director of the Centre for Research Quality and Policy Impact Studies (R-QUEST). Her main research interests include research policy, peer review processes, and research quality and evaluation.

Sara Levander is Senior Lecturer and researcher in Education, Uppsala University. Her research interests are higher education, academic work and faculty evaluation in academic recruitment and promotion. A focal point for her research is the assessment of academic competences in academic recruitment.

Sverker Lindblad is Professor Emeritus of Education, University of Gothenburg, and president of the Swedish Educational Research Association. His ongoing research is in the fields of comparative education and politics of knowledge. A recent book is *Education by the Numbers and the Making of Society: The Expertise of International Assessments* (Lindblad, Pettersson, & Popkewitz, 2018).

Gustaf Nelhans is a theorist of science and senior lecturer at the Swedish School of Library and Information Science (SSLIS), University of Borås, Sweden. His research focuses on the performativity of scientometric indicators as well as on the theory, methodology and research policy aspects of scholarly publications.

Elias Pekkola is a lecturer and Head of Unit at the Faculty of Management and Business at Tampere University (TAU). His publications include several articles and books on academic work, academic profession, careers and HR policy. He has acted in various expert roles in

Finland and in international projects on academic work, careers and higher education policy.

Tine S. Prøitz is a professor at the Department of Business, History and Social Science, University of South-Eastern Norway. Her research interests lie within the fields of education policy and education reform, quality work in higher education, and the practices and procedures of research synthesis. She has recently co-edited the book *Quality Work in Higher Education* (Elken, M., Maassen, P., Nerland, M., Prøitz, T.S., Stensaker, B., Vabø, A. [Eds.]).

Taru Siekkinen is a postdoctoral researcher at the Finnish Institute for Educational Research, University of Jyväskylä (JYU), Finland. Her research interests are academic profession, work and careers, universities and their management. Her latest publication "Change and Continuity in the Academic Profession—Finnish Universities as Living Labs" (2019) was published in *Higher Education*. She is chair of the Consortium of Higher Education Researchers in Finland (CHERIF).

Rudolf Stichweh is Senior Professor of Sociology at the "Forum Internationale Wissenschaft" and the Bonn Center for Dependency and Slavery Studies, University of Bonn. His research fields include comparative study of democratic and authoritarian political regimes; theory of world society and functional differentiation; history and sociology of science; sociology of universities; inequality and asymmetrical dependencies in premodern and modern societies; sociocultural evolution and theory of social systems. His last book publication was *Democratic and Authoritarian Political Systems in Twenty-First-Century World Society. Vol. 1, Differentiation, Inclusion, Responsiveness*. Transcript: Bielefeld 2020.

Malcolm Tight is Professor of Higher Education at Lancaster University (UK). His research interests cover higher education research and social research methodology. His most recent books include *Documentary Research in the Social Sciences* (2019), *Higher Education Research: The Developing Field* (2019), *Syntheses of Higher Education Research: What We Know* (2020) and *Understanding Case Study Research: Small-Scale Research with Meaning* (2017).

Raf Vanderstraeten is director of the Center for Social Theory and a member of the Sarton Center for History of Science, both at Ghent University (Belgium). He is also a visiting professor at the London School of Economics and Political Science (England). His research interests include the sociology of science and the sociology of education.

Tea Vellamo is a PhD student at the Faculty of Management and Business and Chief Specialist at the Faculty of Natural Sciences at Tampere University (TAU). She is writing her doctoral dissertation on technical identity. Recent publications include the chapter "Technical Identity in a Merger Process—Between a Rock and a Hard Place."

Wieland Wermke is Associate Professor in Special Education, Stockholm University. His research interest is in comparative education methodology and teacher/special educator practice at different levels of education. His most recent book is Wermke, W. & Salokangas, M. "Autonomy Paradox. Teachers' Perception of Self Governance Across Europe." Springer 2021.

Don F. Westerheijden is a senior research associate at the Center for Higher Education Policy Studies (CHEPS), University of Twente, the Netherlands. He mostly studies quality assurance in higher education, and its impacts. One of his publications include Jongbloed, B.W.A., Kaiser, F., & Westerheijden, D. (2020). "Improving Study Success and Diversity in Dutch Higher Education Using Performance Agreements," *Tertiary Education and Management, 26,* pp. 329–343.

List of Figures

Fig. 3.1   Average number of authors per article for AERJ and RER   63
Fig. 3.2   Average number of references per article for AERJ and RER   69
Fig. 8.1   Yearly amount in NOK for each Norwegian point   187
Fig. 8.2   Cumulative renegotiated funding by HEI type according to
            the state budget for the years 2010–2016   189
Fig. 11.1   The basic career structure in the Swedish higher education
            system   250
Fig. 11.2   A fourth career track in the case university   251
Fig. 11.3   The process of admitting excellent teachers at the university   252
Fig. 11.4   Requested information to be included in the application
            dossier   254
Fig. 14.1   The process of recruiting full professors in Sweden   334
Fig. 15.1   Schematic overview of temporal focus when evaluating
            research quality   362
Fig. 16.1   Different powers and persons utilising power in university
            recruitment   382

List of Tables

Table 5.1	Classical peer review	110
Table 5.2	Informed and standards-based peer review	112
Table 5.3	Modified peer review	114
Table 5.4	Extended peer review	117
Table 8.1	Three "regimes" of research policy after WWII	181
Table 8.2	Comparison between the Swedish and the Norwegian PRFS	185
Table 8.3	Combined "new" funding and results of renegotiation for the University of Kristianstad in the budget for 2012	189
Table 12.1	The status function of peers in academic publishing and in systematic review	289
Table 13.1	Overview of Studied Proposals and Grants (RCN 1997/98)	304
Table 13.2	Review criteria	307
Table 13.3	The results of different methods for the ranking of applications a-f, of which there is room for three within the budget	317
Table 13.4	Combination of elements of methods 2 and 3, and other methods, in one of the Culture and Society Panels	317
Table 15.1	Characterisation of research fields using Whitley's typology	354
Table 15.2	Judgement devices used to assess the recognition of publications in the discipline (*type of device according to Karpiks typology*)	364

Part I

Peer Review: Introduction and Its Context

1

Peer Review in Academia

Eva Forsberg, Lars Geschwind, Sara Levander,
and Wieland Wermke

Introduction

Over the past few decades, peer review has become an object of great professional and managerial interest (Oancea, 2019) and, increasingly, academic scrutiny (Bornmann, 2011; Grimaldo et al., 2018). Nevertheless, calls for further research are numerous (Tennant & Ross-Hellauer, 2020). This volume is in answer to such interest and appeals. We aim to present a variety of peer-review practices in contemporary academic life as well as

E. Forsberg (✉) • S. Levander
Department of Education, Uppsala University, Uppsala, Sweden

L. Geschwind
Department of Learning in Engineering Sciences, KTH Royal Institute of Technology, Stockholm, Sweden

W. Wermke
Department of Special Education, Stockholm University, Stockholm, Sweden

© The Author(s) 2022
E. Forsberg et al. (eds.), *Peer review in an Era of Evaluation*,
https://doi.org/10.1007/978-3-030-75263-7_1

3

the principled foundation of peer review in scientific communication and authorship. This volume is unique in that it covers many different practices of peer review and their theoretical foundations, providing both an introduction into the very complex field and new empirical and conceptual accounts of peer review for the interested reader. The contributions are produced by internationally recognized scholars, almost all of whom participated in the conference 'Scientific Communication and Gatekeeping in Academia in the 21st Century', held in 2018 at Uppsala University, Sweden.[1] The overall objective of this volume is explorative; framings relevant to the specific contexts, practices and discourses examined are set by the authors of each chapter. However, some common conceptual points of departure may be laid down at the outset.

Peer review is a context-dependent, relational concept that is increasingly used to denote a vast number of evaluative activities engaged in by a wide variety of actors both inside and outside of academia. By peer review, we refer to peers' assessments and valuations of the merits and performances of academics, higher education institutions, research organizations and higher education systems. Mostly, these activities are part of more encompassing social evaluation practices, such as reviews of manuscripts, grant proposals, tenure and promotion and quality evaluations of institutions and their research and educational programmes. Thus, scholarly peer review comprises evaluation practices within both the wider international scientific community and higher education systems. Depending on differences related to scientific communities and national cultures, these evaluations may include additional gatekeepers, internal as well as external to academia, and thus the role of the peer may vary.

The roots of peer review can be found in the assessment practices of reviewers and editors of scholarly journals in deciding on the acceptance of papers submitted for publishing. Traditionally, only peers (also known as referees) with recognized scholarly standing in a relevant field of research were acknowledged as experts (Merton, 1942/1973). Due to the differentiation and increased use of peer review, the notion of a peer

---

[1] Funded by Riksbankens Jubileumsfond (F17-1350:1). The keynotes of the conference are accessible on video at https://media.medfarm.uu.se/play/kanal/417. For more information on the conference, see www.konferens.edu.uu.se/scga2018-en.

employed in various evaluation practices may be extended. Who qualifies as an expert in different peer-review practices and with what implications are empirical issues.

Even though peer review is a familiar phenomenon in most scholarly evaluations, there is a paucity of studies on peer review within the research field of evaluation. Peer review has, however, been described as the most familiar collegial evaluation model, with academic research and higher education as its paradigm area of application and with an ability to capture and judge qualities as its main advantage (Vedung, 2002). Following Scriven (2003), we define evaluation as a practice 'determining the merit, worth or significance of things' (p. 15). Scriven (1980) identifies four steps involved in evaluation practices, which are also frequently used in peer review, either implicitly enacted and negotiated or explicitly stated (Ozeki, 2016). These steps concern (1) the criteria of merit, that is, the dimensions of an object being evaluated; (2) the standards of merit, that is, the level of performance in a given dimension; (3) the measuring of performance relative to standards; and (4) a value judgement of the overall worth.

Consequently, the notion of peer review refers to evaluative activities in academia conducted by equals that distribute merit, value and worth. In these processes of selection and legitimation, issues referring to criteria, standards, rating and ranking are significant. Often, peer reviews are embedded in wider evaluation practices of research, education and public outreach. To capture contemporary evaluations of academic work, we will include a number of different review practices, including some in which the term *peer* is employed in a more extended sense.

The Many Face(t)s of Peer-Review Practices

Depending on the site in which peer review is used, the actors involved differ, as do their roles. The same applies to potential guidelines, purposes, discourses, use of professional judgement and metrics, processes and outcome of the specific peer-review practice. These are all relative to the site in which the review is used and will briefly be commented upon below.

The Interplay of Primary and Secondary Peer Review

It is possible to make a distinction between primary and secondary peer reviews (British Academy, 2007). As stated, the primary role of peer review is to assess manuscripts for publishing, followed by the examination and judgement of grant applications. Typically, many other peer-review practices, so-called secondary peer review, involve summaries of outcomes of primary reviews. Thus, we might view primary and secondary reviews as folded into each other, where, for example, reviews of journal articles are prerequisite to later evaluation of the research quality of an institution, in recruitment and promotion, and so forth (Helgesson, 2016). Hence, the consequences of primary reviews can hardly be overstated.

Traditionally, both forms of primary peer review (assessment of manuscripts and grant applications) are ex ante evaluations; that is, they are conducted prior to the activity (e.g. publishing and research). With open science, open access journals and changes in the transparency of peer review, open and public peer reviews have partly opened the black box of reviews and the secrecy of the process and its actors (Sabaj Meruane et al., 2016). Accordingly, publishing may include both ex ante and ex post evaluations. These forms of evaluation can also be found among secondary reviews, with degree-awarding accreditation an example of the former and reviews of disciplines an example of the latter.

Sites and Reviewer Tasks and Roles

Without being exhaustive, we can list a number of sites where peer review is conducted as part of more comprehensive evaluations: *international, regional and national higher education agencies* conduct accreditation, quality audits and evaluations of higher education institutions; *funding agencies* distribute grants for projects and fellowships; *higher education institutions* evaluate their research, education and public outreach at different levels and assess applications for recruitment, tenure and promotion; *the scientific community* assesses manuscripts for publication, evaluates doctoral theses and conference papers and allocates awards. The evaluation roles are concerned with the provision of human and financial resources, the evaluation of research products and the assessment of future strategies as a basis for policy and priorities. All of these activities are regularly performed by researchers and interlinked in an evaluation spiral in which the same research may be reviewed more than once (Langfeldt & Kyvik, 2015). If we consider valuation and assessment more generally, the list can be extended almost infinitely, with supervision and seminar discussions being typical activities in which valuation plays a central part. Hence, scholars are accustomed to being assessed and to evaluating others.

The role and the task of the reviewer differ also in relation to whether the act of reviewing is performed individually, in teams or in a blending of the two forms. In the evaluation of research grants, the latter is often the case, with reviewers first individually rating or ranking the applications, followed by panel discussions and joint rankings as bases for the final decision made by a committee. In peer review for publishing, there might be a desk rejection by the editor, but if not, two or more external reviewers assess a manuscript and recommend that the editor accept, revise or reject it. It is then up to the editor to decide what to do next and to make the final decision. The process and the expected roles of the involved editor, reviewer and authors may vary depending on whether it is a private publisher or a journal linked to a scientific association, for example. Whether the reviewer should be considered an advisor, an independent assessor, a juror or a judge depends on the context and the task set for the reviewer within the specific site and its policies and practices as well as on various praxes developed over time (Tennant & Ross-Hellauer, 2020).

Power-making in the Selection of Expertise

The selection process is at the heart of peer review. Through valuations and judgements, peers are participants in decisions on inclusion and exclusion: What project has the right qualities to be allocated funding? Which paper is good enough to be published? And who has the right track record to be promoted or offered a fellowship? When higher education institutions and scholars increasingly depend on external funding, peer review becomes key in who gets an opportunity to conduct research and enter or continue a career trajectory as a researcher and, in many systems, a higher education teacher. In other words, peer review is a cornerstone of the academic career system (Merton, 1968; Boyer, 1990) and heavily influences what kinds of scientific knowledge will be furthered (Lamont, 2009; Aagaard et al., 2015).

The interaction involved in peer review may be remote, online or local, including face-to-face collaboration, and it may involve actors with different interests. Moreover, interaction may be extended to the whole evaluation enterprise. For example, evaluations of higher education institutions and their research and education often include members of national agencies, scholarly experts and external stakeholders. Scholarly experts may be internal or external to the higher education institutions and of lower, comparable or higher rank than the subjects of evaluation, and reviewers may be blind or known to those being evaluated and vice versa. Scholarly expertise may also refer to a variety of specialists, for example, to scholars with expertise in a specific research topic, in evaluation technology, in pedagogy or public outreach. A more elaborated list of features to be considered in the allocation of experts to various review practices can be found in a peer-review guide by the European Science Foundation (2011). At times the notion of *peer* is extended beyond the classical idea to one with demonstrated competence to make judgements within a particular research field. Who qualifies as a reviewer is contingent on who has the authority to regulate the activity in which the evaluation takes place and who is in the position to suggest and, not least, to select reviewers. This is a delicate issue, imbued with power, and one that we need to further explore, preferably through comparative studies involving different peer-review practices in varying contexts.

Acting as a peer reviewer has become a valuable asset in the scholarly track record. This makes participating as a reviewer important for junior researchers. Therefore, such participation not only is a question of being selected but also increasingly involves self-election. More opportunities are provided by ever more review activities and the prevalence of evaluation fatigue among senior researchers. The limited credit, recognition and rewards for reviewers may also contribute to limited enthusiasm amongst seniors (Research Information Network CIC, 2015). Moreover, several tensions embedded in review practices can add to the complexity of the process and influence the readiness to review. The tensions involve potential conflicts between the role of the reviewer or evaluator and the researcher's role: time conflict (research or evaluate), peer expertise versus impartiality (especially qualified colleagues are often excluded under conflict-of-interest rules), neutral judge versus promoter of research interests (double expectation, deviant assessments versus unanimous conclusions, peer review versus quantitative indicators, and scientific autonomy versus social responsibility) (Langfeldt & Kyvik, 2015). Despite noted challenges, classical peer review is still the key mechanism by which professional autonomy and the guarding of research quality are achieved. Thus, it is argued that it is an academic duty and an obligation, in particular for senior scholars, to accept tasks as reviewers (Caputo, 2019). Nevertheless, the scholarly exchange value should be addressed in future discussions of gatekeeping in academia.

The Academic Genres of Peer Review

Peer reviews are rooted in more encompassing discourses, such as those concerning norms of science, involving notions of quality and excellence founded in different sites endogenous or exogenous to science. Texts subject to or employed or produced in peer-review practices represent a variety of academic genres, including review reports, editors' letters, applicants' proposals, submitted manuscripts, guidelines, applicant dossiers and curriculum vitae (CVs), testimonials, portfolios and so on. Different genres are interlinked in chains, creating systems of genres. A significant aspect of systems is intertextuality, or the fact that texts within a specific system refer to, anticipate and shape each other. The interdependence of texts is about how they relate to situational and formal expectations—in this case, of the specific peer-review practice. It is also about how one text makes references to another text; for example, review reports often refer to guidelines, calls, announcements or texts in application dossiers. The interdependence can also be seen in how the texts interact in academic communities (Chen & Hyon, 2005): who the intended readers of a given text are, what the purpose of the text is, how the text is used in the review and decision process, and so on. Conclusively, the genre systems of peer review vary depending on epistemic traditions, national culture and regulations of higher education systems and institutions.

Given this diversity, we are dealing with a great number of genre systems involving different kinds of texts and interrelations embedded in power and hierarchies. A significant feature of peer-review texts as a category is the occluded genres, that is, genres that are more or less closed to the public (Swales, 1996). Depending on the context, the list of occluded genres varies. For example, the submission letters, submitted manuscripts, review reports and editor–author correspondence involved in the eventual publication of articles in academic journals are not made publicly available, while in the context of recruitment and promotion, occluded genres include application letters, testimonials and evaluation letters to committees. And for research grants, the research proposals, individual review reports and panel reports tend to remain entirely internal to the grant-making process. However, in some countries (e.g. in Sweden, due to the principle of openness, or *offentlighetsprincipen*), several of these types of texts may be publicly available.

The request for open science has also initiated changes to the occluded genres of peer review. After a systematic examination, Ross-Hellauer (2017) proposed 'open peer review' as an umbrella term for a variety of review models in line with open science, 'including making reviewer and author identities open, publishing review reports and enabling greater participation in the peer review process' (p. 1). From 2005 onwards, there has been a big upswing of these definitions. This correlates with the rise of the openness agenda, most visible in the review of journal articles and within STEM and interdisciplinary research.

Time and space are central categories in most peer-review genres and the systems to which they belong. While review practices often look to the past, imagined futures also form the background for valuation. The future orientation is definitely present in audits, in assessments of grant proposals and in reviews of candidates' track records. The CV, a key text in many review practices, may be interpreted in terms of an applicant's career trajectory, thus emphasizing how temporality and spatiality interact within a narrative infrastructure, for example how scholars move between different academic institutions over time (Hammarfelt et al., 2020). Texts may also feed both backwards and forwards in the peer-review process. For example, guidelines and policy on grant evaluations and distribution may be negotiated and acted upon by both applicants and reviewers. Candidates may also address reviewers as significant others in anticipating the forthcoming reviewer report (Serrano Velarde, 2018). These expectations on the part of the applicant can include prior experiences and perceptions of specific review practices, processes and outcomes in specific circumstances.

Turning to the reviewer report, it is worth noting that they are often written in English, especially ones assessing manuscripts and frequently those on research proposals and recruitment applications as well. Commonly seen within the academic genre of peer review is the use of indirect speech, which can be linked to the review report's significance as related to the identity of the person being evaluated (Paltridge, 2017). Two key notions, politeness and face, have been used to describe the evaluative language of review reports and how reviewers interact with evaluees. There are differences related to content and to whether a report is positive or negative overall in its evaluation. For example, reviewers of manuscripts invoke certain structures of knowledge, using different structures when suggesting to reject, revise or accept and when asking for changes. To maintain social relationships, reviewers draw on different politeness strategies to save an author's face. Strategies employed may include 'apologizing ('I am sorry to have to') and impersonalizing an issue ('It is generally not acceptable to')' (Paltridge, 2017, p. 91). Largely, requests for changes are made as directions, suggestions, clarifications and recommendations. Thus, for both evaluees and researchers of peer reviews, particular genre competences are required to decode and act upon the reports. For beginning scholars unfamiliar with the world of peer review or for scholars from a different language or cultural background than the reviewer, it might be challenging to interpret, negotiate and act upon reviewer reports.

Criteria and the Professional Judgement of Quality

According to the classical idea of peer review, only a peer can properly recognize quality within a given field. Although, in both research and scholarly debate, shortcomings have been emphasized regarding the trustworthiness, efficacy, expense, burden and delay of peer review (Bornmann, 2013; Research Information Network CIC, 2015), many critics still find peer review as the least-worst system, in the absence of viable alternatives. Overall, scholars stand behind the idea of peer review even though they often have concerns regarding the different practices of peer review (Publons, 2018).

Calls for accountability and social relevance have been made, and there have been requests for formalization, standardization, transparency and openness (Tennant & Ross-Hellauer, 2020). While the idea of formalization of peer review refers to rules, including the development of policy and guidelines for different forms of peer review, standardization rather emphasizes the setting of standards through the employment of specific tools for evaluation (i.e. criteria and indicators used for assessment, rating or ranking and decision-making). An interesting question is whether standardization will impact the extent and the way peers are used in different sites of evaluation (Westerheijden et al., 2007). We may add, who will be considered a peer and what will the matching between the evaluator and the evaluation object or evaluee look like?

It is widely acknowledged that criteria is an essential element of any procedure for judging merit (Scriven, 1980; Hug & Aeschbach, 2020). This is the case regardless of whether criteria are determined in advance or if they are explicitly expressed or implicitly manifested in the process of assessment. The notion of peer review has been supplemented in various ways, implicating changes to the practice and records of peer review. Increasingly, review reports combine classical peer review with metrics of different kinds. Accordingly, quantitative measures, taken as proxies for quality, have entered practices of peer review. Today, blended forms are rather common, especially in evaluations of higher education institutions, where narrative and metric summaries often supplement each other and inform a judgement.

In general, quantitative indicators (e.g. number of publications, journal impact factors, citations) are increasingly applied, even though their capacity to capture quality is questioned, especially within the social sciences, humanities and the arts. Among the main reasons given for the rapid growth of demands for metrics, one of the arguments we find is that classic peer review alone cannot meet the quest for accountability and transparency, and bibliometric evaluations may appear cheaper, more objective and legitimated. Moreover, metrics may give an impression of accessibility for policy and management (Gläser & Laudel, 2007; Söderlind & Geschwind, 2019). However, tensions between classical peer review and quantitative indicators have been identified and are hotly debated (Langfeldt & Kyvik, 2011). The dramatic expansion of the use of metrics has brought with it gaming and manipulation practices to enhance reputation and status, 'including coercive citation, forced joint authorship, ghostwriting, h-index manipulation, and many others' (Oravec, 2019, p. 859). Warnings are also issued against the use of bibliometric indicators at the individual level. A combination of peer narratives and metrics is, however, considered a possibility to improve an overall evaluation, given due awareness of the limitations of quantitative data as proxies for quality.

The literature on peer review has focused more on the weighting of criteria than on the meaning referees assign to the criteria they use (Lamont, 2009). Even though some criteria, such as originality, trustworthiness and relevance, are frequently used in the assessment of academic work and proposals, our knowledge of how reviewers ascribe value to, assess and negotiate them remains limited (Hug & Aeschbach, 2020). However, Joshua Guetzkow, Michèle Lamont and Grégoire Mallard (2004) show that panellists in the humanities, history and the social sciences define originality much more broadly than what is usually the case in the natural sciences.

Criteria, indicators and comparisons are unstable: they are situational and dependent on context and a referee's personal experience of scientific work (Kaltenbrunner & de Rijcke, 2020). We are dealing here with assessments in situations of uncertainty and of entities not easily judged or compared. The concept of judgement devices has been used to capture how reviewers delegate the judgement of quality to proxies, reducing the complexity of comparison. For example, the employment of central categories in a CV, which references both temporal and spatial aspects of scholars' trajectories, makes comparison possible (Hammarfelt, 2017). In a similar way, the theory of anchoring effects has been used to explore reviewers' abilities to discern, assess, compare and communicate what scientific quality is or may be (Roumbanis, 2017). Anchoring effects have their roots in heuristic principles used as shortcuts in everyday problem solving, especially when a judgement involves intuition. Reduction of complexity is visible also in how reviewers first collect criteria that consist of information that has an eliminatory function. Next, they search for positive signs of evidence in order to make a final judgement (Musselin, 2002). Dependent on context and situations, reviewers tend to select different criteria from a repertoire of criteria (Hug & Aeschbach, 2020).

On the one hand, the complexity of academic evaluations requires professional judgement: scholars sufficiently grounded in a field of research and higher education are entrusted with interpreting and negotiating criteria, indicators and merits. Still, the practice of peer review has to be safeguarded against the risk of conservatism as well as epistemic and social biases (Kaltenbrunner & de Rijcke, 2020). On the other hand, changes in the governance of higher education institutions and research, as well as marketization, managerialism, digitalization and calls for accountability, have increased the diversity of peer review and introduced new ways to capture and employ criteria and indicators. The long-term consequences of these changes need to be monitored, not least because of how they challenge the self-regulation and autonomy of the academic profession (Oancea, 2019).

How to understand, assess, measure and value quality in research, the career of a scholar or the performances of a higher education institution are complex issues. Turning to the notion of quality in a general sense will not solve the problem, since it has so many facets and has been perceived in so many different ways, including as fitness for purpose, as eligible, as excellent and as value for money (Westerheijden et al., 2007), all notions in need of contextualization and further elaboration to achieve some sense (see also Elken & Wollscheid, 2016).

When presenting a framework to study research quality, Langfeldt et al. (2020) identify three key dimensions: (1) quality notions originating in research fields and in research policy spaces; (2) three attributes important for good research and drawn on existing studies, namely, originality/novelty, plausibility/reliability and value or usefulness; and (3) five sites where notions of research quality emerge, are contested and are institutionalized, comprising researchers, knowledge communities, research organizations, funding agencies and national policy arenas. This multidimensional framework and its components highlight issues that are especially relevant to studies of peer review. The sites identify arenas where peer review functions as a mechanism through which notions of research quality are negotiated and established. The consideration of notions of quality endogenous and exogenous to scientific communities and the various attributes of good research can also be directly linked to referees' distribution of merit, value and worth in peer-review practices under changing circumstances.

The Autonomy of a Profession and a Challenged Contract

Historical analyses link peer review to the distribution of authority and the negotiations and reformulations of the public status of science (Csiszar, 2016). At stake in renegotiations of the contract between science and society are the professional autonomy of scholars and their work. Peer review is contingent on the prevailing contract and is critical in maintaining the credibility and legitimacy of research and higher education (Bornmann, 2011). The professional autonomy of scholars raises the issue of self-regulation. Its legitimacy ultimately comes down to who decides what, particularly concerning issues of research quality and scientific communication (Clark, 1989).

Over the past 40 years, major changes have taken place in many OECD (Organisation for Economic Co-operation and Development) countries in the governance of public science and higher education, changes which have altered the relative authority of different groups and organizations (Whitley, 2011). The former ability of scientific elites to exercise endogenous control over science has, particularly since the 1960s, become more contested and subject to public policy priorities. A more heterogeneous and complex higher education system has been followed by the exogeneity of governance mechanisms, formal rules and procedures, and the institutionalization of quality assurance procedures and performance monitoring. Expectations of excellence, competition for resources and reputation, and the coordination of research priorities and intellectual judgement have changed across disciplinary and national boundaries to varying degrees (Whitley, 2011). These developments can be seen as expressions of the evaluative state (Neave, 1998), the audit society (Power, 1997) and as part of an institutionalized evaluation machinery (Dahler Larsen, 2012).

Changes in the principles of governance are underpinned by persistent tensions around accountability, evaluation, measurement, demarcation, legitimation, agency and identity in research (Oancea, 2019). Besides the primary form of recognition through peer review, the weakened autonomy of academic fields has added new evaluative procedures and institutions. Academic evaluations, such as accreditations, audits and quality assurances, and evaluations of research performance and social impact now exist alongside more traditional forms (Hansen et al., 2019).

Higher education institutions worldwide have experienced the emergence and manifestations of the quality movement, which is part of interrelated processes such as massification, marketization and managerialism. Through organizations at international, national and institutional levels, a variety of technologies have been introduced to identify, measure and compare the performance of higher education institutions (Westerheijden et al., 2007). These developments have emphasized external standards and the use of bibliometrics and citation indexes, which have been criticized for rendering the evaluations more mechanical (Hamann & Beljean, 2017). Mostly, peer review, often in combination with self-evaluation, is also employed in the more recently introduced forms of evaluation (Musselin, 2013). Accordingly, peer review, in one form or another, is still a key mechanism monitoring the flow of scientific knowledge, ideas and people through the gates of the scientific community and higher education institutions (Lamont, 2009).

Autonomy may be defined as 'the quality or state of being self-governing' (Ballou, 1998, p. 105). Autonomy is thus the capacity of an agent to determine their own actions through independent choice, in this case within a system of principles and laws to which the agent is dedicated. The academic profession governs itself by controlling its members. Academics control academics, peers control peers, in order to maintain the status and indeed the autonomy of the profession. Fundamentally, professionals are licensed to act within a valuable knowledge domain. By training, examination and acknowledgement, professionals are legitimated (at least politically) experts of their domain. The rationale of licence and the esotericism of professional knowledge raise the question of how professionals and their work can be evaluated and by which standards. There are rules of conduct and ethical norms, but these are ultimately owned and controlled by the academic profession. From this perspective, we can understand peer review as the structural element that holds academia together.

The increase of peer-review practices in academia can be compared with other professions that also must work harder than before to maintain their status and autonomy. In many cases, their competence and quality must be displayed much more visibly today. Pluralism and individualism in society have also resulted in a plurality of expertise and a decrease of mono-vocational functional systems. A mystique of academic knowledge (as in 'the research says') is not as acceptable in public opinion today as it once was. The term 'postmodern professionals' is suggested to describe experts who expend more effort in the dramaturgy of their competences than people in their positions might have in the past in order to generate trust in clients and in society (Pfadenhauer, 2003). Media makes professional competences, performances and failures much more visible and contributes to trust or mistrust in professions. In a pluralist society, extensive use of peer review may indeed function as a strategy to make apparent quality visible and secure the autonomy of the academic profession, which owns the practice of peer review and knows how to adjust it to its needs.

While most academic evaluations exist across scientific communities and disciplines, the criteria of evaluation can differ substantially between and within communities (Hamann & Beljean, 2017). Thus, research on peer review needs to take disciplinary and interdisciplinary similarities and differences seriously. Obviously, the impact of the intellectual and social organization of the sciences (Whitley, 1984), the mode of research (Nowotny et al., 2001), the tribes and territories (Becher, 1989; Becher & Trowler, 2001; Trowler et al., 2014) and the epistemic cultures (Knorr Cetina, 1999) need to be better represented in future research. Then, examinations of peer review may contribute also to a fuller understanding of the contract between science and society and the challenges directed towards the professional autonomy of academics.

Why Study Peer Review?

As an ideal, peer review has been described as 'the linchpin of science' (Ziman, 1968, p. 148) and a key mechanism in the distribution of status and recognition (Merton, 1968) as well as part and parcel of collegiality and meritocracy (Cole & Cole, 1973). Above all, peer review is considered a gatekeeper regarding the quality of science both in various specialized knowledge communities and in research policy spaces (Langfeldt et al., 2020). Peer review is often taken as a hallmark of quality, expected to both guard and enhance quality. Early on, peer review, or refereeing, was linked to moral institutionalized imperatives. Perhaps most known are those formulated in the *Ethos of Science* by Merton (1942/1973): communism, universalism, disinterestedness and organized scepticism, or CUDOS. These norms and their counter-norms (individualism, particularism, interestedness and dogmatism) have frequently been the focus of peer-review studies. Norms on how scientific work is or should be carried out and how researchers should behave reflect the purpose of science, and ideas of how science should be governed, and are thus directly linked to the autonomy of the academic profession (Panofski, 2010). In short, research into peer review goes to the very heart of academia and its relation to society. This calls for scrutiny.

With changing circumstances, peer review is more often employed, and its purposes, forms and functions are increasingly diversified. Today, academic evaluations permeate every corner of the scientific enterprise, and the traditional form of peer review, rooted in scientific communication, has migrated. Thus, we have seen peer review evolve to be undertaken in all key aspects of academic life: research, teaching, service and collaboration with society (Tennant & Ross-Hellauer, 2020). Increasingly, peer review is regarded as the standard, not only for published scholarship but also for academic evaluations in general. Ideally, peer review is considered to guarantee quality in research and education while upholding the norms of science and preserving the contract between science and society. The diversity and the migration of review practices and its consequences should be followed closely.

In the course of a career, scholars are recurrently involved as both reviewers and reviewees, and this is becoming more and more frequent. As stated in a report on peer review by the British Academy (2007), the principle of *judge not, that ye be not judged* is impossible to follow in academic life. On the contrary, the selection of work for publishing, the allocation of grants and fellowships, decisions on tenure and promotion, and quality evaluations all depend upon the exercise of judgement. 'The distinctive feature of this academic judgement is that it is reciprocal. Its guiding motto is: judge only if you in turn are prepared to be judged' (British Academy, 2007, p. vii).

Indeed, we lack comprehensive statistics on peer review and the involvement of scholars in its diverse practices. However, investigations like the Wiley study (Warne, 2016) and Publons' (2018) *Global State of Peer Review* (2018), both focused on reviews of manuscripts, implicate the widespread and increasing use of peer review. In 2016, roughly 2.9 million peer-reviewed articles were indexed in Web of Science, and a total of 2.5 million manuscripts were rejected. Estimated reviews each year amount to 13.7 million. Together, the continuous rise of submissions and the increase in evaluations using peer reviews expose the system and its actors to ever more pressure.

Peer-review activities produce an incredible amount of talk and gossip in academia. In particular, academic appointments have contributed to the organizational 'sagas' described by Clark (1972). In systems where fierce competition for a limited number of chairs (professorships) is the norm, much is at stake. A single decision, one way or another, can make or break an academic career, and the same is true in relation to recurring judgements and decisions on tenure and promotion (Gunneriusson, 2002). Research on the emotional and socio-psychological consequences of peer rejection or low ratings and rankings is seldom conducted. While rejection may function as either a threat or a challenge to scholarly identities, Horn (2016) argues that rejection is a source of stigmatization pervading the entire academic community. In a similar vein, scholars have to adjust to the maxim of 'publish or perish' and the demands of reviewers, even when these are against the scholars' own convictions. Some researchers consider this a form of 'intellectual prostitution' (Frey, 2003), and reviewer fatigue is spreading through the scientific community. For example, it is widely recognized that editors sometimes have trouble finding reviewers. Obviously, peer review has become a concern to scholars of all kinds and to their identities and everyday practices and careers.

The mundane reality of peer-review practice is quite different from the ideology of peer review, and our knowledge is rather restricted and fragmented (Grimaldo et al., 2018). The roots of peer review can be traced through the seventeenth century and book censorship, the development of academic journals in the eighteenth century and the gatekeeping of scientific communication. As a regular activity, peer review is, however, a latecomer in the scientific community, and it is unevenly distributed across nations and disciplines (Biagioli, 2002). For example, publication practices, discourses and the lingua franca differ between knowledge communities. Traditional peer review is a more prominent feature of the natural sciences and medicine than of the humanities, the social sciences and the arts. This is also reflected in research on peer review. In a similar way, data show that US researchers supply by far the most reviews of manuscripts for journals, while China reviews substantially less. Nevertheless, review output is increasing in all regions and especially so in emerging regions (Publons, 2018).

Even though there are differences, peer review is a fundamental tool in the negotiation and establishment of a scholars' merits and research, of higher education quality and of excellence. Peer review is also considered a tool to prevent misconduct, such as the fraudulent presentation of findings or plagiarism. Thus, peer review may fulfil functions of gatekeeping, maintenance and enhancement. Peer reviews can also be linked to struggles over which form of capital should be the gold standard and over gaining as much capital as possible (Maton, 2005). At stake is, on the one hand, scholastic capital, and on the other hand, academic capital linked to administrative power and control over resources (Bourdieu, 1996).

The introduction of ever new sites for peer review, changing qualifications of reviewers and calls for open science, as well as the increased use of metrics, increase the need for further research. Moreover, the cost and the amount of time spent on different kinds of reviews and their potential impact on the identity, recognition and status of scholars and higher education institutions make peer review especially worthy of systematic studies beyond professional narratives and anecdotes. Peer review has both advocates and critics, although the great majority of researchers are positive to the idea of peer review. Many critics find peer review costly, time consuming, conservative and socially and epistemically biased. In sum, there are numerous reasons to study peer review. It is almost impossible to overstate the central role of peer review in the academic enterprise, and the results of empirical evidence are inconclusive and the research field emergent and fragmented (Bornmann, 2011; Batagelj et al., 2017).

State of the Art of Research on Peer Review

There is a lack of consensus on what peer review is and on its purposes, practices, outcomes and impact on the academic enterprise (Tennant & Ross-Hellauer, 2020). The term *peer review* was relatively unknown before 1970. *Referee* was the more commonly applied notion, used primarily in relation to the evaluation of manuscripts and scientific communication (Batagelj et al., 2017). This lack of clarity has affected how the research field of peer review has been identified and described.

During the past few decades, a number of researchers have provided syntheses of research on peer review in the forms of quantitative meta- and network analyses as well as qualitative configurative analyses. Some are more general in character (Sabaj Meruane et al., 2016; Batagelj et al., 2017; Grimaldo et al., 2018), though the main focus is often research in the natural and medical sciences and peer review for publishing and, to some extent, for grant funding. Others are more concerned with either a specific practice of peer review or different critical topics. Below, we mainly use these recent systematic reviews to depict the research field of peer review, to identify the limits of our knowledge on the subject and to elaborate why we need to study it further.

Academic evaluations, like peer reviews, have been examined from a number of perspectives (Hamann & Beljean, 2017). From a functionalist approach, we can explore how well evaluative procedures serve their purposes—especially those of validity, reliability and fairness—and how well they handle various potential biases. The power-analytical perspective makes critical inquiries into dysfunctional effects of structural inequalities like nepotism and unequal opportunities for resource accumulation. The perspective on the performativity of evaluations and evaluative devices focuses on the organizational impact of the devices, on ranking and on the ways indicators incite strategic behaviour. The social-constructive perspective on evaluation emphasizes that ideas such as merits and originality are socially and historically context dependent. There is also a pragmatist perspective that stresses the situatedness of evaluative practices and interactions (e.g. how panellists reach consensus). More and more frequently used are analytical tools from the field of the sociology of valuation and evaluation, which emphasizes knowledge production as contextualization and the existence and impact of insecurities in the performative situations (Lamont, 2012; Mallard et al., 2009; Serrano Velarde, 2018). Some researchers highlight the variety of academic communities and the intradisciplinary, interdisciplinary and transdisciplinary aspects of research today as significant explanatory factors for evaluative practices (Hamann & Beljean, 2017). We may add changes in the governance of higher education institutions and research and the introduction of new evaluation practices as equally important (Whitley, 2011; Oancea, 2019).

In a network analysis of research on peer review from 1950 to 2016 Batagelj et al. (2017) identified 23,000 indexed records in Web of Science and, above all, a main corpus of 47 articles and books. These texts, which were cited in the most influential publications on peer review, focus on science, scholarship, systematic reviews, peers, peer reviews and quantitative and qualitative analysis. The most cited article allows for an expansion of this list to include the institutionalization of evaluation in science, open peer reviews, bias and the effects of peer review on the quality of research. Most items belonging to the corpus were published relatively early, with only a few published after the year 2000. However, overview papers were published more recently, mainly in the past decade.

The research field of peer review has been described as an emergent field marked by three development stages (Batagelj et al., 2017). The first stage, before 1983, includes seminal work mostly presented in social science and philosophy journals. Main topics include scientific productivity, bibliographies, knowledge, citation measures as measures of scientific accomplishment, scientific output and recognition, evaluations in science, referee systems, journal evaluations, the peer-evaluation system, review processes and peer-review practices. During the second stage, 1983–2002, biomedical journals were influential. Key topics focused on the effects of blinding on review quality, research into peer review, guidelines for peer reviewing, monitoring peer-review performance, open peer review, bias in the peer-review system, measuring the quality of editorial peer review, and the development of meta-analysis and systematic reviews approaches. Finally, in the third stage, 2003–2016, we find research on peer review mainly in specialized science studies journals such as *Scientometrics*. The most frequent topics include peer review of grant proposals, bias, referee selection and links between editors, referees and authors.

Another quantitative analysis (Grimaldo et al., 2018) of articles published in English from 1969 to 2015 and indexed in the citation database Scopus found very few publications before 1970, and fewer than around 100 per year until 2004. Then, from 2004 to 2015 the numbers increased rapidly, 12% per year on average. Half the records were journal articles, books, chapters and conference papers, and the rest were mostly editorial notes, commentaries, letters and literature reviews. Scholars from English-speaking countries, especially the United States, predominated, but authors from prominent European institutions were also found. A fragmented, potentially interdisciplinary research field dominated by medicine, sociology and behavioural sciences and with signs of uneven sharing of knowledge was identified. The research was typically pursued in small collaborative networks. Articles on peer reviews were published mostly by *JAMA*, *Behavioral and Brain Science* and *Scientometrics*. The most important topics were peer review in relation to quality assurance and improvement, publishing, research, open access, evaluation and assessment, bibliometrics and ethics. Among the authors of the top five most influential articles we find Merton, Zuckermann, Horrobin, Bornmann and

Siegelmann. Grimaldo et al.'s (2018) analysis revealed the presence of structural problems, such as difficulties in accessing data, partly due to confidentiality and lack of interest from editorial boards, administrative bodies and funding agencies. More positively, the analysis pointed to digitalization and open science as favourable tools for increases in research, cooperation and knowledge sharing.

In an overview (Sabaj Meruane et al., 2016) of empirical studies on peer-review processes, almost two thirds of the first-named authors had doctoral backgrounds in medicine, psychology, bibliometrics or scientometrics, and around one fifth in sociology of science or science and technology studies. There is definitely a lack of integration of other fields, such as those within the social sciences, the humanities and the arts and education in the study of peer-review processes. The following topics were empirically researched, in descending order: sociodemographic variables (83%), sociometric or scientometric data (47%), evaluation criteria (36%), bias (31%), rates of acceptance/rejection/revision (25%), predictive validity (24%), consensus among reviewers (17%) and discourse analysis of isolated or related texts (14%). The analysis indicates that 'the texts interchanged by the actors in the process are not prominent objects of study in the field' (Sabaj Meruane et al., 2016, p. 188). Further, the authors identified a number of gaps in the research: The field conceives of peer review more as a system than as a process. Moreover, bibliometric studies constitute an independent field of empirical research on peer review. Only a few studies combine analysis of indicators with content or functional analysis. In a similar way, research on science production, reward systems and evaluation patterns rarely includes actual texts that are interchanged in the peer-review process. Discourse analysis, in turn, rarely uses data other than the reviewer report and socio-demographics. Due to ethical issues and confidentiality, discourse studies and text analyses of reviewer reports are less frequent.

It might be risky to state that peer review is an under-studied object of research, considering the vast number of publications devoted to the topic. Nevertheless, it appears that the field of peer-review research has yet to be fully defined, and empirical research in the field has to be more comprehensively done. A common problem the authors consider important to examine is the consequences of the same actor being able to fulfil different roles (e.g. author, reviewer, editor) in various single reviews. Above all, the field requires not only further but also more comprehensive approaches, and in addition, the black box of peer review needs to be fully open (Sabaj Meruane et al., 2016).

Among syntheses focusing on specific topics, those of trustworthiness and bias as well as the employment and negotiation of and the meaning ascribed to criteria in various evaluation practices or in different disciplines are relatively common. In a review of literature published on the topic of peer review, the state of research on journal, fellowship and grant peer review is analysed, focusing on three quality criteria: reliability, fairness and predictive validity (Bornmann, 2011). The interest was directed towards the norms of science, ensuring that results were not incidental, that certain groups or individuals were not favoured or disadvantaged, and that selection of publications and scholars were aligned to scientific performances. Predictive validity was far less studied in primary research than reliability and fairness. Another overview articulates and critiques conceptions and normative claims of bias (Lee et al., 2013). The authors raise questions about existing norms and conclude that peer review is social and that a diversity of norms and opinions among communities and referees may be desirable and beneficial. Bias is also studied in research on who gets tenure with respect to both meritocratic and non-meritocratic factors, such as ascription and social and academic capital (Lutter & Schröder, 2016). These authors show that network size, individual reputation and gender matter.

Epistemic differences point to the necessity of studying peer review within a variety of disciplines and transdisciplinary contexts. An interview study of panellists serving on fellowship grants within the social sciences and humanities shows that evaluators generally draw on four epistemological styles: constructivist, comprehensive, positivist and utilitarian (Mallard et al., 2009). Moreover, peer reviewers employ the epistemological style most appropriate to the field of the proposal under review. In the future, more attention has to be paid to procedural fairness, including from a comparative perspective. In another systematic review of criteria used to assess grant applications, it is suggested that forthcoming research should also focus on the applicant, include data from non-Western countries and examine a broad spectrum of research fields (Hug & Aeschbach, 2020).

As shown in this introductory chapter, the research field devoted to peer review covers a great number of evaluation practices embedded in different contexts. As it is an emergent and fragmented field in need of integration, there are certainly many possible ways to make contributions to the research field of peer review. On the agenda we find issues related to the foundation of science: the ethos of science and the ideology of peer review, the production and dissemination of knowledge, professional self-regulation and open science. There are also questions concerning the development of theoretical framing and methodological tools adapted to the study of diverse review practices in shifting contexts and at various interacting levels. Not least, in response to calls for more comprehensive and integrated research, it is necessary to open the black boxes of peer review and analyse, in empirical studies, the different purposes, discourses, genres, relations and processes involved.

A single book cannot take on all the above-mentioned challenges ahead of us. However, following this brief introduction to the field, the volume brings together research on review practices often studied in isolation. We include studies ranging from the practice of assessing manuscripts submitted for publication to the more recent practice of open review. In addition, more encompassing and general issues are considered, as well as specificities of different peer-review practices. This is further developed below, where the structure of the volume and the contributions of each chapter are presented.

The Structure and Content of the Volume

The structure of the volume falls into three main parts. In the first part, Rudolf Stichweh and Raf Vanderstraeten continue the introduction begun in this chapter. They discuss the term peer review and the contexts of its emergence. In Chap. 2, Rudolf Stichweh explains the genesis of inequalities and hierarchies in modern science. He illuminates the forms and mechanisms of scientific communication on the basis of which the social structures of science are built: publications, co-authorships and multiple authorships, citations as units of information and as social rewards, and peer review as an evaluation of publications (and of projects and careers). Stichweh demonstrates how, in all institutional dimensions of higher education, differences arise between successful and less success-ful participations. Success generates influence and social attractiveness (e.g. as a co-author). Influential and attractive participants are recruited into positions where they assess the achievements of others and thereby limit and control inclusion in publications, funding and careers.

Vanderstraeten, in Chap. 3, puts forward that with the expansion of educational research in the twentieth century, interested 'amateurs' have been driven out of the field, and the scientific community of peers has become the dominant point of orientation. Authorship and authority became more widely distributed; peer review was institutionalized to monitor the flow of ideas within scientific literature. Reference lists in journals demonstrated the adoption of cumulative ideals about science. Vanderstraeten's historical analysis of education journals shows the social changes that contributed to the ascent of an 'imagined' community of expert peers in the course of the twentieth century.

Part II of this volume focuses mainly on how peer-review practices have emerged in many parts of higher education institutions. From being scholarly publication practices in early times, peer review appears to be internationally *the* most significant performative practice in higher edu-cation and research. In this part, the various scholars provide insight into such processes. Don F. Westerheijden, in Chap. 4, revisits the policy issue of the balance between peer review and performance indicators as the means to assess quality in higher education. He shows the paradoxes and unintended effects that emerge when peer review is the main method in the quality assurance procedures of higher education institutions as a whole. Westerheijden argues that attempted solutions of using self-assessments and performance indicators as well as specifically trained assessors increase complaints about bureaucracy from within the aca-demic community.

In Chap. 5, Hanne Foss Hansen sheds light on how peer review as an evaluation concept has developed over time and discusses which roles peer review plays today. She presents a typology distinguishing between classical peer review, informed and standard-based peer review, modified peer review and extended peer review. Peer review today can be found with all these faces. Peter Dahler Larsen argues in Chap. 6 that gatekeepers in institutional review processes who know the future and use this knowledge in a pre-emptive or precautionary way play a key role in the construction of reality, which comes out of Bibliometric Research Indicators, widely used internationally. By showing that human judgement sometimes enhances or multiplies the effects of 'evaluation machineries', this chapter contributes to an understanding of mechanisms that lead to constitutive effects of evaluation systems in research.

In Chap. 7, Agnes Ers and Kristina Tegler Jerselius explore a national framework for quality assurance in higher education and argue that such systems' forms are dynamic, since they change over time. Ers and Tegler Jerselius show how the method of peer review has evolved over time and in what way it has been affected by changes made in the system. *Gustaf Nelhans* engages in Chap. 8 with the performative nature of bibliometric indicators and explores how they influence scholarly practice at macro levels (in national funding systems), meso levels (within universities) and individual levels (in the university employees' practice). Nelhans puts forward that the common-sense 'representational model of bibliometric indicators' is questionable in practice, since it cannot capture the qualities of research in any unambiguous way.

In Chap. 9, Lars Geschwind and Kristina Edström discuss the loyalty of academic staff to their disciplines or scientific fields. They show how this loyalty is reflected in evaluation practices. They elaborate on the extent to which peer reviewers act as advocates for those they evaluate. By doing so, Geschwind and Edström problematize potential evaluator roles. In Chap. 10, Malcom Tight closes *Part II* of this book. Drawing on his extensive review experiences in various areas of higher education institutions, he assesses how 'fit for purpose' peer review is in twenty-first-century academe. He focuses on different practices of peer review in the contemporary higher education system and questions how well they work, how they might be improved and what the alternatives are.

Whereas *Part II* of this volume focuses on the relation and impact of higher education institutions considering education quality and research output, *Part III* illuminates different particular peer-review practices. Eva Forsberg, Sara Levander and Maja Elmgren examine in Chap. 11

peer-review practices in the promotion of what is called 'excellent' or 'distinguished' university teachers. While research merits have long been the prioritized criteria in the recognition of institutions and scholars, teaching is often downplayed. To counteract this tendency, various systems to upgrade the value of education and to promote teaching excellence have been introduced by higher education institutions on a global scale. The authors show that the intersection between promotion, peer review and excellent teaching affects not only the peer-review process but also the notion of the excellent or distinguished university teacher.

In Chap. 12, Tine S. Prøitz discusses how the role of scholarly peers in systematic review is analysed and presented. Peer evaluation is an essential element of quality assurance of the strictly defined methods of systematic review. The involvement of scholarly peers in the systematic review processes has similarities with traditional peer-review processes in academic publishing, but there are also important differences. In systematic review, peers are not only re-judging already reviewed and published research, but also gatekeeping the given standards, guidelines and procedures of the review method.

Liv Langfeldt presents in Chap. 13 processes of grant peer review. There are no clear norms for assessments, and there may be a large variation in what criteria reviewers emphasize and *how* they are emphasized. Langfeldt argues that rating scales and budget restrictions can be more important than review guidelines for the kind of criteria applied by the reviewers. The decision-making methods applied by the review panels when ranking proposals are found to have substantial effects on the outcome. Chapters 14 and 15 focus on peer-review practices in the recruitment of professors. First, Sara Levander, Eva Forsberg, Sverker Lindblad and Gustav Jansson Bjurhammer analyse the initial step of the typecasting process in the recruitment of full professors. They show that the field of professorial recruitment is characterized by heterogeneity and no longer has a basis in one single discipline. New relations between research, teaching and society have emerged. Moreover, the authority of the professorship has narrowed and the amount of responsibilities have increased. Then, Björn Hammarfeldt focuses on discipline—specific practices for evaluating publications oeuvres. He examines how 'value' is enacted with special attention to the kind of tools, judgements, indicators and metrics that are used. Value is indeed enacted differently in the various disciplines.

In the last chapter of the book, Chap. 16, Tea Vellamo, Jonna Kosonen, Taru Siekkinen and Elias Pekkola investigate practices of tenure track recruitment. They show that criteria of this process can exceed notions of individual merits and include assessments of the strategic visions of universities and departments. The use of the tenure track model can be seen as a shift both for identity building related to a university's strategy and for using more managerial power in recruitment more generally.

We dedicate this book to our beloved colleague and friend, professor Rita Foss Lindblad, who was involved in the project but passed away in 2018.

References

Aagaard, K., Bloch, C., & Schneider, J. W. (2015). Impacts of performance-based research funding systems: The case of the Norwegian Publication Indicator. *Research Evaluation, 24*(2), 106–117.

Ballou, K. A. (1998). A concept analysis of autonomy. *Journal of Professional Nursing, 14*(2), 102–110.

Batagelj, V., Ferligoj, A., & Squazzoni, F. (2017). The emergence of a field: A network analysis of research on peer review. *Scientometrics, 113*(1), 503–532. https://doi.org/10.1007/s11192-017-2522-8

Becher, T. (1989). *Academic tribes and territories: Intellectual inquiry and the cultures of disciplines*. Society for Research into Higher Education.

Becher, T., & Trowler, P. R. (2001). *Academic tribes and territories. Intellectual inquiry and the culture of disciplines*. Open University Press.

Biagioli, M. (2002). From book censorship to academic peer review. *Emergences Journal for the Study of Media & Composite Cultures, 12*(1), 11–45. https://doi.org/10.1080/1045722022000003435

Bornmann, L. (2011). Scientific peer review. *Annual Review of Information, Science and Technology, 45*, 197–245. https://doi.org/10.1002/aris.2011.1440450112

Bornmann, L. (2013). Evaluations by peer review in science. *Springer Science Reviews, 1*(1–4). https://doi.org/10.1007/s40362-012-0002-3

Bourdieu, P. (1996). *Homo academicus*. Polity.

Boyer, E. L. (1990). *Scholarship reconsidered: Priorities of the professoriate.* The Carnegie Foundation for the Advancement of Teaching.

British Academy. (2007). Peer review: The challenges for the humanities and social sciences. Retrieved December 1, 2020, from https://www.thebritishacademy.ac.uk/documents/197/Peer-review-challenges-for-humanities-social-sciences.pdf

Caputo, R. K. (2019). Peer review: A vital gatekeeping function and obligation of professional scholarly practice. *Families in Society: The Journal of Contemporary Social Services, 100*(1), 6–16. https://doi.org/10.1177/1044389418808155

Chen, R., & Hyon, S. (2005). Faculty evaluation as a genre system: Negotiating intertextuality and interpersonality. *Journal of Applied Linguistics, 2*(2), 153–184. https://doi.org/10.1558/japl.v2i2.153

Clark, B. R. (1972). The organizational saga in higher education. *Administrative Science Quarterly, 17*, 178–184.

Clark, B. R. (1989). The academic life: Small worlds, different worlds. *Educational Researcher, 18*(5), 4–8. https://doi.org/10.2307/1176126

Cole, J. R., & Cole, S. (1973). *Social stratification in science.* University of Chicago Press.

Csiszar, A. (2016). Peer review: Troubled from the start. *Nature, 532*(7599), 306–308. https://doi.org/10.1038/532306a

Dahler Larsen, P. (2012). *The evaluation society.* Stanford University Press.

Elken, M., & Wollscheid, S. (2016). *The relationship between research and education: Typologies and indicators. A literature review.* Nordic Institute for Innovative Studies in Research and Education (NIFU).

European Science Foundation. (2011). *European peer review guide. Integrating policies and practices into coherent procedures.*

Frey, B. S. (2003). Publishing as prostitution? Choosing between one's own ideas and academic success. *Public Choice, 116*(1/2), 205–223. https://doi.org/10.1023/A:1024208701874

Gläser, J., & Laudel, G. (2007). The social construction of bibliometric evaluations. In R. Whitley & J. Gläser (Eds.), *The changing governance of the sciences. The advent of research evaluation systems.* Springer.

Grimaldo, F., Marušić, A., & Squazzoni, F. (2018). Fragments of peer review: A quantitative analysis of the literature (1969–2015). *PLOS ONE, 13*(2), e0193148. https://doi.org/10.1371/journal.pone.0193148

Guetzkow, J., Lamont, M., & Mallard, G. (2004). What is originality in the humanities and the social sciences? *American Sociological Review 2004, 69,* 190. https://doi.org/10.1177/000312240406900203

Gunneriusson, H. (2002). *Det historiska fältet: svensk historievetenskap från 1920-tal till 1957.* Uppsala: Acta Universitatis Upsaliensis.

Hamann, J., & Beljean, S. (2017). Academic evaluation in higher education. In J. C. Shin & P. Teixeira (Eds.), *Encyclopedia of international higher education systems and institutions.* https://doi.org/10.1007/978-94-017-9553-1_295-1

Hammarfelt, B. (2017). Recognition and reward in the academy: Valuing publication oeuvres in biomedicine, economics and history. *Aslib Journal of Information Management, 69*(5), 607–623. https://doi.org/10.1108/AJIM-01-2017-0006

Hammarfelt, B., Rushforth, D., & de Rijcke, S. (2020). Temporality in academic evaluation: 'Trajectoral thinking' in the assessment of biomedical researchers. *Valuation Studies, 7*(1), 33–63. https://doi.org/10.3384/VS.2001-5992.2020.7.1.33

Hansen, H. F., Aarrevaara, T., Geschwind, L., & Stensaker, B. (2019). Evaluation practices and impact: Overload? In R. Pinheiro, L. Geschwind, H. Foss Hansen, & K. Pulkkinen (Eds.), *Reforms, organizational change and performance in higher education: A comparative account from the Nordic countries.* Palgrave Macmillan.

Helgesson, C.-F. (2016). Folded valuations? *Valuation Studies, 4*(2), 93–102. https://doi.org/10.3384/VS.2001-5992.164293

Horn, S. A. (2016). The social and psychological costs of peer review: Stress and coping with manuscript rejection. *Journal of Management Inquiry, 25*(1), 11–26. https://doi.org/10.1177/1056492615586597

Hug, S. E., & Aeschbach, M. (2020). Criteria for assessing grant applications: A systematic review. *Palgrave Communications, 6*(30). https://doi.org/10.1057/s41599-020-0412-9

Kaltenbrunner, W., & de Rijcke, S. (2020). Filling in the gaps: The interpretation of curricula vitae in peer review. *Social Studies of Science, 49*(6), 863–883. https://doi.org/10.1177/0306312719864164

Knorr Cetina, K. (1999). *Epistemic cultures.* Harvard University Press.

Lamont, M. (2009). *How professors think. Inside the curious world of academic judgment.* Harvard University Press.

Lamont, M. (2012). Toward a comparative sociology of valuation and evaluation. *Annual Review of Sociology, 38*(21), 201–221. https://doi.org/10.1146/annurev-soc-070308-120022

Langfeldt, L., & Kyvik, S. (2011). Researchers as evaluators: Tasks, tensions and politics. *Higher Education, 62*(2), 199–212. https://doi.org/10.1007/s10734-010-9382-y

Langfeldt, L., & Kyvik, S. (2015). Intrinsic tensions and future challenges of peer review. In *RJ Yearbook 2015/2016*. Riksbankens Jubileumsfond & Makadam Publishers.

Langfeldt, L., Nedeva, M., Sörlin, S., & Thomas, D. A. (2020). Co-existing notions of research quality: A framework to study context-specific understandings of good research. *Minerva, 58*, 115–137. https://doi.org/10.1007/s11024-019-09385-2

Lee, C. J., Sugimoto, G. R., Zhang, G., & Cronin, B. (2013). Bias in peer review. *Journal of the American Society for Information Science and Technology, 64*(1), 2–17. https://doi.org/10.1002/asi.22784

Lutter, M., & Schröder, M. (2016). Who becomes a tenured professor, and why? Panel data evidence from German sociology, 1980–2013. *Research Policy, 45*, 999–1013. https://doi.org/10.1016/j.respol.2016.01.019

Mallard, G., Lamont, M., & Guetskow, J. (2009). Fairness as appropriateness: Negotiating epistemological differences in peer review. *Science Technology Human Values*. https://doi.org/10.1177/0162243908329381

Maton, K. (2005). A question of autonomy: Bourdieu's field approach and higher education policy. *Journal of Education Policy, 20*(6), 687–704. https://doi.org/10.1080/02680930500238861

Merton, R. K. (1968). The Matthew effect in science. *Science, 159*(3810), 56–63. https://doi.org/10.1126/science.159.3810.56

Merton R. K. (1973). *The sociology of science: Theoretical and empirical investigations* (Norman W. Storer, Ed.). University of Chicago Press. (Original work published 1942)

Musselin, C. (2002). Diversity around the profile of the 'good' candidate within French and German universities. *Tertiary Education and Management, 8*(3), 243–258. https://doi.org/10.1080/13583883.2002.9967082

Musselin, C. (2013). How peer review empowers the academic profession and university managers: Changes in relationships between the state, universities and the professoriate. *Research Policy, 42*(5), 1165–1173. https://doi.org/10.1016/j.respol.2013.02.002

Neave, G. (1998). The evaluative state reconsidered. *European Journal of Education, 33*(3), 265–284. https://www.jstor.org/stable/1503583

Nowotny, H., Scott, P. B., & Gibbons, M. T. (2001). *Re-thinking science: Knowledge and the public in an age of uncertainty*. Polity Press.

Oancea, A. (2019). Research governance and the future(s) of research assessment. *Palgrave Communications, 5,* 27. https://doi.org/10.1057/s41599-018-0213-6

Oravec, A. (2019). Academic metrics and the community engagement of tertiary education institutions: Emerging issues in gaming, manipulation, and trust. *Tertiary Education and Management.* https://doi.org/10.1007/s11233-019-09026-z

Ozeki, S. (2016). *Three Empirical Investigations into the Logic of Evaluation and Valuing Practices.* Dissertations. 2470. https://scholarworks.wmich.edu/dissertations/2470

Paltridge, B. (2017). *The discourse of peer review. Reviewing submission to academic journals.* Macmillan Publishers.

Panofski, A. L. (2010). In C. J. Calhoun (Ed.), *Robert K. Merton: Sociology of science and sociology as science.* Columbia University Press.

Pfadenhauer, M. (2003). *Professionalität. Eine wissenssoziologische Rekonstruktion institutionalisierter Kompetenzdarstellungskompetenz* [Professionalism. A reconstruction of institutionalized proficiency in displaying competence]. Springer.

Power, M. (1997). *The Audit Society. Rituals of verification.* Oxford University Press.

Publons. (2018). Global state of peer review. Online.

Research Information Network CIC. (2015). Scholarly communication and peer review. The current landscape and future trends. A report commissioned by the Wellcome Trust. Retrieved May 2015, from https://wellcome.org/sites/default/files/scholarly-communication-and-peer-review-mar15.pdf

Ross-Hellauer, T. (2017). What is open peer review? A systematic review (version 2; peer review: 4 approved). *F1000Research, 2017, 6*(588). Last updated: 17 May 2019. Included in Science Policy Research Gateway. https://doi.org/10.12688/f1000research.11369.2

Roumbanis, L. (2017). Academic judgments under uncertainty: A study of collective anchoring effects in Swedish research council panel groups. *Social Studies of Science, 47*(1), 95–116. https://doi.org/10.1177/0306312716659789

Sabaj Meruane, O., González Vergara, C., & Pina-Stranger, Á. (2016). What we still don't know about peer review. *Journal of Scholarly Publishing, 47*(2), 180–212. https://doi.org/10.3138/jsp.47.2.180

Scriven, M. (1980). *The logic of evaluation.* Edgepress.

Scriven, M. (2003). Evaluation theory and metatheory. In T. Kellaghan, D. L. Stufflebeam, & L. A. Wingate (Eds.), *International handbook of educational evaluation* (pp. 15–30). Kluwer Academic Publishers.

Serrano Velarde, K. (2018). The way we ask for money… The emergence and institutionalization of grant writing practices in academia. *Minerva, 56*(1), 85–107. https://doi.org/10.1007/s11024-018-9346-4

Söderlind, J., & Geschwind, L. (2019). Making sense of academic work: The influence of performance measurement in Swedish universities. *Policy Reviews in Higher Education, 3*(1), 75–93. https://doi.org/10.1080/2332296 9.2018.1564354

Swales, J. M. (1996). Occluded genres in the academy. The case of the submission letter. In E. Ventola & A. Mauranen (Eds.), *Academic writing: Intercultural and textual issues*. ProQuest Ebook Central. http://ebookcentral.proquest. com/lib/uu/detail.action?docID=680373

Tennant, J. P., & Ross-Hellauer, T. (2020). The limitations to our understanding of peer review. *Research Integrity and Peer Review, 5*(6). https://doi. org/10.1186/s41073-020-00092-1

Trowler, P., Saunders, M., & Bamber, V. (Eds.). (2014). *Tribes and territories in the 21st century. Rethinking the significance of disciplines in higher education*. Routledge.

Vedung, E. (2002). Utvärderingsmodeller [Evaluation models]. *Socialvetenskaplig tidskrift, 9*(2–3), 118–143.

Warne, V. (2016). Rewarding reviewers—sense or sensibility? A Wiley study explained. *Learned Publishing, 29*, 41–50. https://doi.org/10.1002/leap.1002

Westerheijden, D. F., Stensaker, B., & Joao Rosa, M. (Eds.). (2007). *Quality assurance in higher education. Trends in regulation, translation and transformation*. Springer.

Whitley, R. (1984). *The intellectual and social organization of the sciences*. Clarendon Press.

Whitley, R. (2011). Changing governance and authority relationships in the public sciences. *Minerva, 49*, 359–385. https://doi.org/10.1007/ s11024-011-9182-2

Ziman, J. M. (1968). *Public knowledge*. The University of Chicago Press.

2

Hierarchies and Universal Inclusion in Scientific Communities

Rudolf Stichweh

Tensions and Contradictions in Contemporary Society

This chapter is about a fundamental tension and contradiction in contemporary world society. Society and its function systems such as science are, since the eighteenth-century world, fundamentally based in egalitarian inclusion. But from the operation of egalitarian inclusion arise again and again hierarchical structures in scientific communities and in the system of science that transform this function system into a system with significant and ever-renewing inequalities. These are new inequalities coming from equality—and they are not based in continuities to premodern patterns (Stichweh, 2022).

The argument in this chapter is about the system of science as one of the function systems of society. Functional differentiation is the primary form of social differentiation in contemporary world society. Besides

R. Stichweh (✉)
University of Bonn, Bonn, Germany

© The Author(s) 2022
E. Forsberg et al. (eds.), *Peer review in an Era of Evaluation*,
https://doi.org/10.1007/978-3-030-75263-7_2

science other global function systems crystallize around key social problems: the polity, the economy, religion, law, education, the health complex, the arts and the sports. They shift the 'profile' of society from inequality to heterogeneity. But in the function systems of society new inequalities emerge and therefore the argument of this chapter that is only about science may prove to be paradigmatic for the rise and the forms of inequality in other function systems in society. This will have to be explored in future work.

In looking at science one central interest of this chapter focuses on peer review. Peer review is thought to be a core institution of autonomy and equality in science. It holds the promise that a scientist is judged by those who share his/her interests (autonomy of science) and share the same social status (equality of peers). But just by the selective recruitment for being a peer reviewer and by acquiring scientific influence in becoming a peer reviewer the status of a scientist rises in taking these reviewer roles and therefore the institution of equality contributes to the generation and cumulative expansion of inequalities.

Universal Inclusion

Universal inclusion is a characteristic of all the function systems of society. Inclusion means that there arise possibilities of participation and roles for participation for everyone. Exclusion becomes illegitimate, although it factually is there in numerous variants.

The history of modern society can and should be written as the history of *inclusion revolutions* coming about between the eighteenth and the twenty-first century. These inclusion revolutions are turning points in the differentiation histories of all the function systems. What is meant by this can best be explained in briefly looking at some cases.

In premodern economies the economic well-being of the population was often endangered by population growth. The economies could not absorb the growing populations, and from this condition poverty and hunger, epidemics and loss of population ensued, until on the basis of smaller populations an equilibrium was reestablished. It was for the first time in the second half of the eighteenth century that an economic system, the English economy, succeeded to combine a significant population growth with even faster-growing average incomes. This was the beginning of the inclusion revolution of the modern economy. In the political system the beginnings of democracy (e.g. in France, the United States, Switzerland) started an inclusion revolution. The long-term expansion of voting rights until they included everyone was in this respect the most important process, but to this were clearly added other forms of political inclusion. In the education system the inclusion revolution is coupled to schooling and higher education. Universal schooling already existed in some European regions late in the eighteenth century, and the university transformed itself between 1750 and 2020 from an institution for 1% of the male population to inclusion rates that in some cases approach or even surpass 90%. Religion probably is an especially important and interesting case, as religion is the function system for which arguments claiming the irrelevance and marginality of significant parts of the population would never have made any sense. It is an interesting feature in the history of European Christianity that poor persons and other marginalized groups took central roles in the history of salvation just because of their marginal status. As they had no resources that tied them to this world, poor people were nearer to God than rich people ever could have been and were able to function as mediators and prayed for the salvation of the rich. This is a feature especially prominent in fourteenth- and fifteenth-century Europe, and 100 years later in early modern Europe, confessionalization and its activist and disciplinary demands on the population could be understood as the first inclusion revolution happening before the onset of modernity (Stichweh, 2020a).

We will not look here at all these fascinating cases. Instead we only analyze science. Which are the institutions of universal inclusion in the system of science? The chapter presents the core institutions relevant for our problem (Sections "Publication as the Elementary 'Unit-Act' of the System of Science", "Authorship of Publications as the Form of Inclusion in Science", and "Citations as the Internal Structure of Publications"). And then it analyzes the hierarchies emerging in science on the basis of the operation of these institutions (Sections "Reading and Writing in Scientific Communities: The Hierarchy of Authorship", "The Emergence of Peer Review: The Hierarchy of Readership", "The Two Hierarchies:

Authorship and Readership", and "The Third and Fourth Hierarchy: Hierarchy of Publication Places and Hierarchy of Recruitment for Co-authorship"). The questions we have in view here, in presenting our case, are as much practical questions of the optimal institutional design of a system of science holding to universal inclusion as they are theoretical questions of conceiving a theory of inequality for a functionally differentiated world society.

Publication as the Elementary 'Unit-Act' of the System of Science

Late in the eighteenth century were established the first scientific journals—some of them with disciplinary specializations—('Chemisches Journal' 1778, 'Annales de Chimie et de Physique' 1789, 'Journal der Physik' 1790, 'Philosophical Magazine' 1798) (Hund, 1990; Stichweh, 1984) that are similar to the social and communicative forms that we still use today in communicating science. Journals published scientific papers, which over the next 200 years became an ever more standardized form of the communication of scientific insights. Besides scientific papers in specialized journals there arose the book or the monograph as the second significant form of publication in the system of science. Both publication forms—papers and monographs—then function as the elementary 'unit-acts' (Parsons, 1937) of the communicative and cognitive reproduction of science. 'Unit-acts' are elements; what they say can in principle be reduced to a brief synopsis of their essential insights, and this is even true for long monographs. They share an important property with other elements in other social and natural systems, for example, with atoms. Elements are as well simple as they normally will have an enormous internal complexity. Scientific observers can either focus on the simplicity or on the complexity of elements (i.e. publications or atoms) and the oscillation between the one and the other option is an important part of the practice of science.

Authorship of Publications as the Form of Inclusion in Science

Scientific publications as communicative unit-acts are claimed by authors as their products. The institutionalization of scientific authorship is another core feature of modern science. Authorship is not organizational authorship; a paper is not published by the University of Uppsala or the University of Leiden. And there is no longer a top level of academicians in the major European academies to whom one sends the report of one's discoveries and who decide if these informations are printed (as a letter to the respective academicians) in the pages of the academy journal. Instead of these hierarchical or organizational solutions there now is individual authorship that at the same time is inclusive authorship as everybody who is able to write a paper can now publish a paper under his or— later—her name. Therefore, it can be claimed that the genesis of the specialized scientific journal is at the same time the starting point of an inclusion revolution in the system of science that over time significantly expands the author space of the science system.

Around 1800 it can safely be said that authorship is nearly always individual authorship. There are some cases of co-authorship even at this early point in time—perhaps 2% of all papers in 1800 and still not more than 7% in 1900 (Beaver & Rosen, 1978, 1979)—but the dominant pattern is individual publication by authors who enter science by this act of individual publication. When this changes again, in the twentieth and twenty-first centuries, behind these changes are transformations in the social structure of scientific communities. There are two major changes, the normalization of co-publication by at least two authors and added to this an escalation of the number of authors per paper that in our days may include significantly more than two authors (the most frequent number of authors today is three) or even dozens of authors and in some cases (in high energy physics and clinical medicine) hundreds and thousands of authors (Adams et al., 2019).

Some sociological characteristics of this process have to be mentioned. (1) It is still individual authorship. The system of science never opted for the substitution of collective or organizational authorship for individual authorship. There are some cases of collective authorship among whom the collective of French mathematicians called Bourbaki may be the most famous. But Bourbaki was primarily established for the production of mathematical textbooks. (2) The rise of co-authorship and then of multi-authorship reflects the emergence of cooperation and division of labor as the normal modus operandi of doing scientific research. (3) Co-authorship expands once more the author space, as it opens the way to publication for all those who could not produce a paper alone. Or, in the ironic formulation of De Solla Price, it allows publication for those who only have half a paper in them at the present time (Price, 1986). (4) But co-authorship is not only about cooperation and division of labor; it implies an expansion of the number of cognitive perspectives integrated into one scientific paper. There are more methods, more theories, more subdisciplines and disciplines that are integrated into one scientific paper. This expansion of the number of cognitive perspectives drives the growth of multi-authorship. (5) Co-authorship changes the relation of authorship and writing. Not everyone who is one of the authors of a paper has been participating in the writing of the paper. On the other hand, writing a paper may become a relevant competence in its own right and may become for some persons the major contribution they made to the paper. (6) Over time there arise ever more social roles and statuses and contributions that may be accepted as legitimate claims for authorship. There are places for senior scientists, guest authors and honorific authors, reciprocal offers of 'free tickets' on one's papers exchanged between two scientists, authorship for departmental heads and for other positions in organizational hierarchies (Adams et al., 2019). (7) There is, finally, the question of international co-authorship and its fast expansion. Partially, it results from the same forces just mentioned: the division of labor, the need for ever more theories and methods and for knowledge from other disciplines. But there are additional reasons, too. In many projects one needs data from other countries, one has to stay and to work in these countries, and these things in many situations can't be done if one does not include authors from these countries. Often this is even a political imperative. A good example is a recent very interesting paper on the physiological and genetic adaptations to extreme diving to be observed in one of the last remaining populations of sea nomads (people living on boats and spending hours every day in and under water to catch and collect fish and plants from the sea). The paper (Ilardo et al., 2018) has 17 authors, with institutional addresses from six countries. One of these addresses is from Indonesia. The author is from a Department of Education. In a short note on author contributions it is said in the paper that this author contributed logistical support to the project (obviously a strange claim for authorship). Shortly after the publication of the paper in *Cell* objections were raised in Indonesia that the researchers had violated Indonesian rules by not sufficiently consulting with Indonesian institutions and researchers (Rochmyaningsih, 2018; Van Groenigen & Stoof, 2020) and not getting permission for the transport of DNA material out of the country. (8) There are other strong reasons for the international extension of the recruitment of coauthors. International coauthors clearly enhance the visibility of scientific papers. Adding a further country demonstrably has a stronger effect on future citations of a scientific paper than simply adding one more author from a country that is already represented by an author, and this is true up to the eighth country (Adams et al., 2019).

The scientific paper becomes an extremely flexible instrument for the inclusion in science. The list of authors is a very simple list of names, with footnotes added to the individual names that point to organizational addresses. In some cases in our days, the list of names is longer than the paper. The list is nearly never alphabetical. It is bidirectionally rank-ordered, with positions at the beginning and the end especially prominent. But nonetheless the list suppresses hierarchy more than it makes hierarchy visible. It symbolizes science as a collective endeavor. But the collectivity is represented as a collection of individuals, and the point is incessantly made by every scientific paper that every individual counts in the production of science.

Citations as the Internal Structure of Publications

In the nineteenth and twentieth centuries another core structure slowly arose in science. Science invariably became second-order observation. Scientific observers observe reality but they always do this in relating their observations to the observations other scientists have made before.

From this arises a core obligation for every scientific paper: It has to review the insights proposed by other scientific papers and it has to relate the novelties it claims to these anterior insights. These relations between the present paper and earlier publications have to be documented by precise citations to these publications.

Citations are a microstructure of the publications in which they occur. If two papers make use of the same citations the papers are seen as cognitively similar, as belonging to a network of papers who are related to one another by cognitive neighborhoods. But the most remarkable property of citations is that they combine two heterogeneous functions. They are units of cognitive information. They inform readers of a paper where further relevant information is to be found. For a scientist to read a specific paper is often primarily motivated by the hope to get access to the population of papers that are relevant for work on a specific scientific problem. But besides being of informational relevance citations are at the same time social rewards for the authors of the papers that are cited. In the social dimension citations are acts of recognition. They certify that the authors of the cited publication have done something worthwhile. They have contributed to science. Even if the citing scientist(s) try to refute the citing paper and its cognitive claims, the social function of the citation remains intact. It is still said that the respective paper is a relevant part of science and that it is useful for the progress of science to refute its cognitive claims. As we know since Karl Raimund Popper (Popper, 1963), science deals in a symmetrical way with affirmations and refutations of the cognitive claims of other publications.

For the cited authors it can be said that a new atom of reputation is added to their balance sheet by the act of citation. Among other things citations are acts of inclusion. As long as one has only published, there remains a fundamental insecurity: Has my paper ever been read by anybody? After the first atom of reputation created by the first citation, careers can begin and inequality can start to arise. There is a cumulation of citations over time—and this happens on the basis of 'preferential attachment' (Newman, 2001) and 'cumulative advantage' (DiPrete & Eirich, 2006; Merton, 1988) as mechanisms of the production of inequality.

Reading and Writing in Scientific Communities: The Hierarchy of Authorship

In most of the function systems of society there is a split that distinguishes performance roles and observer roles (Ahlers et al., 2020; Stichweh, 2016). There are professionals and clients, doctors and patients, professional artists and their public, and so on. In scientific communities there are authors and readers. One is included in scientific communities as an author (of papers and monographs) and as a reader (of papers and monographs). Role-taking is in both cases based on self-selection, although the decision to write a paper does not guarantee that the presumptive author is able to publish the paper.

There is a strong preference toward authorship in scientific communities. One enters a scientific community by authorship, by contributing publications, not by reading publications. Science is a community of publishing authors, not a community of readers. The fact of reading (scientific papers) becomes visible and relevant not as a creative act in itself (as is the case in literature) (Moretti, 2013) but by citations in publications that document the readings of authors. The hierarchy of science is not a hierarchy of perceptive readers but a hierarchy of authors who are highly cited by other scientific authors in their publications.

But there is an outer fringe of participants in scientific disciplines who only read publications and who do not and mostly cannot contribute publications to the respective discipline. These participants in most cases are visible as authors in other disciplinary communities. Therefore, this phenomenon is akin to interdisciplinarity and is related to the learning processes of which interdisciplinarity consists (Stichweh, 2017).

The Emergence of Peer Review: The Hierarchy of Readership

The self-selection for doing research and for publishing the research one has done that was for a long-time characteristic of modern science is strongly changed in twentieth-/twenty-first-century science by the emergence of peer review (Cole & Cole, 1981; Cole et al., 1978; Squazzoni et al., 2020).

Peer review means the institutionalization of a new class of readers in science who decide on the research that can be done (by preparing funding decisions) and who decide on the papers and books that will be published (by preparing publication decisions for journals and for book publishers). The readings of these readers do not enter the public communication processes in science. They are mostly private (private to the journals and publishers they work for), invisible readings. But they are very influential. And they imply the rise of a new type of reader roles (readers who do not channel their readings into publications) and a new hierarchical level of especially influential readers in science that establishes a supervenient level of control in science that wasn't there before.

The Two Hierarchies: Authorship and Readership

In the modern system of science the inclusion in authorship is the primary mode of inclusion. It is universal (only demanding the capability to write a scientific paper) and it demonstrates the primacy of performance roles in science. Science is about doing science and not about knowing science by reading scientific papers. Only when reading is part of a production process it is integrated into this understanding of science.

Peer review creates a new kind of reader role in science. The access to these new reader roles presupposes previous success as a scientific researcher and author. Therefore, these reader roles are highly selective and are mostly accessible only at later points in one's career. When these roles are offered, the persons to whom they are offered know that they are advanced in their careers and participate in science not only as researchers and authors but additionally as reviewers who decide on the quality of the research and authorship of other scientists.

The semantic term for this activity is 'peer review' and this suggests that one is judged upon by one's equals. But these peers are a little bit more equal than others. Peer review creates a level and forms of influence that differs from the influence derived from publishing papers. It creates a new hierarchical level of influence.

This hierarchy of influential readers prominent in peer review restructures the inclusion in research and publication. Reviewers as readers decide who can do research (funding decisions in funding agencies) and who can publish (as reviewers for journals and publishing houses)—and they decide on scientific careers by reviewing publications that are counted for advancement, and by reviewing suggestions for hiring decisions.

The most influential readers as reviewers are often no longer authors and researchers themselves. Their readings have enormous weight. But these readings do not enter the scientific discourse and they do not enter the ongoing cumulation of scientific knowledge.

The inclusion in research and publication is drastically restructured by the emergence of readership roles. In principle, science is still characterized by universal inclusion. But there are ever new control levels added (for a comparative perspective on other systems (Power, 1997)). One needs funding, one's papers have to be accepted, for a career one needs calls to professional positions, one needs recommendations and reviews for fellowships and other stays at places relevant for research and publication, teaching reviews become a part of a university career and the curriculum one teaches has to be audited, the research institute that is the place of work needs regular evaluations. The university one works for wants to be excellent and is ranked. All this is structured by two hierarchies that are strongly linked: the hierarchy of authors, in which the individual scientist climbs on the basis of publications and citations, and finally gets access to the most influential positions and then becomes a professional reader of the publications of others and does no longer do this as a preparation for one's own publications. Instead one becomes ever more important in a hierarchy of readers (= evaluators, auditors) that is the highest level of control in organizing the system of science.

The Third and Fourth Hierarchy: Hierarchy of Publication Places and Hierarchy of Recruitment for Co-authorship

Over time, there are further hierarchies built into the scientific production and communication processes. As publications are the major products of the processes of research defining the core of science and as the citation of publications and the cumulative aspects of citations become the simplest and most basic reward for the cognitive achievements documented in publications, new hierarchies emerge around publication and the authorship of publications.

Besides the hierarchy of authors and the hierarchy of readers (reviewers, evaluators, auditors) nested into one another, there comes about a hierarchy of publication places (journals, publishing houses). It is no accident that this hierarchy is defined by levels and forms of peer review, by the probability of citations (impact factors) and by rejection rates.

The same self-referential intensification of hierarchy is to be observed in the fourth hierarchy establishing itself: the hierarchy in selecting and recruiting coauthors for publication. Scientists who search for coauthors are looking for other scientists who are identified by numerous publications in highly ranked journals and by a great number of citations they succeeded to cumulate over a publication career.

In this argument it is easily to be seen how the reciprocal intensification of the four hierarchies characteristic of the communication system of present-day science transforms science as a system based on universal inclusion into a social system with extreme inequalities.

Two Modalities of Quality Control in Science

Anticipatory, Centralized Control by Scientific Elites

Cumulative rewards for successful authors, their promotion to influential readers/reviewers who are installed as central agents of quality control in science, the intensification of these patterns by a steep hierarchy of ranked journals, and the recruitment of coauthors on the basis of advanced positions in the other three hierarchies—all these patterns create a remarkable system of quality control by scientific elites. A major property of this system of control is that it is 'anticipatory' control. Papers are rejected or printed before they have been examined by a significant number of members of the relevant scientific communities and projects are funded on the basis of prognoses regarding their probable scientific success. To believe in the rationality of these decisions demands a strong belief in the superior knowledge and wisdom of the scientific elites who practice this anticipatory control. It is a mode of control that is very conservative, as it concentrates control in the hand of elites whose individual members may have been active for decades and who may have a prejudice against innovation, newcomers, outsiders and heterodoxies.

Post-hoc, Decentralized Market Control Based on Universal Inclusion

There is one alternative control modality that is based on institutional alternatives that have already been practiced at some places. It substitutes post-hoc control of research and publications for anticipatory control by elites. This implies liberal standards for self-selected research (that is mostly done with basic funding available for everyone, a funding level that may be adapted on the basis of successes) and the publication of results on liberal publication platforms. Evaluation mostly happens after the research has been done and after the results have been published. But this post-hoc evaluation is entrusted to the decentralized expertise of diversified communities emerging on the basis of universal inclusion.

Concluding Remarks

It is probable that the two modalities of control will coexist in the foreseeable future of science. The first modality, 'anticipatory centralized control', is connected to stable hierarchies of established elite researchers who control the access to careers, research funds, co-authorship options and possibilities of publication in high-status journals. This is a very conservative model that may hinder scientific innovation.

The second modality of control is compatible with publication of unreviewed papers on platforms such as arXiv. Peer review may be 'open peer review' (Ross-Hellauer, 2017) after publication. Reviews will often be based on self-selection for reviewing and may be published together with the papers reviewed. The whole process of publishing, reviewing and revising papers on the basis of reviews becomes an open process visible to everyone and accessible (liberalization of publication, accessibility of reviewing) in a universal way. This modality, 'post-hoc, decentralized market control', has a higher compatibility with the self-professed universalism of modern science. Even under these circumstances, inequalities will arise (as differences in success between papers will always be considerable). But the hierarchies will be much less stable, as most forms of influential writing and reading (as a reviewer) will be available to everyone.

References

Adams, J., Pendlebury, D., Potter, R., & Szomszor, M. (2019). *Global Research Report: Multi-authorship and research analysis*. Institute for Scientific Information.

Ahlers, A. L., Krichewsky, D., Moser, E., & Stichweh, R. (2020). *Democratic and authoritarian political systems in 21st Century World Society. Vol. 1— Differentiation, inclusion, responsiveness*. Transcript.

Beaver, D. d B., & Rosen, R. (1978). Studies in scientific collaboration part I.: The professional origin of scientific co-authorship. *Scientometrics, 1*, 65–84.

Beaver, D. d B., & Rosen, R. (1979). Studies in scientific collaboration part II: Scientific co-authorship, research productivity and visibility in the French scientific elite, 1799–1830. *Scientometrics, 1*(2), 133–149.

Cole, J. R., & Cole, S. (1981). *Peer review in the National Science Foundation: Phase II of a study*. National Science Foundation.

Cole, S., Rubin, L., & Cole, J. R. (1978). *Peer review in the National Science Foundation. Phase one of a study.* National Science Foundation.

DiPrete, T. A., & Eirich, G. M. (2006). Cumulative advantage as a mechanism for inequality. *Annual Review of Sociology, 32,* 271–297.

Hund, F. (1990). Die "Annalen der Physik" im Wandel ihrer Aufgabe. *Physikalische Blätter, 46*(6), 172–175.

Ilardo, M. A., Moltke, I., Korneliussen, T. S., Cheng, J., Stern, A. J., Racimo, F., de Barros Damgaard, P., Sikora, M., Seguin-Orlando, A., Rasmussen, S., van den Munckhof, I. C. L., Ter Horst, R., Joosten, L. A. B., Netea, M. G., & Saling, S. (2018). Physiological and genetic adaptations to diving in sea Nomads. *Cell, 173,* 569–580.

Merton, R. K. (1988). The Matthew effect in science, II. Cumulative advantage and the symbolism of intellectual property. *ISIS, 79,* 606–623.

Moretti, F. (2013). *Distant reading.* Verso.

Newman, M. E. J. (2001). Clustering and preferential attachment in growing networks. *Physical Review E, 64,* 025102.

Parsons, T. (1937). *The structure of social action.* Free Press (of Glencoe).

Popper, K. R. (1963). *Conjectures and refutations. The growth of scientific knowledge.* Routledge and Kegan Paul.

Power, M. (1997). *The audit society: Rituals of verification.* Oxford University Press.

Price, D. J. D. S. (1986). *Little science, big science … and beyond.* Columbia University Press.

Rochmyaningsih, D. (2018). Study of 'sea nomads' under fire in Indonesia. *Science, 361,* 318–319.

Ross-Hellauer, T. (2017). What is open peer review? A systematic review. *F1000Research, 6,* 588. https://doi.org/10.12688/f1000research.1369.1

Squazzoni, F., et al. (2020). Unlock ways to share data on peer reviews. *Nature, 578,* 512–514.

Stichweh, R. (1984). *Zur Entstehung des modernen Systems wissenschaftlicher Disziplinen. Physik in Deutschland 1740–1890.* Suhrkamp.

Stichweh, R. (2016). *Inklusion und Exklusion. Studien zur Gesellschaftstheorie.* Transcript.

Stichweh, R. (2017). Interdisziplinarität und wissenschaftliche Bildung. In H. Kauhaus & N. Krause (Eds.), *Fundiert forschen. Wissenschaftliche Bildung für Promovierende und Postdocs* (pp. 181–190). Springer VS.

Stichweh, R. (2020a). Der Beitrag der Religion zur Entstehung einer funktional differenzierten Gesellschaft. In M. Pohlig & D. Pollack (Eds.), *Die*

Verwandlung des Heiligen: Die Geburt der Moderne aus dem Geist der Religion (pp. 173–187). Berlin University Press.

Stichweh, R. (2022). How Do Divided Societies Come About? In: Stichweh, R., *Functional Differentiation of Society.* Transcript: Bielefeld.

Van Groenigen, J. W., & Stoof, C. R.. (2020). Helicopter research in soil science: A discussion. *Geoderma* 373. https://doi.org/10.1016/j.geoderma.2020.114418

3

"'Disciplining' Educational Research in the Twentieth Century"

Raf Vanderstraeten

Introduction

The research programme in sociology of science, which Robert K. Merton began to envisage in the mid-twentieth century, focused on the normative structure of science. Echoing broader democratic concerns, Merton depicted the peer review system, developed for scientific journals, "as crucial for the effective development of science" (1973, p. 461). As an evaluation mechanism, it provided an "institutionalised form for the application of standards of scientific work" (1973, p. 469). Despite its many imperfections, "the structure of authority in science, in which the referee system occupies a central place, provides an institutional basis for the comparative reliability and cumulation of knowledge" (1973, p. 495).

The original version of the chapter has been revised. A correction to this chapter can be found at https://doi.org/10.1007/978-3-030-75263-7_17

R. Vanderstraeten (✉)
Ghent University, Ghent, Belgium

London School of Economics and Political Science, London, UK
e-mail: Raf.Vanderstraeten@UGent.be

© The Author(s) 2022, corrected publication 2023
E. Forsberg et al. (eds.), *Peer review in an Era of Evaluation*,
https://doi.org/10.1007/978-3-030-75263-7_3

In the view of Merton and his collaborators, the scientific system was largely self-organizing and self-policing, and the scientific literature with its peer review system was largely where that happened (see also Hollinger, 1990; Jacobs, 2002; Baldwin, 2015; Csiszar, 2018, pp. 1–21).

One crucial problem with this approach is that the "structure of authority in science" is presented as a "natural" feature of how the scientific system is supposed to operate. This approach builds upon the idea that the publication norms and practices have remained more or less constant throughout their existence—from the first scientific journals in the seventeenth century to its successors in the early twenty-first century. The idea, however, that there is essential stability from the first, early modern scientific journals to their contemporary counterparts has encouraged sociologists and historians of science to project back onto earlier epochs' contemporary sensibilities about what journals are for, and how scientific communities ought to operate. It has not encouraged them to analyse more closely how peer review has become a *sine qua non* of scholarly journals and publication practices, and how this evaluation mechanism has changed the scientific system itself (for discussions of the state of the art, see Hirschauer, 2004; Bornmann, 2011; Pontille & Torny, 2015).

In this chapter, an analysis is presented of relevant changes in publication and evaluation practices in one field of research, namely education, and more particularly in the journals published by the largest association in this field, namely the American Educational Research Association (AERA). Founded in 1916, this association was originally known as the National Association of Directors of Educational Research (NADER). Shortly after World War I, however, it opened active membership to anyone who displayed the ability to conduct research: "the criterion for inclusion became demonstrated competence as a researcher—and the primary indicator of that competence was written work … that the members of the policy-making Executive Committee could assess" (Mershon & Schlossman, 2008, p. 319). More inclusive names were adopted to reflect this shift: first Educational Research Association of America (ERAA), and shortly afterwards American Educational Research Association (AERA).

In the course of its history, this association has launched several scientific journals. In 1930, it started with the publication of the *Review of Educational Research* (RER). Although *RER* was AERA's only journal for about three decades, the association expanded rapidly in the course of the

1960s and 1970s. The *American Educational Research Journal* (AERJ) first appeared in 1964, and *Educational Researcher* (ER), emanating from AERA's member newsletter, was published in 1972. One year later, the annual *Review of Research in Education* (RRE) started to appear. Two other, more specialized journals came out in the latter half of the 1970s: the *Journal of Educational and Behavioral Statistics* (JEBS), in 1976, and *Educational Evaluation and Policy Analysis* (EEPA), in 1979. More recently, in 2015, the association also launched *AERA Open,* an open-access online journal. All of these journals rank among the most influential publication outlets in the field of education.

The following analyses, which build on previously published work (Vanderstraeten et al., 2016), make use of two types of material. On the one hand, quantitative material on all the articles published in RER and AERJ is presented. Because the coverage of the content of the older volumes of the AERA journals is often incomplete in the existing bibliographical databases, the data were hand-checked and cleaned with the help of the content pages of all journal issues themselves. On the other hand, all editorial documents and guidelines that have appeared in AERA journals were analysed. Despite the fact that I did not have access to the journals' archives, the editorial documents allow me to provide a sociological history of the evolution of the publication and evaluation practices in the field of education. Because the journals are used as source materials, their contents are hereafter cited by referring to the journal, publication year and page numbers. In order to avoid overburdening the reader, particular attention is paid to the publication and evaluation practices in AERA's oldest journals, *RER* and *AERJ*, although it is worth noting that the data gathered for the other AERA journals confirm the analyses based on these two (Vanderstraeten et al., 2016).

The focus of this chapter thus is on the changing publication and review practices in the AERA journals. The field of education research allows for an interesting case study, not only because it is perceived to be interdisciplinary oriented, with close ties to psychology, philosophy and sociology, but also because it generally does not enjoy high status, and therefore seems quite receptive to changes in other fields of research (Vanderstraeten, 2011; Jacobs, 2013, pp. 100–120). From this perspective, the discussion first focuses on changing expectations regarding editorship and authorship, as well as changing forms of authority and inclusion in authorial roles. Next, attention is paid to how publication pressures ("publish or perish") and evaluation mechanisms delimit what is valued in the scientific system, namely peer-reviewed papers. Afterwards, the focus is on changing citation cultures, and thus on the question of how authors are expected to incorporate and build on the arguments developed in other publications, which have gone through the process of peer review. In the more general reflections, with which this chapter concludes, I try to illustrate that my analyses not only shed light on the ways the scientific system organizes itself, but also help in imagining ways in which improvements can be made.

Reviewers and Authors

Initial Expectations

Of course, scientific journals have never been characterized by any truly unified format. Even during the last century, journals have varied widely in the nature of their contents, their size, frequency, and submission and acceptance procedures. Papers have varied not only in their length, from short notes or letters to more extended memoirs, but also in the genre expectations of diverse research fields (see Bazerman, 1988; Gross et al., 2002). Despite such variations, however, it is widely accepted that scientific journals constitute a special class of publications that can be demarcated from other forms of literature. Evaluation mechanisms, based on peer review, are often understood to protect the integrity of this corpus. These evaluation mechanisms can also be seen to separate a small body of legitimate scholarly work from other, unscientific enterprises. Editors or experts called on to judge whether a paper ought to be published are imagined as doing their duty not only to a journal's reputation and prestige, but to science as a whole (e.g., Merton, 1973).

The evolution of the AERA journals shows, however, that the review mechanisms also display much historical variation. The ways in which editors and reviewers are able to understand or define their own role have changed quite considerably. How we conceive of authority in the system of science is the outcome of a series of attribution and evaluation processes. How editors and reviewers position themselves and their journals, and how authors can take both credit and responsibility for particular publication output, also is the result of a series of historically contingent choices. At the same time, the perceived scientific eminence of the editors and the authors, as well as the representation of the various interest groups to which the journals intend to direct themselves, also seems instrumental in establishing and maintaining the authority of the journals and their association.

Overall, RER was in its first decades not what we would now call a "traditional" journal: it did not publish original research papers. It was rather conceived as a periodical reference work, regularly summarizing recent research on "the whole field" of education (RER, 1931, p. 2). It was to appear five times per year, with each issue devoted to a specific topic. In the first issue, the editors presented a cycle of 15 topics to be addressed over a three-year period. RER's first volumes dealt with topics such as the curriculum, teacher personnel, school organization, finances, intelligence and aptitude tests, and so on. The last topic of the first cycle was "methods and technics of educational research." For each issue (and thus for each topic), the idea also was to assign an issue editor and a committee of experts, who were to solicit and review all manuscripts. As it turned out, these designated editors and experts would frequently author several review articles themselves.

The original aim of the journal was to disseminate the results of scientific research to a broader audience: "to review earlier studies" and "to summarize the literature" for an audience of "teachers, administrators, and general students of education" (RER, 1931, p. 2). But this editorial strategy was characterized by a hierarchical structure. It is quite clear that authority and authorship were closely connected: Issue editors and authors were chosen because of their authority on the topics, but inclusion in RER also granted the issue editors and authors considerable authority.

Interestingly, some authorship problems appeared. Authorship was held to be exclusive; it was not easily extended beyond a small group of specialists. Co-authorship, in the strict sense of two names listed alongside one another at the front of a text, was not self-evident. Several authors of early RER articles were aided by "assistants." Sometimes authors published "in cooperation with" others—but neither the assistants nor the "cooperating" contributors were identified as full co-authors. In 1935 and 1936, moreover, errata had to be published to add co-authors to reviews that had appeared in print in previous issues (see Excerpts 1+2). Although the inclusion of these errata illustrates that the attribution of authorship could be contested (no other errata appeared in the early volumes), RER did, in the first decades of its existence, entrust only a few scholars with reviewing the relevant research. The journal entrusted and *authorized* only a few scholars to summarize and review what was considered to be the relevant research and hence to speak to the broader community of people interested in education and the results of education

---

Excerpts 1+2: Authorship corrections in early RER issues

Erratum

Owing to an unfortunate oversight, the name of Richard Wilkinson as co-author of the chapter entitled "Recent Developments in the Written Essay Examination" was omitted from the manuscript of the *Review of Educational Research,* Volume V, Number 5, December, 1935.—W. J. Osburn.

Erratum

Through an unfortunate circumstance, the following names were omitted from the manuscript of the chapters entitled "School Marks" and "Recording and Reporting" in the *Review of Educational Research,* Volume VI, No. 2, April, 1936:

- William A. McCall, *Teachers College, Columbia University, New York, New York*
- Florence Mapes, *Paterson, New Jersey*
- B. Duke Small, *Board of Education, Atlanta, Georgia*
- Adelaide P. Bostick, *New York, New York*
- Dorothea Walsh, *New York, New York*
- Freda Smulker, *Long Island City, New York.*

—Arch O. Heck, *Chairman,
Committee on Pupil Personnel,
Guidance, and Counseling.*

research. The editors and experts appointed by the journal often filled the pages of the journal with their own contributions.

Further Expansion

For almost four decades, the editors of RER stayed close to their ambition to treat "the whole field" by means of a cyclical coverage of all important topics in education. Already in the 1930s, however, questions emerged as to the proper readership of RER. The interests of education practitioners, on the one hand, and education researchers, on the other, proved difficult to align. In 1938 and 1939, for example, the editorial board adopted five new topics to be covered in three-year cycles. In an editorial foreword, it was underlined that the new strategy would allow focussing on instruction and therefore be of benefit to practitioners in schools instead of to researchers in universities. As no scholars specialize in such instructional areas, "they are much more difficult to prepare," but, as the editors added, "it is hoped that they will render a larger service to a greater number of users and thus justify the increased effort that they call for" (RER, 1940, p. 75).[1] In the following decades, however, AERA would increasingly orient itself to the growing and influential community of education researchers instead of to education practitioners.

Prompted by the rapid expansion of education research, especially in the decades after World War II, RER adopted, beginning in 1970, a new editorial policy in which each issue was expected to include unsolicited reviews on topics of the authors' choice. The incoming editor, Gene V Glass, stated "the new editorial policy" as follows: "The purpose of the *Review* has always been the publication of critical, integrative reviews of published education research. *In the opinion of the Editorial Board, this goal can now best be achieved by pursuing a policy of publishing unsolicited reviews of research on topics of the contributor's choosing* ... The reorganization of the *Review of Educational Research* is an acknowledgment of a

---

[1] At the same time, more emphasis was put on research methods to help researchers cope with a proliferation of both quantitative and qualitative techniques (e.g., RER, 1939, p. 451, 1956, pp. 323–343). Clearly, some inconsistencies were part of the editorial strategies of the AERA journals.

need for an outlet for reviews of research that are initiated by individual researchers and shaped by the rapidly evolving interests of these scholars" (RER, 1970, p. 323). The last issue that reflected the old editorial policy appeared in 1971.

At that time, the landscape of scholarly publishing in the field of education had already changed. In 1964, AERA began publishing AERJ, with a mission to publish "original reports of experimental and theoretical studies in education." In the rapidly expanding field of scientific journals, AERJ was a "traditional" journal that put emphasis on the presentation of novel findings. Its establishment was an indication of the fact that AERA aspired to a more active, innovative role at the level of scholarly communication about education (see AERJ, 1966, pp. 211–221, 1968, pp. 687–700). In the same period of time, moreover, the RER editors put forward their new expectations regarding the content and orientation of articles and submissions. RER shifted its emphasis from summaries or reviews to critical evaluations; it now explicitly required its authors to provide an overview of the strengths and shortcomings of the existing knowledge base. Articles now had to advance research on the topics they discussed. Glass wrote: "It is hoped that the new editorial policy of the Review, with its implicit invitation to all scholars, will contribute to the improvement and growth of disciplined inquiry on education" (RER, 1970, p. 324). No doubt, these new expectations corresponded with changes in the composition of AERA's membership and RER's readership base. Its readership came to consist mainly of specialists, who did not need a "review" to learn about developments in their field of research. The *raison d'être* of RER—as well as of the other AERA journals that were established in the 1960s and 1970s—now lay in the presentation of findings that were relevant primarily to other researchers. Seen in this light, the new editorial policy expressed by RER *disqualified most of the journal's own early educational publications as either unoriginal or not properly scientific.*

In the same editorial, Glass also indicated that "the role played by the *Review* in the past [would] be assumed by an *Annual Review of Research in Education*, which AERA [was] planning" (RER, 1970, p. 323). The first volume of the *Review of Research in Education* appeared only three years later. RRE again solicited reviews in particular research areas. In this regard, the "Statement From the Editor" accompanying the first issue of the *Review of Research in Education* was reminiscent of the old editorial policy of RER: "The more important areas will appear periodically but not necessarily regularly. Some areas, relatively dormant or unproductive, may not appear for years" (RRE, 1973, p. vii; see also ER, 1976/11, p. 10). However, the RRE editor also took pains to underline that the new venue would orient itself towards scholars, who would read it to inform themselves about ongoing education research. "Summaries of research studies are valuable and appropriate, but too much summary distracts from criticism and perspective" (RRE, 1973, p. vii). And the RRE editor added: "Many conceive of reviewing as the summarizing of research studies and trends in order to inform readers and keep them abreast of their fields. Such an annotated bibliographic approach can have little impact, however" (RRE, 1973, p. vii). Although it thus proved difficult to give up the idea that the research field could be authoritatively surveyed by a few leading scholars, the expectations regarding the role of editors and reviewers changed around 1970. Instead of filling the pages of the journal with their own contributions, the editors and reviewers became increasingly engaged as *gatekeepers* of scientific communication channels (Vanderstraeten, 2010, 2011).

Authors and Reviewers

Community of Peers

The expression "publish or perish," which became widely used in the 1960s and 1970s, can be seen to signal the institutionalization of a "communication imperative" in science (see also de Solla Price, 1963). Publications have not only become increasingly perceived as indices of full membership in the scientific community, but peer-reviewed papers

---

[2] Within AERA, the differential value attributed to peer-reviewed journal papers also became evident. While presentations at annual meetings were valued, more value was attached to what could, after peer review, be published in the AERA journals. "[In the 1950s] … members who proposed a paper for the program were generally assured that it would be accepted" (ER, 1982, p. 9).

have also become a base unit for sizing up careers, with publication lists a significant factor in decisions about hiring, tenure and grants.[2] In the process, changing expectations emerged for journal editors, reviewers and authors.

Underlying this evolution were important demographic changes within the academic system. As already mentioned, the field of education research was a clear beneficiary of the expansion of the American system of higher education in the 1950s and 1960s. In his presidential address presented at the AERA 1966 Annual Meeting, which was published in the first issue of AERA's new journal, AERJ, the educational psychologist Benjamin Bloom provided a short overview of this rapid expansion. "From the level of support of 1960," Bloom estimated, the growth in federal funding of education research and development had been "of the order of 2,000 per cent" (AERJ, 1966, p. 211). The number of education researchers had also increased substantially during that period; Bloom noted that in the previous five years, membership in AERA had grown "at the rate of about 25 per cent per year" (AERJ, 1966, p. 213). The growing number of journals devoted to education was another factor in (and indicator of) the expansion and "academization" of this field. If the 1960s constituted a "Renaissance" in education research, the expansion and ensuing professionalization of research drove the "amateurs" out of the association (ER, 1982/9, pp. 7–10). As a result of the growth of the scholarly community, researchers had to direct their communications to other researchers instead of to "those off campus" (see AERJ, 1973, pp. 173–177; RER, 1999, pp. 384–396). New forms of competition and/or collaboration between potential authors also emerged.

To clarify the extent of these changes, it is interesting to point to developments at the level of the authorial roles. Figure 3.1 displays the evolution of the number of authors or co-authors per published article in RER and AERJ. It is clear that single-authored articles were the norm for a relatively long time. In 1931, all but two RER articles were single-authored (although "assistants" contributed to four of these articles). Forty years later, the majority of the articles in RER were still written by single authors. But the expectations and conventions quickly changed after that. In the case of *RER*, which adopted a new editorial policy in the 1970s, the average number of authors per article increased from 1.05 in

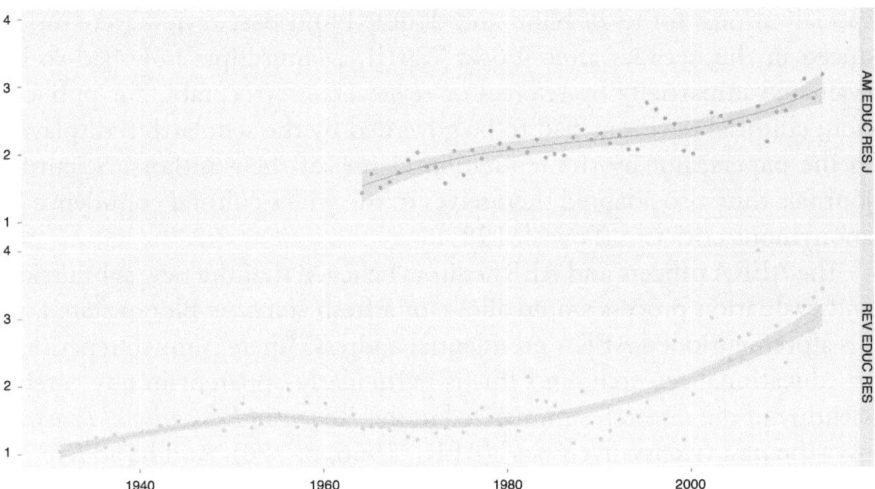

Fig. 3.1   Average number of authors per article for AERJ and RER

1931 to 1.21 in 1970 and 3.61 in 2018 (with a standard deviation of 2.75). In the case of AERJ, there was a relatively steady increase in the number of co-authored articles; the average changed from 1.42 in 1965 to 2.30 in 1990 to 2.66 in 2018 (with a standard deviation of 2.03). In 2018, only about 1 in 6 RER and 1 in 5.5 AERJ articles were single-authored. Co-authored, if not multiple-authored, publications have become the norm. For sure, the rise of "big science" has influenced this evolution (de Solla Price, 1963). But the rise of co-authored publications also implies that forms of peer review become incorporated into the publications themselves. More and more peers now are (co-)authors, involved in the production—and not just the evaluation—of papers. For many scholars, collaboration with peers has become part of their research and publication strategies.

Blind Peer Review

It is also interesting to direct attention to the new evaluation mechanisms that were expected to replace the former system of invited submissions. Not just in the field of education, but in a broad variety of scientific

specializations, forms of blind and double-blind peer review were introduced in the decades after World War II. Manuscripts now had to be
evaluated impartially by referees or reviewers as acceptable for publication; editorial decisions had to be governed by the scholarship displayed
in the papers, not by the reputation claims of their authors. Scientific
journals thus also adapted themselves to the wider cultural confidence in
anonymous criticism (Powell, 1985).

The AERA officers and AERJ editors believed that the new submission
and evaluation process would allow for a fresh start. As Bloom stated, in
his aforementioned AERA presidential address, "there is much repetition
in educational research, and this is particularly apparent in any careful
scrutiny of the research summarized in the *Review of Educational Research*
over the past twenty-five years" (AERJ, 1966, p. 220). And he added: "It
is this redundancy that in part explains why there are so few examples of
crucial research in the period under consideration" (ibid.). But Bloom
also believed that major improvements could be realized in education
research, provided that some structural changes were implemented. In
part, his plea reminds of free market ideologies. The rapid communication of research findings had to be facilitated; journals had to focus on the
publication of new, innovative findings, instead of on summarizing existing research. Priority had to be given to submissions based on the initia-
tive of individual researchers, but some form of invisible hand (peer
review) was thought to be necessary. In this way, the system would benefit
the entire scientific community.

In order to maintain authority and trust in the field, the journal editors
were also forced to take a distanced stance on all decisions that could be
perceived as injurious to others (such as rejections of individual contributions). To maintain authority, they could not be perceived as exercising it
(see also Pontille & Torny, 2015). They rather assigned editorial responsibility to others. In 1973, the AERJ editors appointed two "Reviewers-
at-Large ... [to] serve as a regulatory agent over the editorial process"
(AERJ, 1973, p. 174). At the same time, they promised to protect the
diversity of the publication output. They strived for a "corporate identity"
that could represent the field as a whole, and rely on the expertise available in the field as a whole. Although "the basic mechanism for maintenance of high standards remains to be a peer review carried out

anonymously and in good professional taste," they promised to call upon "all the expertise in the AERA" (ibid.). A longer citation may illustrate the editors' prevailing concerns: "In the past, it has been customary for the editors to appoint a board of consulting editors and, after screening out some manuscripts for policy and load reasons, to refer all the rest to the board for review and recommendation. For instance, the 1971–1972 board had 35 members on it, many providing their precious service for six long years ... (Incidentally, and unfortunately, only one of the 35 was female.) This system of a fixed body of readers works well in a monolithic professional organization which, alas, the AERA is not. Though unintentionally, it is easy for this sort of board to become homogeneous in composition, narrow in focus, and dogmatic in judgment. To avoid these dangers and to allow readers a greater share of responsibility for *their* magazine, we have done away with the arrangement and, instead, decided to rely upon a large number of consultants selected from the general AERA membership and, if deemed necessary or desirable, even from outside the organization" (AERJ, 1973, p. 174–175).

By stressing the decisive role of the assessments of the various expert reviewers, the AERJ editors also tried to respond to "some irate colleagues" (AERJ, 1973, p. 176). The editors of all AERA journals, they stated in their somewhat unconventional "Message From the Editors," do "*not* meet or work as a group, even though all are doing what they can to contribute to the production of fine, worthwhile publications. They certainly do *not* 'conspire' for or against any authors, subjects, or types of study" (ibid.). Moreover, "frequent phone calls or letters to the editorial office do *not* facilitate the review process. Once a manuscript has been sent out to consultants, editors do not have any further information until the reviews and recommendations are back" (ibid.). They added, moreover, that "the editors are *not* monsters with sinister motives, out to get this author or insult that scholar ... [They make mistakes but] they are not so bad as to justify unbridled invectives and tirades on the part of some of our fellow educational researchers" (AERJ, 1973, p. 177). In short, the development of the discipline required discipline of all its members. The new evaluation mechanisms built on the institutionalization of different judging instances, but also required some difficult

> Excerpt 3: A list of "anonymous" referees and editorial consultants included in RER[3]

Acknowledgment of Editorial Consultants

The *Review of Educational Research* is made possible in large part by the generosity of the following individuals who served during the period January 1, 1970 to August 1, 1970 as editorial consultants on the publication of submitted manuscripts.

Marvin C. Alkin *Univ. of California, Los Angeles*	Robert M. Gagné *Florida State Univ.*
Richard C. Anderson *Univ. of Illinois*	John I. Goodlad *Univ. of California, Los Angeles*
Alexander W. Astin *American Council on Education*	William L. Goodwin *Univ. of Colorado*
Daniel E. Bailey *Univ. of Colorado*	Arden D. Grotlueschen *Univ. of Illinois*
Frank B. Baker *Univ. of Wisconsin*	A. Ralph Hakstian *Univ. of Alberta*
Albert Bandura *Stanford Univ.*	Dale B. Harris *Pennsylvania State Univ.*

socialization processes on the part of editors, reviewers, (would-be) authors and readers.

Involvement in peer review could be legitimated in terms of membership of the research community. By involving an increasing number of education researchers in the role of referee, the editors could hope for a better understanding of the complexities of the decision-making processes they were involved in. By adopting various role perspectives, especially those of author and of referee, and thus quite literally taking the part of the other, researchers could be expected to understand and accept the expectations of the other. By taking up the role of referee, they could learn to meet the demands of referees and editors. Being asked to act as referee thus could also be seen to constitute a privilege that would bring

---

[3] Of course, the annual publication of lists of consulted referees is also a way to give credit to the scholars on whose expertise the editors relied. Databases, such as Publons, now also allow reviewers to get credit for work that would otherwise remain invisible. On the other hand, a small but growing group of periodicals have turned to open peer review, to give—among other things—recognition to the efforts of their reviewers.

its own rewards. This psycho-social integration into the entire process of scientific communication could be presented as acting as accumulation of advantage that accrues to scholars, who are perceived to be successful in their field of expertise, just as much as the more tangible advantages of research grants and large labs (Merton, 1973, pp. 439–459; Bazerman, 1988, p. 146). Full membership of the scientific community seemed to involve individuals in the roles of author and referee, but its ideological roots of this line of thinking are obvious. The structures of an "audit culture" became gradually visible (Power, 1997).

Another remark may be added. While the invention of new editorial positions to handle issues of general policy and of referees to handle issues concerning individual contributions may have helped the editors and their journals maintain authority and trust, the underlying concerns also led to a somewhat paradoxical strategy. The AERA journals, like many other scientific journals, started to publish—mostly annual—lists of scholars who served as referees. Displaying the identity of their (anonymous) reviewers seems necessary to enhance the journals' prestige in the field. In the light of the institutionalization of more complicated procedures of double-blind peer review, the journals obviously can no longer only build on the visibility and scientific eminence of their editors.

Papers and References

Suggestions for Contributors

As already mentioned, the shifting editorial strategies had an impact on the publication formats of the journals. It has been suggested that the introduction of double-blind peer review has gone along with the standardization of publication output (Bazerman, 1988; Grafton, 1997; Gross et al., 2002). Standardization of publication formats can also be observed in the AERA journals in the course of the 1970s. Shortly before the introduction of RER's new editorial policy, for example, broad editorial guidelines were communicated: "There are no restrictions on the size of the manuscripts nor on the topics reviewed" (e.g., RER, 1969, inside cover). One decade later, much more detailed instructions were common in all AERA journals. Not only were strict page limitations introduced, but prospective authors were also referred to the publication manual of the American Psychological Association, which included (and includes) detailed guidelines on manuscript structure and content, writing styles, referencing methods and so forth. Manuscripts now also needed to be accompanied by an abstract of 100–150 words. To enable blind review, the list of authors had to be typed on a separate sheet (e.g., RER, 1980, p. 201; AERJ, 1980, pp. 1, 125). As more emphasis was placed on individual scholarship, and as more scholars were pushed to submit manuscripts to peer-reviewed journals, the publication formats also became increasingly regulated and predefined. To make fair comparisons of the scientific quality of different manuscripts possible, and thus to enable fair editorial decisions about acceptance or rejection, standardization seemed imperative.

More detailed "suggestions for contributors," which pertained to the content and orientation of the articles that could be considered for publication, were also put forward.[4] At the time that RER shifted its emphasis from summaries or reviews to critical evaluations, it started to require that all submitted manuscripts would provide an overview of the strengths and shortcomings of the existing knowledge base. Articles now had to advance research on the topics they discussed; would-be authors had to display familiarity with the existing body of specialized knowledge and present their own work as a new, innovative contribution to this body (see also ER, 2006/6, pp. 33–40). It should thereby be taken into account that, in the case of RER, individual articles now often had to be placed in an issue without any substantive relation to the topics being discussed in the other articles of the same issue.

Citation Consciousness

Following the shift of attention towards the published paper, as the accredited product of research, it has increasingly become expected that

---

[4] A related discussion concerns the rapid diffusion of the IMRAD (Introduction, Methods, Results, Analysis and Discussion) structure for scientific papers. In the health sciences, it became in the 1980s the only pattern adopted in original papers. In education research, more diversity remained possible, although standards for reporting the findings of empirical research were also imposed (see ER, 2006/6, pp. 33–40).

papers build upon, and refer to, other publications. They are expected to build upon the authority of other publications, of publications which have gone through double-blind peer review themselves. At the same time, they are expected to invite responses, that is, become cited, and thereby further advance research (Stichweh, 2001). The readership of the journals at present predominantly consists of potential authors of new journal papers. The focus on the paper hence supports the image of science as a cumulative endeavour; it also supports the image of a self-regulating social system with the scientific literature and its gatekeepers at its core.

In all AERA journals, the reference lists have over time gained much weight. As Fig. 3.2 shows, there was a significant rise in the number of references per article over the last five decades. (For the articles published before 1956 no citation data have been collected by the *Web of Science* [WoS].) For AERJ, the average number of references per article per year

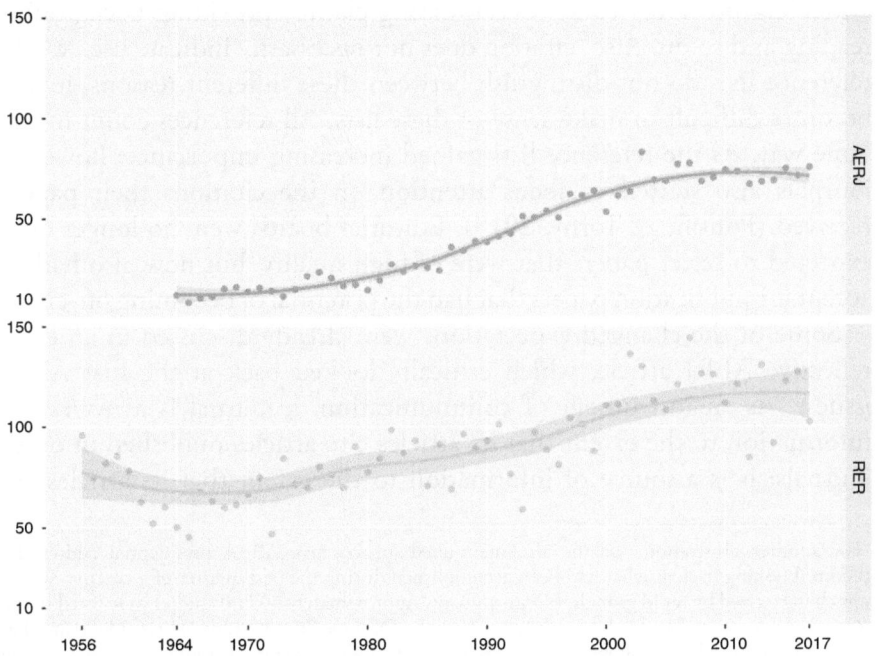

Fig. 3.2 Average number of references per article for AERJ and RER

multiplied by a factor of 7.5 in a period of half-a-century, from an average of about 10 references during the mid-1960s and early 1970s to an average of 75 references in the most recent years. Most of the increase took place between the 1980s and the 2010s, thus in 30 years' time. For RER, it should come as no surprise that the historical change is somewhat different, as this journal was traditionally focused on summarizing and reviewing a broad body of literature. In the course of the last 50 years, however, the average number of references per article doubled within this journal. There is more variation in RER than in AERJ, but RER articles now list on average some 120 publications in their reference sections.

In a broad sense, a "citation consciousness" is thought to be an essential part of good scholarly practice; scholars have to build on and refer to the scholarly work that is relevant to their topic. By citing particular work, they add their voices to already-published papers and to the journals which validated these papers. But, of course, authors can use citations for many different reasons: giving credit to related publications, criticizing previous work, substantiating claims, providing background reading and so on. Also, citation does not necessarily indicate use. While reference lists do not distinguish between these different reasons, it also becomes difficult to make sense of these lists. All references count in the same way. As the reference lists gained increasing importance, however, journals also started to focus attention on the citations their papers received (Pontille & Torny, 2015). Editorial boards were no longer only expected to select papers that were of high quality, but now also had to accept for publication papers that had the potential of becoming oft-cited.

Some of the changed expectations were already discussed in an early reflective AERJ article, which critically looked back at the first AERJ issues: "As an instrument of communication, a journal is a *receiver* of information to the extent that its articles cite articles published in other journals; it is a source of information to the extent that its articles are

---

[5] For another illustration, see the aforementioned, provocative AERA presidential address by Bloom. Looking back at what had been accomplished during the past quarter of a century, which was characterized by rapid growth, Bloom argued: "Approximately 70,000 studies were listed in the *Review of Educational Research* over the past 25 years. Of these 70,000 studies, I regard about 70 as being crucial for all that follows. That is, about 1 out of 1,000 reported studies seem to me to be crucial and significant, approximately 3 studies per year" (AERJ, 1964, p. 218). He thus also questioned the review practices that had prevailed within RER.

cited as bibliographical references in other journals. Assuming that a journal should serve more than an archival function, the latter is the more important index of a journal's impact" (AERJ, 1968, p. 694). Already in the 1960s, the journals were prompted to reflect on the impact they could have on education and education research.[5]

In the same period of time, Eugene Garfield had already started to market science citation indexes and impact factors with his *Institute for Scientific Information* (see also Garfield, 2004). The *Web of Science* impact factors indicate that the AERA journals occupy central positions within the field of education research (Vanderstraeten et al., 2016). However, the databases of the *Web of Science* also show that the relatively high impact factors of the AERA journals are the result of the visibility of a small number of papers. In the case of AERJ, a few papers are now highly referenced (nine are cited more than 500 times, two more than 1000 times), but half of the referenced AERJ papers are cited ten times or less, and one-third are cited three times or less. In the case of RER, seven papers are now cited more than 1000 times, but two-third of the referenced RER papers are cited ten times or less, while more than half of them are cited three times or less. For all AERA journals (but most pronounced for RER), there is a major gap between a tiny core of highly cited papers and the vast majority of the other work, which is barely referenced at all. The overall impact of these journals is very much dependent on the visibility of a few papers.

The highly skewed visibility of the RER and AERJ papers displays that the ways in which referees and editors evaluate submissions strongly differ from the ways in which published papers are referenced or valued in other publications.[6] Merton was aware of such divergences, but they did not lead him question his faith in the peer review system of scientific journals. He rather seemed to believe that improvements in editorial decision-making procedures could bring both forms of evaluation in line with one another (1973, p. 476, note 18). However, given the fact that

---

[6] The Gini indexes of the distributions for both journals are quite similar (>0.90). A comparison of the results with the indexes for wealth distributions within countries (for which the Gini index is commonly used) is telling. The distribution of citations to the AERA journals is worse than the figure for the distribution of wealth in the most unequal countries in the world, like Haiti, South Africa or Botswana (±0.65).

journal rankings and impact factors have become incorporated into the everyday decision-making routines of (would-be) authors, editors and science administrators alike, it no longer makes sense to conceive of peer review as the epitome of legitimate scientific assessment (see Sïle & Vanderstraeten, 2019). Other mechanisms are now increasingly used to value and measure the products of scientific work. Neither theoretically nor historically, there are good reasons to attribute privileged status to the system of double-blind peer review. In the course of the past century, the structural units of science have been more fluid than they might seem. We may therefore also question whether we still need to depict science as a self-regulating social system with the scientific literature and its gate-keepers at its core.

Conclusion

As we have seen, RER was initially conceived of as a journal that had to compile and review research findings, which in most cases were not readily available to its subscribers. The role which this journal initially fulfilled was mainly one of critically reporting on developments in the field of education and education science. For the editors, to publish in their own journal or their special issue was no abuse of privilege. Setting the tone by including work of their own and thereby making public judgements about the work of others was rather understood as the prerogative and even the duty of the editors. In the early decades of RER, the boundaries between the roles of author, reviewer and editor were blurry. The author was a reviewer, while the editor or reviewer also was an author! (Moreover, some editors/authors believed that they could withhold authorship credit from collaborators.)

With the expansion and increasing specialization of education research, structural changes in the publication process took place. The *raison d'être* of RER shifted: from summarizing existing research to presenting new, original findings. Like other scientific journals (including AERJ), RER came to rely on unsolicited papers. As a consequence, the distinction between different roles, especially roles for reviewers and for authors, became more pronounced. The journals' editors and reviewers became gatekeepers. As the journals started to rely on (double-blind) peer review as the way to organize and legitimate the editorial selection process, the article itself also acquired increasing importance. The prestige of these evaluation structures delimited the type of output that is valued in the scientific system. Since the latter part of the last century, the peer-reviewed paper has become widely used to identify who counts as a legitimate scientific practitioner and as a qualified expert in particular fields of research (Csiszar, 2018). In education research, as in other fields of research, individual careers have become dependent on publication lists, on authorship of peer-reviewed articles.

The focus on journals, journal articles and double-blind peer review has led to new conceptions of science. More democratic ideas about scientific communities have gained acceptance. Following the expansion and rapid specialization of different fields of research, small elite networks could no longer be entrusted with assessing the claims made in unsolicited submissions. Authorship and authority became more widely distributed; peers have become expected to monitor the flow of ideas within the scientific literature. Likewise, publications have to incorporate cumulative ideals about science. Authors have to highlight their reliance on other authors through citations and references. As our analysis of the education journals shows, the ascent of an "imagined" community of expert peers was the result of changes which took place in the course of the twentieth century. It was not a relatively stable social structure that made it possible to govern scientific activities.

The credit that comes from publishing papers in peer-reviewed journals has privileged certain kinds of scientific activity. The value ascribed to publishing research papers, and the general expectations about the format that those papers ought to take, has a strong influence on the types of projects scholars choose to pursue, the modes of collaboration that they are apt to engage in and the kinds of knowledge that make it into print. Among the ironic consequences of this focus on journals is the legitimation of short, standardized articles as equal to or even preferable to longer texts and books. Short articles, especially when they are stripped of materials that present the broader context but instead focus on standardized presentations of research results, are typically of use only to the most informed inner circle of experts (Johns, 1998). In this sense, the structure of science has contributed to a specialized notion of science that set it apart from other forms of knowledge exchange. The focus on peer review certainly contributed to securing and strengthening this orientation.

The way we nowadays conceive of scientific exchange—with authors, editors, reviewers, papers, reference lists and so on—is the outcome of a series of historical contingencies. The formation and institutionalization of these basic units made it relatively easy to speak of a self-organizing and self-regulating system of scientific research. If this mode of self-organization provides the basis for the reliability and the cumulative structure of science, as, for example, R. K. Merton put it, it does not seem advisable to call any of its basic units into question. Analyses of the historical contingencies of peer review, however, make it possible to shed light on the very social structure that made modern science possible. As we have seen, the social structure of science has been much less stable than it still seems to be. When the historical contingencies underlying this structure are taken into account, it should not be too difficult to imagine alternatives (see Vanderstraeten, 2019).

References

Baldwin, M. (2015). *Making "Nature": The history of a scientific journal.* University of Chicago Press.

Bazerman, C. (1988). *Shaping written knowledge: The genre and activity of the experimental article in science.* University of Wisconsin Press.

Bornmann, L. (2011). Scientific peer review. *Annual Review of Information Science and Technology, 45*(1), 197–245.

Csiszar, A. (2018). *The scientific journal: Authorship and the politics of knowledge in the nineteenth century.* University of Chicago Press.

de Solla Price, D. J. (1963). *Little science, big science.* Columbia University Press.

Garfield, E. (2004). The intended consequences of Robert K. Merton. *Scientometrics, 60*(1), 51–61.

Grafton, A. (1997). *The footnote: A curious history.* Harvard University Press.

Gross, A. G., Harmon, J. E., & Reidy, M. (2002). *Communicating science: The scientific article from the 17th century to the present.* Oxford University Press.

Hirschauer, S. (2004). Peer Review Verfahren auf dem Prüfstand. *Zeitschrift für Soziologie, 33*(1), 62–83.

Hollinger, D. A. (1990). Free enterprise and free inquiry: The emergence of laissez-faire communitarianism in the ideology of science in the United States. *New Literary History, 21*(4), 897–919.

Jacobs, J. A. (2013). *In defense of disciplines: Interdisciplinarity and specialization in the research university.* University of Chicago Press.

Jacobs, S. (2002). The genesis of 'scientific community'. *Social Epistemology, 16*(2), 157–168.

Johns, A. (1998). *The nature of the book: Print and knowledge in the making.* University of Chicago Press.

Mershon, S., & Schlossman, S. (2008). Education, science, and the politics of knowledge: The American Educational Research Association, 1915–1940. *American Journal of Education, 114*(3), 307–340.

Merton, R. K. (1973). *The sociology of science: Theoretical and empirical investigations.* University of Chicago Press.

Pontille, D., & Torny, D. (2015). From manuscript evaluation to article valuation: The changing technologies of journal peer review. *Human Studies, 38*(1), 57–79.

Powell, W. W. (1985). *Getting into print: The decision-making process in scholarly publishing.* University of Chicago Press.

Power, M. (1997). *The audit society: Rituals of verification.* Oxford University Press.

Sïle, L., & Vanderstraeten, R. (2019). Measuring changes in publication patterns in a context of performance-based research funding systems: The case of educational research in the University of Gothenburg (2005–2014). *Scientometrics, 118*(1), 71–91.

Stichweh, R. (2001). History of scientific disciplines. In N. J. Smelser & P. B. Baltes (Eds.), *International encyclopedia of the social and behavioral sciences* (Vol. 20, pp. 13727–13731). Elsevier.

Vanderstraeten, R. (2010). Scientific communication: Sociology journals and publication practices. *Sociology, 44*(3), 559–576.

Vanderstraeten, R. (2011). Scholarly communication in education journals. *Social Science History, 35*(1), 109–130.

Vanderstraeten, R. (2019). Systems everywhere? *Systems Research & Behavioral Science, 36*(3), 255–262.

Vanderstraeten, R., Vandermoere, F., & Hermans, M. (2016). Scholarly communication in AERA journals, 1931 to 2014. *Review of Research in Education, 40*, 38–61.

Part II

Peer Review and the Higher Education Evaluation Machinery

4

Gatekeepers on Campus: Peer Review in Quality Assurance of Higher Education Institutions

Don F. Westerheijden

Which Gates Do Peers Keep?

An internet search of 'gatekeepers on campus' leads on the first page to the guards of the physical gates on campus, and also to the university[1] services to prevent students from committing suicide. The gatekeepers in this chapter are concerned with less physical gates, and although they may themselves lead to stress and anxiety on campus, it is not often dramatic, though I will show that the anxiety has pernicious consequences. I mean the members of external evaluation committees that form the core of practically every quality assurance and accreditation procedure, and the gates are those of academic existence, because the characteristic outcome of accreditation is a decision to accept a university as legitimate (Adelman,

---

[1] For brevity, I tend to use 'university' or 'institution' rather than 'higher education institution', even though I mean all types of higher education and research institutions.

D. F. Westerheijden (✉)
University of Twente, Enschede, the Netherlands
e-mail: d.f.westerheijden@utwente.nl

© The Author(s) 2022
E. Forsberg et al. (eds.), *Peer review in an Era of Evaluation*,
https://doi.org/10.1007/978-3-030-75263-7_4

1992; Schwarz & Westerheijden, 2004; Sursock, 2001; Westerheijden, 2001)—whether it is a traditional campus with gates, or an open university that teaches online. In quality assurance without accreditation, a positive judgement by external evaluators may not decide about the survival of the institution, but still guards the gates of academic repute.

'Gatekeepers of science' has been a term applied to the persons who perform peer review, that is, who control access to journal space for publications, and to money through competitive research grants. 'Gatekeepers of science' is found in article titles since the late 1960s (Crane, 1967), while some years earlier the term appeared already in the text of an article on 'The Reception System of Science' (de Grazia, 1963). Peer review had been transposed from early-medieval legal contexts ('the lawful judgment of his peers') to the emerging world of science in the seventeenth and eighteenth centuries to safeguard validity of submissions for publication (Benos et al., 2007), although 'prior to the Second World War the process was often quite uncodified, and editors frequently made all decisions themselves with only informal advice from colleagues. Only quite recently has the paradigmatic "editor plus two referees" system become widespread' (Rowland, 2002, p. 248). Peer review has also been used for many decades in decisions about awarding competitive research grants (Marsh et al., 2008). Peer review was transposed to a new context again when quality assurance appeared in the 1980s as a policy tool to safeguard the quality of higher education institutions and their study programmes. In this chapter, issues of the latter transposing of peer review will be addressed: first, what is 'quality' of universities; second, what are peers in the process of quality assurance of universities; and third, how do these quality assurance processes affect the university?

Quality of Higher Education Institutions

Quality is unavoidably a contested and multi-faceted concept (Harvey & Knight, 1996). According to my Latin dictionary, the word is connected with *qualis*, asking 'how'. Hence, quality asks about 'how-ness', characteristics. Characteristics cannot exist without an object, hence the often-used quote of Pirsig's 1970s novel *Zen and the Art of Motorcycle Maintenance*, that 'when you try to say what the quality is, apart from the things that have it, it all goes *poof!* There's nothing to talk about' (Pirsig, 1984, p. 163). Accordingly, there is good sense in the definition agreed upon in the International Organization for Standardization (ISO) of quality as the 'degree to which a set of inherent characteristics ... of an object ... fulfils requirements' (ISO, 2015). In higher education, the objects whose characteristics are of interest might include universities, the persons making up those institutions (teachers, researchers, support staff, and perhaps students), or their activities (in particular, their research or their involvement in education processes). A high-quality university then is one with a positive image of, for example, delivering good graduates to society, of having good professors, or of being famous for discoveries and inventions. This loose illustration of what a high-quality university might mean points to several important points: first, it incorporates several 'objects' that may have quality; second, quality is a matter of image (it is 'in the eye of the beholder'); third, specifically for the class of objects in which higher education fits, the quality of the objects is not immediately obvious to most beholders, which explains why methods to assess the quality of higher education are so important; a fourth interesting point about the ISO definition is its connecting quality judgements to requirements, which might lead to debates about the roles of higher education in society (i.e. requirements to do what?) and to the political question, which stakeholders have or ought to have the power to define those roles and requirements. In this chapter, I cannot even begin to tease out most of the argumentations for and against potential answers to most of these questions, even though I will touch upon all four questions mentioned. A complication, which may make my contribution less legible than I would wish, is that the four issues are interconnected, which makes it difficult to treat one without making assumptions about the others.

A single higher education institution encompasses many objects that have quality: the 'primary processes' in a university are education and research,[2] which 'produce' graduates and scientific results—that defines the first two objects. Then, there are supporting processes taking place, as

---

[2] For simplicity, I ignore the third mission of universities. Much of the third mission is predicated on research (in knowledge exchange and innovation) or on education (e.g. in post-experience training), so much of the argument in this chapter applies to the third mission as well, but I miss out on parts of the increasingly relevant area of community engagement. For further simplicity, most of my attention will be on quality in relation to education rather than research.

in any other organisation: support for teaching staff in their education and research work, support for students in their learning and living, provision and management of the facilities, and management of the organisation as a whole. But the primary processes must be divided further: each separate study programme has its own quality, just like every research project. That makes universities multi-product firms. Multi-product firms are common—this alone does not set higher education apart in the quality debate. When one considers that the 'production technology' in universities can be characterised as a professional-based service, the complexity increases. This means that education (to focus on that primary process) relies largely on the behaviour of teachers, who generally act separately, based on their long training in a certain field of knowledge. It also means that, like every service, learning as the 'product' of education only appears in interaction with the students. The final complication to the education process that needs to be considered here comes from the insight of Harvey that education is not just a service for unchanging clients, but aims to transform and empower clients, that is, students (Harvey & Knight, 1996).

The other primary process, research, also relies largely on the knowledge and skills embodied in the professionals, even if there may be more reliance on machinery (laboratories, computers) in the research process than in education (although digitalisation makes technology more important there as well) and even if producing new knowledge is not necessarily a service for or with a client. Notably, the organisational units that 'produce' education may differ from those that 'produce' research: in delivering study programmes, teachers from several departments may be involved, who—at least in research universities—may do their research in other constellations (e.g. laboratories or research institutes).

Mintzberg has elaborated the consequences from the professional production technologies of education and research for the organisation of universities, which he subsumed under his category of professional bureaucracies (Mintzberg, 1983): the power in such organisations lies with the 'operational core', that is, the teachers and researchers, more than with the central management; hence, universities are very decentralised, both horizontally and vertically. Kogan saw more of an exchange relation, leading to an (almost) equal balance of power between the operational core and institutional managers: 'without active academics securing the reputation of the institution the managers would have nothing worthwhile to manage. There is, therefore, a process of exchange between those who manage and those who provide the main academic outputs of the institutions. They provide the expertise upon which the institution thrives or fails. The institution provides the resources enabling them to perform their academic tasks' (Kogan, 1984, p. 64).

In the following, I will first consider the primary processes—specially education—as the object of quality judgements, and then turn to the organisation that houses the primary (and secondary) processes as the object.

Quality of Performance and Performance Indicators

Out of the different conceptions of quality distinguished by Harvey and Green (1993), the one traditionally held in academia is that of distinctiveness, which in Harvey & Green's view translated to the more modern conception of quality as excellence, surpassing the highest standards. Excellence sounds alluring; who would not want to excel? However, in most 'naïve' debates about quality, the question is never asked, at what one wants to excel: fundamental research to gain a Nobel Prize? Being an excellent educator for first-generation undergraduates? Gaining a top-ten position in the Shanghai ranking? Leaving the object of excellence unspecified eases agreement in conversation but does not really help to evaluate or enhance quality. Moreover, '[e]xcellence, by definition, is a normative concept, i.e. not everyone can be excellent' (Elton, 1998, p. 4). Excellence is a *position good*, in economic terms, and its wide—often unconsciously self-evident—acceptance in academia may explain why university league tables with their explicit ascription of positions found so much (albeit grudgingly) recognition.

The other conception often quoted in higher education is that of 'fitness for purpose', which comes close to the ISO definition of quality: do what you are supposed (or required) to do. However, in higher education, the 'supposed' is more often linked to the institution's mission than to external requirements—although in state-controlled higher education systems, missions may have been defined externally as well.

Anyhow, the process view on quality has as object of valuation the performance of a university: at stake is the excellence or fulfilling the purpose by research 'products' and by education 'products'. This view easily leads to quantitative indicators: numbers of publications and citations for research quality, and numbers of graduates or employment statistics for education quality. Accordingly, it fits the use of performance indicators in quality assurance, which in turn fits the neoliberal turn (Harvey, 2005) in managing universities and in higher education policy. In fact, the rise of the neoliberal notion of New Public Management (NPM) spurred the introduction of quality assurance to higher education in the first place, around the 1980s (Paradeise et al., 2009; Pollitt, 1990; Pollitt & Bouckaert, 2011). The contrast between quality assurance through performance indicators and through peer review will be a *leitmotiv* in this chapter.

At this place, I draw attention to one consequence of focusing on products or outputs, namely that it leads to an analytical view on quality in higher education: the quality of output A may differ from that of output B. And once one starts analysing, the different outputs in both education and research multiply quickly: by faculties and by disciplines, by levels of education, by study programmes and by course units. The overview over quality gets lost in a multitude of different qualities of different objects within the university, and that makes assessing or communicating quality exceedingly difficult (Barefoot et al., 2016; Branco et al., 2016; Cremonini et al., 2007). The information costs for prospective students—but also for other stakeholders interested in quality—to learn about quality become prohibitive in this way. Another solution must be found for quality assurance to remain practicable.

Another issue hinted at before is that quality of higher education is not like the economy textbook case where customers in advance know the quality of the good they intend to buy. The standard case is called a *search good*. Higher education, however, is an *experience good* or maybe even *a credence good*. Of an experience good the quality can only be judged during or after consumption: services are all of this kind, such as going to the movies (Henze et al., 2015; Reinstein & Snyder, 2005; von Ungern-Sternberg & von Weizsacker, 1985). With credence goods, even afterwards, consumers do not know the quality of the good (Bonroy & Constantatos, 2008; Dulleck & Kerschbamer, 2006): doctors' consults, computer repairs, and education are given as standard examples of which it remains all but impossible to know if beneficial outcomes of the services can be ascribed to the service or are caused by other circumstances. No matter whether the balance tilts towards higher education being an experience good or a credence good, asymmetry of information exists beforehand, and that is what quality assurance (but also university rankings or labelling) aims to address in its function of informing external stakeholders (Baksi et al., 2016; Morphew & Swanson, 2011; Westerheijden, 2009). That is why it is relevant not to drown in a multitude of performance indicators about all possible objects of quality in a university.

Referring back to the argument that education aims to be transformative, a further complication is that by the time they graduate, students' ideas of what they wanted to achieve with their study, that is, their quality criteria, may have changed. This too argues against education being a search good. As an aside, it also implies that current students' satisfaction with their education is not necessarily correlated with their opinions on quality of the study once they have graduated.

Strengths and Weaknesses of Performance Indicators

As mentioned above, policies to assess quality in higher education arose with neoliberal NPM, around 1980 in early-adopter countries. NPM brought a new approach to the public sector, much more geared towards 'producing output', serving the customers (newspeak for citizens), and for that reason much more focused on efficiency instead of legality and legitimacy, through service units (newspeak for government agencies) led by powerful executive managers. In short, private sector management became the ideal for the public sector. In the wake of this movement, new methods in the public sector included quality assurance (Pollitt & Bouckaert, 2011; van Vught & Westerheijden, 1995), which had engulfed industry in the 1970s, when the rise of Japan's industry showed the leading industries in the western world that methods of production and management needed much improvement to keep up with East Asian quality levels (Deming, 1993; Dill, 1999). The industrial, managerial approach to quality assurance relied on the use of numeric performance indicators for fact-based decision-making.

The main criticism of performance indicators when they were first introduced to higher education in the 1980s remains true today: each indicator captures a small aspect of the performance of a university, and it is often geared towards efficiency, far removed from what the actors in higher education themselves understand by its quality. Already in the early days of quality assurance in higher education, Elton drew attention to this as an instance of Goodhart's law: 'any observation of a social system affects the system both before and after the observation, and with unintended and often deleterious consequences' (Elton, 2004).

For instance, graduation rates and employment rates were (and are) popular performance indicators. The obvious response to maximise such rates—incentivised by politicians and managers—easily led to churning out as many graduates as possible in the shortest time possible, with readily applicable job skills. Whether these graduates gained much deep learning, whether they became critical thinkers capable of bringing an analytical attitude to bear on their first job, further career and social life, whether they gained competences to become valuable, engaged citizens in an open, democratic society, was not measured by such indicators. Such performance indicators imply a short-term utilitarian view on higher education that was embedded in neoliberalism: higher education should train the country's workforce here and now. If one holds the view that higher education has the role (also) to transmit long-term values, to educate the next generation of leaders and thinkers in society—the view of Humboldt as much as of cardinal Newman (Labrie, 1986; Rothblatt, 1997; van Vught, 1994)—such short-term performance indicators invited goal displacement for all of the higher education sector.

In sum, the weaknesses of performance indicators as a tool for assessing quality of higher education are: they are (often distant) proxies to quality as the concept is understood by actors in higher education; they measure and promote efficiency which may lead to goal displacement. Moreover, they empower managers (both inside and outside universities) rather than teachers, which may be deleterious to quality enhancement in a professional production technology such as higher education, because to achieve improvement the professionals need to adjust their behaviour 'in the classroom', which is precisely what managers cannot control.

Nevertheless, performance indicators have strengths as well. First, they provide objective measurements. Objectivity is important in the bureaucratic and legal contexts in which (especially public, but to a large extent also private) higher education operates. It is a solid foundation for decision-making by politicians, civil servants, quality assurance agencies, institutional leaders or managers, and even—if, for example, accreditation decisions are disputed to the highest levels—by judges. Second, as a consequence of their fit with bureaucracy (in its original, objective meaning of rule-based organisations), performance indicators make higher education institutions more manageable. Better insight into the different processes in the institution makes it possible to control and improve those processes from a managerial perspective. Managers would do well, though, to remember what early quality 'evangelist' in higher education, Herb Kells, said about them: an indicator puts a question mark rather than an exclamation mark (Kells, 1992); that is, performance indicators can show that something unexpected happens, but they do not define the solution for any problem they may uncover.

Quality of the Organisation and Peer Review

The closing statements in the previous paragraph turn attention again from the role of objective, analytical performance indicators back to the managers or leaders in higher education institutions who have to take action based on the indicators' information about unexpected performances. This brings me to the alternative approach to quality assurance, that is, not the analytical view of the university's performances, but a synthetic look at the organisation that 'produces' quality: research groups, faculties or—in the spotlight in this chapter—whole institutions. In that perspective, the organisation's quality is a capacity to act, a potential, rather than a performance.

From an information or communication point of view the organisation perspective has as major advantage that in a single bit (or at most a few bytes—depending on the number of different values in the beholder's scale of quality), stakeholders get informed about the quality of the institution. The flipside is, obviously, that a single quality judgement for the whole institution hides potentially many and very diverse qualities (van Vught & Ziegele, 2012).

Another drawback of the organisational perspective is that good organisation and good processes do not guarantee good results; it is a potential or capacity, as I just stated, and the 'production function' of good education is not known—if there is such a thing as a single function across the diversity of courses, educational goals, teachers and, most importantly, students (Eisner, 1976; Hanushek, 2007; Scheerens, 1987). The perception of failure of the old Weberian, bureaucratic governance of higher education with its focus on regulating and funding capacity (inputs, facilities) had instigated the NPM approach to the public sector, including quality assurance as a new policy instrument, in the first place. On the positive side, creating conditions (staff, facilities) and processes (teaching), as well as implementing those processes, is precisely what a university does. It defines the extent of the institution's contribution to the quality of education and the outcome is (at least partly) out of its control: students' learning, employment, citizenship and so on depend on active participation by the students themselves. Good inputs and good design of education processes can influence the occurrence of desired outputs: High-Impact Practices (HIPs) in teaching for student engagement and deep learning (Kuh, 2008), the right mix of theoretical and practical elements in study programmes to engender academic competence, but also other key competences such as entrepreneurialism, social competences, transferable skills (European Council, 2018) and so on.

In a way the turn towards the quality assurance of the organisation is a return to more traditional ways of envisioning the governance and management of the university: a focus on inputs and process. It is also more comfortable to teaching staff and administrators alike than external scrutiny of the results of their efforts, the actual performances, because it is perceived as less of an inroad on academic freedom and institutional autonomy than detailed examination of the teaching and learning itself.

What may be less comfortable to the teaching staff, especially in traditional universities, where the power of Mintzberg's 'operational core' has been augmented by the ideology of academic freedom—for good reasons, but that lies outside the scope of this chapter—is that the introduction of quality assurance and other NPM policies necessarily has strengthened the grip of administrators on the university. The old jokes about the president of the university only being there to settle disputes about parking space, to make sure that the lawn got mown on campus, or that 'almost his only duty was to preside at banquets' like the Mayor of Michel Delving (Tolkien, 1966, p. 19), do not raise a smile any longer. The locus of autonomy in the university has shifted to managers—who no longer resent being called 'managers' (Westerheijden, 1997).

In briefest summary, autonomy is about 'who decides what?' Ever since higher education studies emerged, this has been a question of interest. Clark (1983) invented the triangle of hierarchical state coordination, academic collegial decision-making and price-regulated market coordination to show that in every country a balance was reached among hierarchy, market, and the peculiarly academic coordination of collegial decisions among peers. In a more recent analysis, de Boer et al. (2010) distinguished five coordination mechanisms in their 'governance equaliser': state regulation, stakeholder guidance, managerial self-regulation, academic self-regulation, and competition. Moreover, they emphasised that in any system at any moment, each coordination mechanism could be present at different levels of intensity, thus distinguishing low-governance from high-governance balances. Introducing quality assurance was one of many changes in the decades since the late 1970s that introduced NPM to higher education (since Reagan in the US and Thatcher in the UK) and that implied that managerial self-regulation has become stronger in universities (de Boer et al., 2010; Paradeise et al., 2009; Thoenig & Paradeise, 2014). This changed the balance of autonomy (Westerheijden, 2018), even if there would have been no changes to the previous level of academic self-regulation. However, for any given decision item, autonomy is a zero-sum game in the university: if the management decides about the pedagogical approach of a module (e.g. by prescribing problem-based learning), the professor cannot do it anymore.

The previous is not to state that professors' academic freedom was necessarily the best way towards high-quality education: management may act upon more advanced (and maybe even evidence-based) views on education than the academic staff who—at least in traditional research-oriented universities—have been trained in research methods rather than in teaching methods. (This has changed to some extent in some countries, in recent decades.) There may be collective benefits to the increased managerial self-regulation, and I would venture to state that the advent of quality assurance has led to a redefinition of academic freedom: teaching has become less of a private affair in the minds of many teaching staff. Team teaching and a much more student-centred view on teaching have become commonplace. Already 10–15 years after the introduction of quality assurance in the Netherlands, the discourse among academics and administrators in research universities had embraced terms and categories of thought that would have been unthinkable before. Institutional managers but also academics invariably found the teaching and research processes legitimate objects of management (Westerheijden, 1997)—whether due to quality assurance itself or to the broader permeation of society by neoliberal ideology, does not really matter (similarly: Kolsaker, 2008; Leišytė, 2016). At the same time, this cultural turnaround did not lead to quality management becoming fully systematised: after 25 years of experience with quality assurance, external evaluations still lead to a scramble for dispersed or absent data, and a state of light panic in most higher education institutions. Quality cultures have changed, but have not become mature in most European higher education institutions; studies into how to establish a positive, pervasive quality culture are still deemed necessary to spread the quality 'gospel' (Bendermacher et al., 2016; Brennan et al., 2017; Harvey & Stensaker, 2008; Kottmann et al., 2016; Sursock, 2011). It still appears difficult to ensure that universities adopt a positive quality culture with a widely shared set of values and norms in most universities, underpinned by effective structures and processes such as centres for educational excellence, that gives teachers a sense of engagement with the quality of their education work (Kottmann, 2017).

Besides the shift of power within the university from the professionals (teaching and research staff) to managers, external stakeholders also gained influence over the decades of neoliberal and neo-Weberian steering of higher education, at least in Europe (de Boer et al., 2010; Pollitt & Bouckaert, 2011; Westerheijden, 2018). Instead of accepting higher education, and in particular research universities as a value in themselves, as something perhaps quirky but anyway accepted in society, it was especially the strong stakeholder of the Thatcher government in the UK in the late 1970s that demanded that higher education be useful in the short term. Polytechnics were held up as better examples of 'useful knowledge' (a term dating back to the Bentham and other Victorians, even to eighteenth-century Enlightenment (Berg, 2007)) than British universities in those days (Goedegebuure et al., 1990). Taking higher education serious as a sector of national economy and security had started around World War II, with rocket science developing in Nazi Germany and the atom bomb in, eventually, the USA. The US 'endless frontier' report engraved pure research into the realm of public policy in times of peace as well, since 1945 (Bush, 1945; Pielke, 2010). In the 1960s and 1970s, critical voices in society sparked off protest movements that emphasised universities' critical and democratising roles. In a sort of sedimentary process, all these movements from Enlightenment to neoliberalism have left traces in the catalogue of at least partly conflicting demands that are put on universities, which translate into different and partly conflicting requirements that define what universities are expected to achieve. Consequently, what counts as quality depends on who defines quality; quality is not a matter of objective indicators, but an inherently political issue, considered at this level (Brennan, 1999; Morley, 2003; Ramirez, 2013; Skolnik, 2010; Westerheijden, 1990a). It then becomes evident that different actors in society may hold different views of universities' quality, and if one accepts a pluralist view on society, that holding different views is legitimate.

Traditionally, the providers of education and research, the academics, have predominated in this process, in collusion with the government—in Europe the main provider of funds for higher education and research. In Clark's 'triangle' (Clark, 1983), the market used to be largely absent as a coordination mechanism until the 1980s in Europe. However, since the

1980s, many governments have changed the old coalition of state officials and academic oligarchy into a principal–agent *quid pro quo* relation in the wake of NPM ideas. The role of students as 'consumers' of a 'service' was stressed and even stimulated in the same vein. Viewing higher education as a 'public good' rather than as a private, marketable service, returned on the agenda in reaction to the debates for and against the General Agreement on Trade in Services (GATS) of the early 2000s (EUA and ESIB, 2002; Vlk et al., 2008), but only slowly and again adding to the sediment layers of demands, without eradicating all of the previous developments.

These considerations point us towards the political, institutional embedding of quality assurance of higher education in quality assurance agencies, ministries, professional associations (e.g. lawyers, engineers or chartered accountants) and so on, but also towards who actually make the quality judgements, that is, the evaluators in quality assurance.

Strengths and Weaknesses of Peer Review in Quality Assurance

The term peer review was transposed to quality assurance from the quality judgements in journal publications, as stated at the beginning of this chapter. When advising about publication of a submitted article, peer review enables a holistic judgement of the article's quality 'in a way that defies strict logic but has won popular acceptance over the centuries' (Robertson, 2015). Peer reviewers are expected to judge the combination of originality of the research question and if its solution would contribute to advancement of the field and/or social relevance; correct, up-to-date, and inventive use of the theory in the field(s); correct and imaginative use of methodologies; and interesting, ingenious discussion of the findings and conclusions. The combination is what makes the judgement holistic—with different weights for different elements depending on the submission and with a case-specific, perhaps only semi-conscious way to combine the different elements. Peer evaluation is a matter of 'connoisseurship', which requires not only flexible application of knowledge of the field, but also—exemplified in the next quote for educational science—'to have a background sufficiently rich in educational theory, educational philosophy, and educational history to be able to understand the values implied by the ongoing activities and the alternatives that might have otherwise been employed' (Eisner, 1976, p. 145). The holistic judgement is a major strength of peer review.

Closely related is another strength: the human connoisseurship in peer review can be applied in different contexts, quite different from the stringent requirements of exactly the same definitions of data needed for performance indicators. One only has to look at the multitude of footnotes in OECD's annual *Education at a Glance* reports to see that comparison of data is quite difficult across different contexts with different data collection processes and definitions of the data. Within countries, similar problems may exist at a smaller scale between higher education institutions that collect their own data. Connoisseurs can handle such ambiguity—not perfectly from a data point of view perhaps, but good enough to get a sense of the quality profile of a university. They have deep insight into academic life as a consequence of their own long training in academia—it is the academic professional production model reflexively applied to itself.

The flip side of holistic and flexible judgement is that it is impossible to objectify it completely, which makes it less legitimate in a rule-bound, bureaucratic, or legal context. Peer review requires trust in the peers' expertise and honesty—but all of these three conditions are questionable in the context of quality assurance: trust, expertise, and honesty.

Lack of trust in the public sector was a basic assumption in neoliberalism (Harvey, 2005), which at the individual level translates into reduced trust in the expertise of peers. This sentiment has had real consequences even in the heart of scientific communication, that is, in the practices of journal submission decision-making. The debates around the functioning of peer review in scientific journals, experiments with either double-blind or more transparent forms of it seem instigated by the neoliberal turn. However, even if it may have strengthened in recent decades, criticism of peer reviewers' expertise predates the widespread influence of neoliberalism: already in the 1960s–1980s, studies appeared showing that peer review involved a large degree of *random error* (disagreement among reviewers, republication experiments). Moreover, there could be *intellectual bias* against non-majority, consensual theories and methodologies in a field, against negative findings (Epstein, 1990, 2004), even though falsification of a hypothesis teaches us much more than corroboration (Popper, 1980), and *social bias* against non-majority scientists (by gender, by colour, by institution) (Cislak et al., 2018; Hopkins et al., 2013; earlier overview: Westerheijden, 1991).

Random error and bias may occur unconsciously, but when the interests associated with a decision increase, honesty may become jeopardised as well. This issue may already arise in publication of discoveries and inventions (secrecy, stalling competitors' publications, stealing their ideas, etc.), may be more visible in competitive grant reviews, but may reach its peak in institutional evaluation for quality assurance. Especially if in a review accreditation is at stake—hence the legitimate existence of an institution—the interest in a positive decision may threaten peers' honesty either through their own anticipation of the consequences of their decision or through subversive actions by the university under scrutiny. Universities have been known to rig data or to respond to questionnaires strategically—and sometimes to deny such reports vehemently—in the case of university ranking (Jaschik, 2018a, 2018b; Lederman, 2009a, 2009b), so one can imagine that the temptations are still larger when it comes to the existence of the institution.

A crucial issue with peer review when applied to quality assurance is that the situation of the peers differs from peer reviewers of journal submissions. In terms of principal–agent theory, in journal submission reviews, peers are agents on behalf of the academic field as an abstract principal, who work for the benefit of the field, whose own (career) interest runs largely parallel with the task of reviewing, and who can use the state of knowledge in the field as a temporarily stable base for their decisions: the submitted paper does or does not add to the current knowledge in the field. In institutional quality assurance, the peers are agents on behalf of very real external principals, i.e. quality assurance agencies, which are usually either (quasi-)governmental agencies or agents of the profession-outside-universities (law firms, engineering bureaus, hospitals etc.). Besides, reviewers' judgements in quality assurance are made against criteria that are defined by the quality assurance agencies—perhaps with input from academic peers, but in any case translated into the bureaucratic discourse of standards and requirements in a legal context (Baumann & Krücken, 2018; Harman, 1998; Langfeldt et al., 2010). Admittedly, quality assurance agencies regularly encountered difficulties in making academic *prima donna* peer reviewers apply the bureaucratic standards and criteria rather than tacit knowledge about their field, especially in the early years of quality assurance when 'the majority of those who judge the teaching excellence of their colleagues have undergone little if any professional development as teachers and none as assessors of excellence' (Elton, 1998, p. 4). Similarly, in Germany, where accreditation several years after its introduction only enjoyed limited legitimacy among academics, the 'role separation' between being teachers and being reviewers on behalf of an agency outside the academic field 'was only obtained to a very limited extent' and the reviewers insisted on their primary role as peers, criticising the accreditation system while taking part in it (Baumann & Krücken, 2018). Yet, in most countries in the twenty-first century, quality assurance agencies engage in fairly extensive training of the peer reviewers, and have developed more control capacity of the process through guidelines, report templates, and the presence of quality assurance agency staff members as 'secretary', 'coordinator', or 'auditor' in the external review team, so that the agency's externally defined standards are applied; peer-specific field knowledge is only allowed to substantiate vague norms like 'up-to-date' textbooks, 'adequate' teaching facilities, and the like.

Since the external reviewers in quality assurance do not primarily apply (tacit) criteria from within the field, as they do in journal review, they are taken out of their field in this task and it becomes questionable if they still act as respected equals (i.e. peers) while they are agents for quality assurance agencies as principals. It may be a matter of convenience to call external review by a committee in quality assurance processes peer review, but it stretches the meaning of the term 'peer'. In defence of quality assurance agencies, it must be added that some (but I do not know what proportion) try to instil what might be called a 'field identification' in the reviewers rather than a 'government identification'. Thus, in the US context, where institutional accreditation has a century-long tradition, institutional ('regional') accreditation agencies give as much room as possible to evaluation from a 'fitness for purpose' conception of quality, that is, mission-based evaluation, so that the institution's mission defines part of the quality criteria, which embeds the agents more in the field than when they have to evaluate according to purely externally defined criteria. For instance, one agency admonishes: 'Evaluators are encouraged to approach their assignments as colleagues rather than as auditors' (NWCCU [Northwest Commission on Colleges and Universities] 2017: p. 9). Nevertheless, especially in recent years, also in the US, governmental requirements have come to define a substantial part of the institutional evaluations. Still, the British Quality Assurance Agency for Higher Education (QAA) states even regarding England—where there is more emphasis on complying with externally imposed standards than in Scotland or Wales—that institutional audit operates on the 'principle of peer review' (QAA, n.d.).

In the reasoning above, I focused on the external reviewers' task of passing a judgement, a summative evaluation on behalf of the principal so that they can decide about an institution's legitimacy, funding and so on. Most quality assurance processes have, however, a second aim, that is, to assist quality enhancement in the university—the formative part of evaluation. In that perspective, it is important that the reviewers be acknowledged by the evaluees as peers from whom they accept advice, as respected colleagues in the field, fellow professionals who by their training and status in the field have expertise of how higher education works. If this is successful, peer review is very effective, because 'one of the strongest pressures on any group of academics is the prospect of being judged by senior peers in the discipline' (Harman, 1998, p. 354). This is a tenable thesis when the evaluation concerns teaching or research in a specific field of knowledge—the field is what defines the professional community of teachers and researchers. The peer concept gets stretched again, however, once the object of evaluation is not a recognised field of knowledge within a university, but the institution as a whole. Who are then the peers? In a 1990s internationally comparative publication, Harman mentioned 'panels of experts, usually involving at least some "external" members' (Harman, 1998, p. 353), but I am not aware of more recent or more precise studies on this question. A quick scan for this chapter of nine US regional accreditors and European quality assurance agencies with institutional audits showed varied practices of composing evaluation teams.

The commonality is, however, that all agencies choose reviewers largely from within higher education institutions, who must have several years of experience of working in university. First, all but one of the agencies emphasise that evaluation teams must include 'educators', and one quality assurance agency adds that such persons 'typically will represent one of the types of teaching disciplines at the institution being visited'. Another agency prescribes that experienced teaching staff usually constitute the majority of the visiting team—at least three out of at least five—although most agencies only mention requiring at least one teaching representative. The one agency that does not include educators defines peers exclusively as fellow leaders of institutions: 'former rectors and vice-rectors'. Most other quality assurance agencies, too, mention senior institutional administrators as the second major category of evaluators. Implementing their mission-based approach, some US agencies specify that the administrative contingent includes financial officers and 'specialists whose expertise is related to the known areas of concern of the institution, (e.g. assessment, student personnel, finance, planning, etc.)'. Specialists on quality assurance or institutional research also are mentioned. The third major category of reviewers are external stakeholders, which are included in a minority of quality assurance agencies' review teams, for example, 'members of boards of trustees of accredited institutions, legal counsel, state education or system employees, representatives of the business community, public members'. Finally, European quality assurance agencies include a student representative among the reviewers. In sum, the composition of external review teams for institutional evaluations and accreditations indeed broadens the concept of peers further. The inclusion of experienced institutional administrators but also of specialists in managing a university testifies to what has been called the rise of 'third space' professions in between traditional teachers and traditional university leaders (Whitchurch, 2013). Quality assurance thus has contributed to the evolution of a new species in higher education, it would seem—an unintended consequence—first by coercing universities to professionalise quality assurance with teaching excellence centres and quality assurance offices, and secondly by then promoting the new species of professionals into institutional evaluation teams. Remarkably, the spread of external evaluators seems to be less pronounced than might have been expected under an NPM regime: not all quality assurance agencies in my, admittedly, small convenience sample include externals, and if they do, there often is a certain degree of connection between them and higher education; for example, they may be lay members of a university's board of trustees, or they are student representatives—if students can be called external stakeholders rather than internal ones.

Persistent Dilemmas: Instead of a Conclusion

The general argument in this chapter holds that the traditional way of assessing quality in higher education is through peer review. This method has strengths, namely holistic and flexible judgement, but also weaknesses especially when applied to quality assurance of higher education institutions, because it is prone to random error, to intellectual and social biases, and honesty is threatened as a result of the high stakes involved, in particular when accreditation of the university might be at risk. Moreover, trust in inside, human judgements has waned—NPM makes that explicit in its stress on transparency and accountability through performance indicators (Westerheijden, 1990b). Strengths and weaknesses of performance indicators mirror those of peer review: they are objective but partial and often distant proxies to the concept of quality. Moreover, reliance on performance indicators leads to goal displacement and concomitant distortions of behaviour of and in universities, even to fraud in creating data for indicators.

Actual quality assurance processes use a combination of peer review and performance indicators in the hope to balance the weaknesses of one with the strengths of the other. The capacity to collect data on the primary processes in university—my focus in this chapter has been on education rather than on research—and to analyse them has increased over the course of the recent decades after the introduction of quality assurance in higher education systems. We know much more about the processes, and this information, often in the form of indicators, is used as input for evaluation processes based on human judgement, which is still called peer review, even if the term has grown to carry a new, broader meaning than before.

Peer review seems to be used in institutional quality assurance also to emphasise the 'soft' side of it, the quality enhancement that needs the non-threatening yet authoritative voice of respected peers. Some of the quality assurance agencies whose criteria for reviewer status I scanned seem to downplay the administrators' character for just this purpose; for example, the QAA states: 'Most of our reviewers are academics with postgraduate qualifications, many with doctorates. Some hold senior roles such as Vice Chancellor, Principal or Pro-Vice-Chancellor' (QAA, n.d.)—as if they just happened to have held that position for a while and accidentally. Yet, the UK is the most managerial, hard-NPM country in Europe, where more than in other countries, being an institutional administrator is a full-time occupation and a career in its own right. The rhetoric cannot hide that the external reviewers come in from a position of power—power to affect the institution's reputation, its leaders' careers (e.g. in the UK and the US), or even the existence of the institution (in countries where accreditation is a condition for legitimate operation). Realising this, I once formulated the following dilemma (Westerheijden, 1990b):

Dilemma I—Without (the threat of) serious consequences, quality assurance is not taken seriously in academe and turns into an administrative burden, yet with (the threat of) serious consequences, quality assurance turns into a game to gain positive outcomes, not to assure or enhance quality.

Considering that the peer review side of institutional reviews is stressed to make them more accepted among the evaluees, with the further aim to strengthen the quality enhancement function of the evaluation, another dilemma arises from the above as well (Westerheijden, 2013):

Dilemma II—Quality enhancement demands evidence-based decision-making, where the evidence usually consists of performance indicators, but performance indicators threaten quality enhancement through the goal displacement that they induce.

Having studied quality assurance in higher education for more than three decades, these dilemmas continue to puzzle me. Assuring and enhancing quality remains a balancing act involving all who work in a higher education institution, spurred on by external review, yet simultaneously hindered by its distorting effects, whatever the mix of peer review and performance indicators it employs. Peer reviewers in quality assurance, the gatekeepers of prestige and even of the existence of the campus, make tightrope walkers out of their peers on campus, the teachers as much as the leaders.

References

Adelman, C. (1992). Accreditation. In B. R. Clark & G. Neave (Eds.), *The encyclopedia of higher education* (Vol. 1, pp. 1313–1318). Pergamon.

Baksi, S., Bose, P., & Xiang, D. (2016). Credence goods, misleading labels, and quality differentiation. *Environmental and Resource Economics.* https://doi.org/10.1007/s10640-016-0024-4

Barefoot, H., Oliver, M., & Melar, H. (2016). Informed choice? How the United Kingdom's key information set fails to represent pedagogy to potential students. *Quality in Higher Education, 22*(1), 3–19.

Baumann, J., & Krücken, G. (2018). Debated legitimacy: Accreditation in German Higher education. *Higher Education Policy.* https://doi.org/10.1057/s41307-018-0120-x

Bendermacher, G. W. G., Oude Egbrink, M. G. A., Wolfhagen, I. H. A. P., & Dolmans, D. H. J. M. (2016). Unravelling quality culture in higher education: A realist review. *Higher Education.* https://doi.org/10.1007/s10734-015-9979-2

Benos, D. J., Bashari, E., Chaves, J. M., Gaggar, A., Kapoor, N., LaFrance, M., Mans, R., Mayhew, D., McGowan, S., Polter, A., Qadri, Y., Sarfare, S., Schultz, K., Splittgerber, R., Stephenson, J., Cristy, T., Grace Walton, R., & Zotov, A. (2007). The ups and downs of peer review. *Advances in Physiology Education, 31*(2), 145–152. https://doi.org/10.1152/advan.00104.2006

Berg, M. (2007). The genesis of "Useful Knowledge". *History of Science, 45*(2), 123–133. https://doi.org/10.1177/007327530704500201

de Boer, H. F., Enders, J., & Schimank, U. (2010). On the Way towards New Public Management? The governance of university systems in England, the Netherlands, Austria, and Germany. In D. Jansen (Ed.), *New forms of gover-*

nance in research organizations—disciplinary approaches, interfaces and integration (pp. 137–152). Springer.

Bonroy, O., & Constantatos, C. (2008). On the use of labels in credence goods markets. *Journal of Regulatory Economics, 33*(3), 237–252.

Branco, F., Sun, M., & Miguel Villas-Boas, J. (2016). Too much information? information provision and search costs. *Marketing Science, 35*(4), 605–618. https://doi.org/10.1287/mksc.2015.0959

Brennan, J. (1999). Evaluation of higher education in Europe. In M. Henkel & B. Little (Eds.), *Changing relationships between higher education and the state* (pp. 219–235). Jessica Kingsley.

Brennan, J., Cremonini, L., King, R., Lewis, R., Wells, M., & Westerheijden, D. (2017). *Cultures of quality: An international perspective—final report of phase 2.* Quality Assurance Agency for Higher Education.

Bush, V. (1945). *Science, the endless frontier.* United States. Office of Scientific Research and Development.

Cislak, A., Formanowicz, M., & Saguy, T. (2018). Bias against research on gender bias. *Scientometrics, 115*(1), 189–200.

Clark, B. R. (1983). *The higher education system: Academic organization in cross-national perspective.* University of California Press.

Crane, D. (1967). The gatekeepers of science: Some factors affecting the selection of articles for scientific journals. *American Sociologist, 2*, 195–201.

Cremonini, L., Westerheijden, D. F., & Enders, J. (2007). Disseminating the right information to the right audience: Cultural determinants in the use (and misuse) of rankings. *Higher Education, 55*, 373–385. https://doi.org/10.1007/s10734-007-9062-8

Deming, W. E. (1993). *The New economics: For industry, government, education.* Massachusetts Institute of Technology Center for Advanced Engineering Study.

Dill, D. D. (1999). Through Deming's eyes: A cross-national analysis of quality assurance policies in higher education. In CHEPS/QSC (Ed.), *Quality management in higher education institutions: Reader, Institutional management and change in higher education* (pp. 309–328). Lemma.

Dulleck, U., & Kerschbamer, R. (2006). On doctors, mechanics, and computer specialists: The economics of credence goods. *Journal of Economic Literature, 44*(1), 5–42.

Eisner, E. W. (1976). Educational connoisseurship and criticism: Their form and functions in educational evaluation. *The Journal of Aesthetic Education, 10*(3/4), 135–150.

Elton, L. (1998). Dimensions of excellence in university teaching. *International Journal for Academic Development, 3*(1). https://doi.org/10.1080/1360144980030102

Elton, L. (2004). Goodhart's law and performance indicators in higher education. *Evaluation and Research in Education, 18*(1–2), 120–128.

Epstein, W. M. (1990). Confirmational response bias among social work journals. *Science, Technology & Human Values, 15*, 9–38.

Epstein, W. M. (2004). Confirmational response bias and the quality of the editorial processes among American social work journals. *Research on Social Work Practice, 14*(6), 450–458. https://doi.org/10.1177/1049731504265838

EUA, and ESIB. (2002). *EUA and ESIB joint declaration: Students and universities: An academic community on the move.* Paris.

European Council. (2018). *Council recommendation of 22 May 2018 on key competences for lifelong learning (2018/C 189/01).* Official Journal of the European Union.

Goedegebuure, L. C. J., Maassen, P. A. M., & Westerheijden, D. F. (1990). *Peer review and performance indicators: Quality assessment in British and Dutch higher education.* Lemma.

de Grazia, A. (1963). The scientific reception system and Dr. Velikovsky. *American Behavioral Scientist, 7*(1), 38.

Hanushek, E. A. (2007). Education production functions. In *Palgrave encyclopedia.* s.l.

Harman, G. (1998). The management of quality assurance: A review of international practice. *Higher Education Quarterly, 20*, 345–364.

Harvey, D. (2005). *A brief history of neoliberalism.* Oxford University Press.

Harvey, L., & Green, D. (1993). Defining quality. *Assessment & Evaluation in Higher Education, 18*(1), 9–34.

Harvey, L., & Knight, P. T. (1996). *Transforming higher education.* Society for Research into Higher Education & Open University Press.

Harvey, L., & Stensaker, B. (2008). Quality culture: Understandings, boundaries and linkages. *European Journal of Education, 43*(4), 427–442.

Henze, B., Schuett, F., & Sluijs, J. P. (2015). Transparency in markets for experience goods: Experimental evidence. *Economic Inquiry, 53*(1), 640–659. https://doi.org/10.1111/ecin.12143

Hopkins, A. L., Jawitz, J. W., McCarty, C., Goldman, A., & Basu, N. B. (2013). Disparities in publication patterns by gender, race and ethnicity based on a survey of a random sample of authors. *Scientometrics, 96*(2), 515–534. https://doi.org/10.1007/s11192-012-0893-4

ISO. (2015). ISO 9000:2015. *ISO 9000:2015(En) Quality management systems—Fundamentals and vocabulary.* Retrieved October 17, 2020, from https://www.iso.org/obp/ui/#iso:std:iso:9000:ed-4:v1:en:term:3.6.2

Jaschik, S. (2018a, August 24). Eight more colleges identified as submitting incorrect data for "U.S. News" Rankings. *Inside HigherEd.*

Jaschik, S. (2018b, July 10). Temple ousts business dean after report finds online M.B.A. program for years submitted false data for rankings. *Inside HigherEd.*

Kells, H. R. (1992). An analysis of the nature and recent development of performance indicators in higher education. *Higher Education Management, 4*, 131–138.

Kogan, M. (1984). The political view. In B. R. Clark (Ed.), *Perspectives on higher education*. University of California Press.

Kolsaker, A. (2008). Academic professionalism in the Managerialist Era: A study of English universities. *Studies in Higher Education, 33*(5), 513–525. https://doi.org/10.1080/03075070802372885

Kottmann, A. (2017). *Unravelling tacit knowledge: Engagement strategies of centres for excellence in teaching and learning.* Working paper

Kottmann, A., Huisman, J., Brockerhoff, L., Cremonini, L., & Mampaey, J. (2016). *How can one create a culture for quality enhancement? Final Report.* s.l. [Enschede, Gent]: CHEPS; CHEGG.

Kuh, G. (2008). *High-impact educational practices: What they are, who has access to them, and why they matter.* Association of American Colleges and Universities.

Labrie, A. (1986). *'Bildung' En Politiek 1770–1830: De 'Bildungsphilosophie' van Wilhelm von Humboldt Bezien in Haar Politieke En Sociale Context.* Universiteit van Amsterdam.

Langfeldt, L., Stensaker, B., Lee, H., Huisman, J., & Westerheijden, D. F. (2010). The role of peer review in Norwegian quality assurance: Potential consequences for excellence and diversity. *Higher Education, 59*(4), 391–405. https://doi.org/10.1007/s10734-009-9255-4

Lederman, D. (2009a, June 3). "Manipulating," Er, Influencing "U.S. News". *Inside Higher Ed.*

Lederman, D. (2009b, June 4). Rankings Rancor at Clemson. *Inside Higher Ed.*

Leišytė, L. (2016). Bridging the duality between universities and the academic profession: A tale of protected spaces, strategic gaming, and institutional entrepreneurs. In L. Leišytė & U. Wilkesmann (Eds.), *Organizing academic work in higher education: Teaching, learning and identities* (pp. 55–67). Routledge.

Marsh, H. W., Jayasinghe, U. W., & Bond, N. W. (2008). Improving the peer-review process for grant applications: Reliability, validity, bias, and generalizability. *American Psychologist, 63*(3), 160–168. https://doi.org/10.1037/0003-066X.63.3.160

Mintzberg, H. (1983). *Structure in fives—designing effective organizations*. Prentice Hall.

Morley, L. (2003). *Quality and power in higher education*. Open University Press; Society for Research into Higher Education.

Morphew, C. C., & Swanson, C. (2011). On the efficacy of raising your university's rankings. In J. C. Shin, R. K. Toutkoushian, & U. Teichler (Eds.), *University rankings: Theoretical basis, methodology and impacts on global higher education* (pp. 185–199). Springer.

NWCCU (Northwest Commission on Colleges and Universities). (2017). *Handbook for peer evaluators*. Redmond, WA.

Paradeise, C., Reale, E., Bleiklie, I., & Ferlie, E. (2009). *University governance: Western European comparative perspectives*. Springer.

Pielke, R. (2010). In retrospect: Science—The endless frontier. *Nature, 466*(7309), 922–923. https://doi.org/10.1038/466922a

Pirsig, R. M. (1984). *Zen and the art of motorcycle maintenance*. Bantam Books.

Pollitt, C., & Bouckaert, G. (2011). *Public management reform: A comparative analysis: New public management, governance, and the Neo-Weberian state. 3rd rev. and upd.* Oxford University Press.

Pollitt, C. (1990). Measuring university performance: Never mind the quality, never mind the width? *Higher Education Quarterly, 44*(1), 60–81.

Popper, K. R. (1980). *The logic of scientific discovery (rev)*. Hutchinson.

QAA. (n.d.) Our reviewers. Retrieved October 5, 2020, from https://www.qaa.ac.uk/reviewing-higher-education/our-reviewers.

Ramirez, G. B. (2013). Studying quality beyond technical rationality: Political and symbolic perspectives. *Quality in Higher Education, 19*(2), 126–141.

Reinstein, D. A., & Snyder, C. M. (2005). The influence of expert reviews on consumer demand for experience goods: A case study of movie critics. *Journal of Industrial Economics, 53*(1), 27–51. https://doi.org/10.1111/j.0022-1821.2005.00244.x

Robertson, G. (2015). Magna Carta and Jury Trial. *Magna Carta*. Retrieved October 6, 2020, from https://www.bl.uk/magna-carta/articles/magna-carta-and-jury-trial.

Rothblatt, S. (1997). *The modern university and its discontents: The fate of Newman's legacies in Britain and America*. Cambridge University Press.

Rowland, F. (2002). The peer-review process. *Learned Publishing, 15*(4), 247–258. https://doi.org/10.1087/095315102760319206

Scheerens, J. (1987). *Het Evaluerend Vermogen van Onderwijsorganisaties*. Universiteit Twente.

Schwarz, S., & Westerheijden, D. F. (2004). Accreditation in the framework of evaluation activities: A comparative study in the European higher education area. In S. Schwarz & D. F. Westerheijden (Eds.), *Accreditation and evaluation in the European higher education area* (pp. 1–41). Kluwer Academic Publishers.

Skolnik, M. L. (2010). Quality assurance in higher education as a political process. *Higher Education Management and Policy, 22*(1), 79–98.

Sursock, A. (2001). *Towards accreditation schemes for higher education in Europe? Final Project Report.* CRE Association of European Universities.

Sursock, A. (2011). *Examining quality culture part II: Processes and tools— Participation, ownership and bureaucracy.* European University Association.

Thoenig, J.-C., & Paradeise, C. (2014). Organizational governance and the production of academic quality: Lessons from two Top U.S. research universities. *Minerva, 52,* 381–417. https://doi.org/10.1007/s11024-014-9261-2

Tolkien, J. R. R. (1966). *The fellowship of the ring.* George Allen & Unwin.

von Ungern-Sternberg, T., & von Weizsacker, C. C. (1985). The supply of quality on a market for "experience goods". *The Journal of Industrial Economics, 33*(4), 531. https://doi.org/10.2307/2098391

Vlk, A., Westerheijden, D., & van der Wende, M. (2008). GATS and the steering capacity of a nation state in higher education: Case studies of the Czech Republic and the Netherlands. *Globalisation, Societies and Education, 6*(1), 33–54. https://doi.org/10.1080/14767720701855584

van Vught, F. A. (1994). The Humboldtian University under pressure: New forms of quality review in Western European Higher Education. In P. A. M. Maassen & F. A. van Vught (Eds.), *Inside academia: New challenges for the academic profession* (pp. 185–226). de Tijdstroom.

van Vught, F. A., & Westerheijden, D. F. (1995). Quality measures and quality assurance in European higher education. In C. Pollitt & G. Bouckaert (Eds.), *Quality improvement in European public service* (pp. 33–57). Sage.

van Vught, F. A., & Ziegele, F. (2012). *Multidimensional ranking: The design and development of U-Multirank.* Springer.

Westerheijden, D. F. (1990a). A political view of quality assessment in higher education. In T. W. Banta (Ed.), *Proceedings 2nd International Conference on Assessing Quality in Higher Education.* Knoxville, TN.

Westerheijden, D. F. (1990b). Peers, performance, and power: Quality assessment in the Netherlands. In L. C. J. Goedegebuure, P. A. M. Maassen, & D. F. Westerheijden (Eds.), *Peer review and performance indicators: Quality assessment in British and Dutch higher education* (pp. 183–207). Lemma.

Westerheijden, D. F. (1991). Promises, problems and pitfalls of peer review: The use of peer review in external quality assessment in higher education. In T. W. Banta (Ed.), *Proceedings of the third international conference on assessing quality in higher education* (pp. 130–142). University of Tennessee.

Westerheijden, D. F. (1997). A solid base for decisions: Use of the VSNU research evaluations in Dutch universities. *Higher Education, 33*(4), 397–413.

Westerheijden, D. F. (2001). Ex Oriente Lux? National and multiple accreditation in Europe after the fall of the wall and after Bologna. *Quality in Higher Education, 7*(1), 65–75.

Westerheijden, D. F. (2009). Information of quality and quality of information to match students and programme. In J. Newton & R. Brown (Eds.), *The future of quality assurance* (pp. 29–46). EAIR.

Westerheijden, D. F. (2013). Achieving the focus on enhancement? In R. Land & G. Gordon (Eds.), *Enhancing Quality in higher education: International perspectives, International Studies in Higher Education* (pp. 39–48). Routledge.

Westerheijden, D. F. (2018). University governance in the United Kingdom, the Netherlands and Japan: Autonomy and shared governance after new public management reforms. *Nagoya Journal of Higher Education, 18*, 199–220.

Whitchurch, C. (2013). *Reconstructing identities in higher education: The rise of 'third space' professionals*. Routledge.

5

The Many Faces of Peer Review

Hanne Foss Hansen

Introduction

Evaluation is an ongoing activity in most parts of contemporary societies. Terms such as "the evaluative state" (Neave, 1998) and "the evaluation society" (Dahler-Larsen, 2012) have been used to describe this.

In academia evaluation organized as peer review dates back to the eighteenth century (Benos et al., 2007). Since then peer review practices have been discussed and developed. Today peer review processes take many forms. Almost all aspects of scientific enterprise rely on evaluation done by peers (Bornmann, 2011). Peers evaluate doctoral dissertations, applicants to academic positions, applications for promotion, applications for research grants, manuscripts submitted for publication, scholars proposed to receive awards and prizes as well as research organizations in the form of groups, departments, programs, institutes and universities (Langfeldt & Kyvik, 2011, 2015). In some countries peers even evaluate

H. F. Hansen (✉)
University of Copenhagen, Copenhagen, Denmark
e-mail: hfh@ifs.ku.dk

© The Author(s) 2022
E. Forsberg et al. (eds.), *Peer review in an Era of Evaluation*,
https://doi.org/10.1007/978-3-030-75263-7_5

disciplines and interdisciplinary fields at the national level across universities and other research organizations. In some research areas they also critically evaluate newly published books in order to assess whether they contribute to new knowledge to the field. Furthermore, peers sometimes assess research in order to decide on which knowledge to be used as input to policy-making and regulation.

The scientific enterprise is in this way permeated with evaluation activities (Hamann & Beljean, 2017). The same goes for higher education. Here peer panels and other types of panels, including peers, experts on educational leadership, labor market representatives and students, evaluate the quality of educational programs and educational quality assurance systems (Hansen, 2009a, 2014). All in all, peers thus undertake many evaluator roles.

Alongside the development and expansion of peer review activities, also other governance structures in academia have been transformed. In the last decades, university systems in the Nordic countries have experienced continuous change. Funding arrangements, accountability measures as well as institutional management and organizational structures have been transformed (Geschwind et al., 2019; Hansen et al., 2019a). Universities have increasingly developed into what observers have termed "corporate actors" with organizational traits such as identity, hierarchy and rationality (de Boer et al., 2007), and "complete organizations" where an authoritative center coordinates and controls actions through the hierarchy (Brunsson & Sahlin-Andersson, 2000; Seeber et al., 2015).

Changes in the institutional context in which peer review activities are embedded seem to have influenced the role and use of peer review. Whereas peer review historically was an opportunity for peers to exercise academic power in a professional self-regulated system, it today often plays the role as giving managerial advice and ensuring accountability. Using the concepts of Johan P. Olsen (2007), the importance of peer review practices exercised in the context of the universities as self-governing communities of scholars has diminished while it has increased in the contexts of the universities as instruments for political agendas as well as service enterprises embedded in competitive markets.

This chapter sheds light on this development. Focus is on two research questions: (1) How has peer review as an evaluation concept been developed over time? (2) What is the role of peer review today? Focus is both on research evaluation on the scientific side of universities and on the educational side. The analysis is mostly conceptual, aiming at mapping different types of peer review. The mapping of peer review is supplemented with examples from Denmark and the other Scandinavian countries.

The analysis is based on documentary material and self-experience from working in the university sector for more than 35 years. During that time I have been doing research in the development of the university sector and in evaluation practices (Borum & Hansen, 2000; Hansen, 2009a, 2009b; Hansen & Borum, 1999; Hansen et al., 2019a, 2019b). Further, I have been acting in the role as peer in Denmark and abroad in many contexts, among others, in the publication system, in assessment of dissertations, in assessing applicants for positions and promotion as well as in assessing project applications in research councils. Moreover, I have served as peer in a national discipline evaluation (Political Science in Norway, initiated by the Norwegian Research Council in 2002) and participated in panels in educational quality assurance (University of Tromsø, 2006; NLA University College, 2005; Norwegian Academy of Music, 2004; all under the auspices of the Norwegian Agency for Quality Assurance in Education [NOKUT]). I have also worked with evaluation authorities in both the research and educational fields in Denmark, Norway and Sweden as advisor and board member. I draw on these experiences in the analysis, knowing of course that other scholars may have other experiences from other scientific fields and other organizations.

The chapter is structured in seven sections. The next section presents classical peer review as an evaluation model. The following three sections discuss how peer review related to the evaluation of research has developed over time into other types of evaluation models, termed informed and standards-based peer review, modified peer review and extended peer review. Then, the next section is dedicated to the role of peer review in the higher educational field. The final section holds the conclusion and some personal reflections on the challenges which peer review practices meet these years.

Classical Peer Review as One of Several Evaluation Models

The concept of evaluation can briefly be defined as assessment or appraisal. In the specialized evaluation literature, the concept is defined as systematic assessment of the merit, worth and value (Scriven, 1991) of evaluands (= objects for evaluation). As mentioned above, there are multiple evaluands in relation to research and higher education. The literature on evaluation offers a range of evaluation models. Peer review can be characterized as a professional or collegial evaluation model (Vedung, 1997; Hansen, 2005) where the evaluation criteria are defined by the peers and not by other stakeholders, for example, users. The fundamental idea is that members of the profession are trusted to evaluate other members' activity and results. Evaluation may be summative, assessing against a standard or benchmark, for example, is the article worth publishing or is the applicant qualified for a professorship, or it may be formative, focusing on whether activities are in progress.

Peer review is the classical type of evaluation in research. In classical peer review evaluation recognized researchers read and assess other researchers' contributions, focusing either on project ideas, as in, for example, research councils; on manuscripts submitted for publication, as in relation to journals and publishing houses; or on CVs and publications handed in in relation to applications for appointment and promotion. The content and context of classical peer review is characterized in Table 5.1 (Hansen, 2009b). Classical peer review evaluation is the corner stone in gatekeeping in academia aiming at ensuring quality control and the best possible distribution of scarce resources.

Table 5.1 Classical peer review

Task	Assessment of research quality of products and individuals.
Process	Reading first-order material, assessing, nominating and sometimes ranking.
Peer panel composition	Homogeneous. Mono-professional.
Evaluation approach	Summative, clear decision focus.

Classical peer review is based on reading what can be termed first-order material, for example, publications, manuscripts and project proposals. Classical peer review is most often organized with a number of peers. Peers are either working in parallel, as in review processes of manuscripts submitted for publication, or as a panel, as in assessment of applicants for a position. The assessment process is relational in the sense that it is made in a context. A dissertation, for example, is assessed in relation to the research area to which it seeks to contribute, just as an applicant to a position is assessed in relation to the job description. Most often, the process includes a form of cross-control. The assessment of a manuscript is passed on from the reviewers to the editor(s) and the assessment of a dissertation from the assessment panel to, for example, the dean.

Classical peer review is still of great importance first and foremost in the publication system and in relation to assessing research applications. With country variations, classical peer review in the Nordic countries is however being transformed (Hansen et al., 2019b). In a Danish context classical peer review sometime back constituted the corner stone in the recruitment system. All applicants for a position were assessed by a peer panel reading their enclosed publications, nominating whether the individual applicant was qualified for the position and finally ranking the qualified applicants. The assessment document was passed on to the university management, which normally acted according to the proposed ranking. Only in cases of disagreement in panels there was more leeway for management. Further, the assessment document was distributed to all applicants for the position, a process often initiating and supporting a discussion in the research environment on the attributes of research quality. As I will return to in below, peer review has been transformed into applicant assessment processes, which, these years, are organized very differently.

Informed and Standards-Based Peer Review

In the Danish university system peer review in relation to recruitment and promotion has changed considerably. Peer influence has been reduced, managerial power increased. Applicants are assessed as either qualified or not qualified, no longer ranked. Publicness is reduced, confidentialism is the norm. In still more contexts peers in assessment committees only look at the qualifications of the applicants shortlisted. Criteria for shortlisting are blurred.

Further, the reading of first-order material which are at the core of classical peer review has increasingly become supplemented by the use of metrics, and a range of other assessment criteria have become increasingly important. We can term this informed and standards-based peer review. The content and context of this type of peer review is characterized in Table 5.2.

In recent years still more easy accessible metrics have been developed. Some are global systems. Examples are Scopus and Google Scholar counting citations and presenting indexes such as, for example, the h-index as well as journal metrics such as the journal impact factors (JIF). Others are national systems such as the Danish bibliometric research indicator (BFI) developed upon inspiration from the Norwegian model and dividing journal and publishers into two, in some disciplines three, quality categories.

Although metrics are contested, it is my experience that they move into use in peer review processes, for example, in assessments of applicants for positions and promotion. Often, it is applicants themselves referring to metrics. Sometimes it is peers arguing that metrics are relevant. Sometimes it is managers arguing that panels are expected to or even demanded to include metrics. And this happens, even though bibliometric experts, for example, in the Leiden Manifesto (Hicks et al., 2015) argue that metrics should not be used at the level of individuals. It

Table 5.2 Informed and standards-based peer review

Task	Assessment of research quality of products and individuals.
Process	Reading first-order material and assessing, nominating and sometimes ranking by including metrics such as publication ranking, citation metrics, etc. Further giving priority to multiple evaluation criteria.
Peer panel composition	Homogeneous. Mono-professional.
Evaluation approach	Summative, clear decision focus.

seems that accessible metrics increasingly are used also for purposes they were not prepared to be used for.

Standardization seems to be another development tendency at least in Denmark. In relation to recruitment processes, departments increasingly specify the dimensions applicants have to be assessed upon, sometimes termed a scholarly qualification matrix, the SQM. For a professorship a SQM may, for example, outline four dimensions, all with several sub-dimensions, to be assessed: Research (internationally recognized, proven ability to engage in new areas of research, frequent publications, good track record, acquiring external grants, experience with leading roles in networks), education (ample experience and good results with supervision, solid experience and good results with course development, solid experience of PhD supervision), service to society (proven ability to engage with stakeholders outside the university,) and personal (evidence of active contributions to the administrative and managerial tasks, evidence of active mentorship, including co-authorships with junior colleagues). In addition to such performance goals there may be further criteria such as the ability to establish cross-disciplinary networks at faculty level, good results with study program management and evidence of how own research has had non-academic impact. Compared to classical peer review related to recruitment, criteria these years have become very multidimensional.

This development of peer review becoming more standardized and metric based may reflect an aim to make assessments more transparent and fair, but it transforms peer review practices from being discretion-norm-based practices to standards-based practices. Further, it reflects an increasing individualization. In the thinking of the SQM, departments are not entities which on a group or collective basis have to fulfill multi-dimensional performance goals. Rather, every individual in the department has to fulfill every imaginable performance goal.

Another area where peer review becomes more standardized is in assessments of applications for research projects, centers and so on. In some contexts, research councils demand peers to use multiple specified criteria and sometimes also to put grades on all criteria. Again, this development may reflect an aim of transparency and fairness, but it enables decision-makers to make mechanical decisions.

Modified Peer Review

In addition to scientific floor-level evaluands such as research products and scholars, more evaluands such as research groups, departments and universities have been introduced. Evaluation is no longer only going on at the micro level. In the meta- and macro-level evaluation cases, the reading of first-order material (publications) either is less important or it has been replaced by the reading of other types of materials such as self-evaluation reports, a range of available metrics and, in some systems, impact cases. Also site visits and interviews have become important evaluative information. We can term this modified peer review. The content and context of modified peer review is characterized in Table 5.3.

Modified peer review is implemented both on the individual department, faculty or university level, and, in some countries, in national evaluation systems. In all three Scandinavian countries modified peer review has been on the agenda for some years.

The Research Council of Norway in the last 20 years has carried through several rounds of research quality evaluation based on modified peer review (https://www.forskningsradet.no/Statistikk-og-evalueringer/evalueringer/). Most evaluations have had the focus on disciplines, one panel evaluating all departments in the discipline across the universities. In recent years, focus has been broadened. In 2011 an evaluation of biology, medicine and health research was carried out, in 2017 one of the humanities and in 2018 one of the social sciences. The broad evaluations have been organized with several panels. In the case of the biology,

Table 5.3 Modified peer review

Task	Assessment of research quality at the organizational levels (groups, departments, disciplines, universities).
Process	Reading second-order material, doing site visits, presenting assessments by reporting. Sometimes rating.
Peer panel composition	Heterogeneous. Panel members having different fields of specialization as to cover the organizational level in question.
Evaluation approach	Summative.

medicine and health evaluation, seven independent panels looked into subfields and a principal evaluation committee integrated the findings, conclusions and recommendations from the seven panels in a joint evaluation report. In the context of the national evaluation system, Norwegian universities have been reluctant to initiate modified peer review evaluation at the university level.

The opposite situation is found in Denmark and Sweden, where the universities have been the primary agenda setters in relation to modified peer review evaluation. In the Scandinavian context Copenhagen Business School was a frontrunner in implementing modified peer review at university level in the 1990s (Hansen & Borum, 1999; Borum & Hansen, 2000). After internal discussions followed by voluntary trials, an evaluation program over a number of years passed all departments through evaluation processes. The individual departments had significant room for maneuver in regard to how to organize and which material to produce to the disposal for peer review panels. The possibilities for local adaption reduced conflicts but also resulted in variations in the value of the processes.

In Sweden, Uppsala University has been a frontrunner in implementing modified peer review. In 2006/2007 Uppsala University organized a process, called the KoF07, taking 75 departments and units through peer review conducted by 24 peer panels with a total of 176 panel members. The aim of the evaluation was to find and display the "gold nuggets" in the university's basic production units, both those which could already provide evidence of success and those that appeared to have significant potential for the future. Later, other Swedish universities followed in the footsteps of Uppsala, and Uppsala University itself repeated the evaluation exercise in 2011.

Other universities in Scandinavia has also been inspired by Uppsala and other European universities. At my own university, University of Copenhagen, modified peer review of departments some years ago was taken up at the Faculty of Social Science. Later, a common concept for the whole university was worked out and implemented across all faculties, the concept being considerably more standardized than the prior one used at the business school.

Like in Uppsala University, the Copenhagen approach was first and foremost a summative approach. In addition to inspiration from Uppsala, the University of Copenhagen also looked to the UK experiences, where modified peer review evaluation has been used as a method to grade research institutions and subsequently use the grades as a basis for distributing resources. The University of Copenhagen wanted to use the UK grades in the form of numbers in its approach. This proposal met resistance in the organization. Instead of asking the panels to grade through numbers, the panels were asked to use the prose version of the UK grades.

The Copenhagen evaluations have not been used for distributing resources but have served as input into a strategy and planning process. Panel reports were used as leadership information in the management hierarchy. Department heads had to work out plans as to how to act upon the assessment and advice in the panel reports. Deans had to work out a faculty report on the basis of the individual department reports, and the vice-chancellor, a report to the board on the basis of the reports from the deans. In this way the summative evaluation approach turned out to be used in a more formative learning-oriented process. This seems also to be the case at the department level, where department heads characterize the self-evaluation part of the process as more valuable than the panel reports.

Contrary to Uppsala University, where evaluation reports are public and easily accessible (see: https://uu.se/en/about-uu/quality/evaluation/evaluation-of-research/), the University of Copenhagen chose not to go public, but treat evaluation reports as internal documents. Only the short vice-chancellor report to the board is publicly accessible. The decision to not go public was anchored in discussions about whether publicness would turn self-evaluations into beautification as well as restrict the panels in presenting honest critique.

The Scandinavian comparison shows that different actors may set the scene and the agenda for research quality evaluation based on modified peer review. The comparison between Uppsala University and the University of Copenhagen further shows that the modified peer review evaluation concept is spacious. It is possible to adapt it to local organizational values and agendas.

Extended Peer Review

Peer review at organizational levels may also be implemented in a formative approach. We can term this extended peer review. The content and context of extended peer review is characterized in Table 5.4. In extended peer review focus is not on the quality of research results and contributions to knowledge production, but instead on whether the way of organizing, strategies and management processes support research quality development.

Recently, Uppsala University organized a third round of peer review evaluation. In the last round, called the KoF17, focus was less on the quality of research and more on whether the research environments were well-functioning, with a special emphasis on conditions for and processes contributing to research quality and renewal. In this round the evaluation approach was thus formative, whereas the approaches in the two first rounds were summative. In this third round 130 external peers organized into 19 panels participated. Before peers came on site visits, the university conducted a survey of research staff aiming at investigating their view on the quality of the research environments. As a consequence of the evaluation, several development initiatives have been launched among these initiatives related to the development of clear career paths. The Uppsala example illustrates how peer review evaluation at organizational levels across time can take different directions.

Modified and extended peer-review-based research evaluation initiated by universities aims at securing and developing quality in research and

Table 5.4 Extended peer review

Task	Assessment of aspects related to research quality at organizational levels (e.g. organization, structure, management, strategy).
Process	Reading second-order material, doing site visits and interviews with stakeholders, presenting assessments by reporting.
Peer panel composition	Heterogeneous. Multi-professional.
Evaluation approach	Formative.

research organizations. It also serves other purposes. One important purpose is to act as a shield toward national authorities and political initiatives. By taking ownership to quality assurance of research the universities try to protect their autonomy. In this perspective it is interesting that the Association of Swedish Higher Education Institutions, SUHF, which is an interest organization for Swedish universities and university colleges, has worked out a joint framework for quality assurance and quality development in research (https://suhf.se/gemensamt-ramverk-for-larosatenas-kvalitetssakring-och-kvalitetsutveckling-av-forskning/). The idea is to support quality work at individual institutions. Further, it is an attempt to influence the Swedish Higher Education Authority's (UKÄ), which has recently been asked by the Swedish government to develop a national system for the scrutiny of higher education institutions' (HEIs) quality assurance of research.

Peer Review in the Educational Field

The use of peer review is also important in the field of education, where quality assurance systems and accreditation procedures in the wave of the Bologna process have become part of the daily life in higher education institutions. In this sphere the professional and collegial peer review evaluation model has become mixed with other actor models, including the user evaluation model inviting, among others, labor market representatives and students into the evaluation process. In this sphere there are also several evaluands. The quality of educational programs (their curriculum content, pedagogical principles and, sometimes, even student learning outcomes) may be in focus and/or institutional quality assurance procedures and systems.

Sweden was in the late 1960s and early 1970s the frontrunner country in Scandinavia in relation to developing a national system for quality assurance in higher education. Pedagogical research and development projects delivered the ideational raw material for system development (Gröjer, 2004). Across time, different agencies have developed and used different concepts, moving back and forth between giving priority mostly to assessing quality in programs and assessing quality work at institutions.

These years the Swedish Higher Education Authority, UKÄ, has regained focus on scrutinizing quality work at institutions (https://english.uka.se/quality-assurance/quality-assurance-of-higher-education.html). Reviews focus on how well institutional quality systems help to improve the quality of programs. Six assessment areas are in focus: governance and organization, preconditions, design, implementation and outcomes, student and doctoral student perspective, working life and collaboration, and, finally, gender equality.

The review process includes several elements, among these: asking the institutions to work out self-evaluations, inviting student associations to give input to how they experience their influence on institutional quality work, panels doing site visits, posing questions to the self-evaluation and examining one or two specific quality assurance processes, so-called audit trails. Panels consist of student and doctoral student representatives, employer and working life representatives, and experts/peers from the higher education sector. The overall judgment is given on a three-point scale: approved, approved with reservation and quality assurance processes under review. The panel makes a preliminary judgment, which serves as the basis for UKÄ's decision. All review processes have follow-ups, but the form differs according to the overall judgment given.

In Denmark educational evaluation has also been organized differently across time. Ad hoc initiatives saw the light of the day in the late 1980s. Inspiration came from, among other countries, the Netherlands. In 1992 the Danish Centre for Evaluation of Higher Education was established. In 1999, the center was reorganized into the Danish Evaluation Institute. In 2007 a large re-organization was undertaken and an accreditation agency called ACE Denmark was introduced (Hansen, 2009a, 2014). In 2013 one more reform followed and the name of the agency was changed to the Danish Accreditation Institution (https://akkr.dk/akkreditering/).

Along with all the organizational changes, evaluation approaches have also changed. In the early years, evaluations were formative and learning oriented. When accreditation was introduced, the approach became summative. Every higher educational program at bachelor's and master's level, old and new ones, had to go through an accreditation process in order to become approved.

With the 2013 reform the system has gradually switched from accreditation of individual programs to accreditation of entire education institutions. The model for institutional accreditation gives the individual institution a free hand to organize its own quality assurance system as long as it lives up to the five criteria for quality and relevance laid down in the ministerial order. Two of the five criteria concern the quality assurance system, while the remaining three criteria concern the quality and relevance of the educational programs.

In all the shifting evaluation regimes, expert panels have been set up to perform the evaluation process. In institutional accreditations, panels typically have five members, including peers, one labor market representative and one student. The panel visits the institution to interview management, teachers, quality assurance employees and other relevant stakeholders. The accreditation institution works out a report gathering the panel's assessment.

It is emphasized to panels that they have no decision power. This rests with the Accreditation Council (https://akkrediteringsraadet.dk), comprised of nine members, including experts on higher education, labor market representatives and one student. On the basis of the panel report and other sources of information, the council takes decisions on whether to approve, conditionally approve or reject institutional accreditation.

There are examples of panels experiencing that their assessments are somewhat overruled. Peer discretion may be restricted by a more authoritative council culture giving priority to fairness and equal treatment across institutions. The question is whether overruling experiences in the long run have consequences for peer recruitment.

As a consequence of a decision of an institution being conditionally approved or rejected, the institution is faced with demands for accreditation of individual educational programs. The evaluation system is in this way an arena for negotiation about university autonomy parallel to the dynamics related to research evaluation.

In Norway, Norgesnettrådet was established in 1997, one of the tasks being to draw up guidelines for quality work within higher education. Some years later in 2002, the Norwegian Agency for Quality Assurance in Education (NOKUT) was established (https://www.nokut.no/en/).

NOKUT was given the task to review the quality assurance systems at all higher education institutions and this has been done in several rounds.

Contrary to Denmark, the universities in Norway have authority to establish educational programs. Other types of higher educational institutions have to apply for accreditation of programs. Rules for this differ across institutional types. Institutions may also apply for accreditation in another institutional category. In this way institutions can follow a path to become university, move up the hierarchy so to speak, thereby obtaining increased autonomy. And several former university colleges have become universities across time. The Norwegian system is in this respect very different from the Danish system, with the Swedish system in between. Like in Denmark and Sweden, panels including peers are important in the assessment processes. There are different rules for panel composition according to the assessment task.

NOKUT itself was evaluated in 2007/2008 by a panel composed by researchers assigned to the task by the Norwegian Ministry of Higher Education. As part of the evaluation a survey was sent to institutional leaders, administrative staff, students and academic staff investigating their experience of the impact of different forms of external evaluation. One of the findings of the study was that although NOKUT worked with different types of assessments with different purposes, impacts were perceived as quite similar regardless of the evaluation approach (Stensaker et al., 2011). One explanation discussed was that "formal procedures, rules and regulations are 'softened' during practice, underlining the classical distinctions between 'talk' and 'action', and between formal rules and more pragmatic practices" (Stensaker et al., 2011, p. 475). The authors further state that this possibility may occur due to the fact that all evaluations include peers who seem to contribute to change the (formal) focus of the process. Peers thus appear to have some discretion to translate the authoritative point of departure to a professional practice. Another interesting finding was that the institutional leadership and the administration were the groups identifying most positive effects of the schemes. Positive effects seem not to trickle down to academic staff.

Comparing across countries, both similarities and differences can be observed. In all three countries quality assurance systems are in place and steadily further developed. Also, peers are important members in the panels carrying through the assessments. Further, hitherto the national systems have been specialized in evaluating the educational side of universities. This may be changing in the Swedish context, where UKÄ has been given the task of doing pilots in evaluation of institutional quality assurance of research. In relation to how agencies are organized, there are considerable differences. Structures are different and there seems to be differences in the relations between panels and the formal decision-makers, with the Danish Accreditation Council going further than the decision-makers in Norway and Sweden in restricting the discretion of panels.

Conclusion and Discussion

Focus in this chapter has been on two research questions: (1) How has peer review as an evaluation concept been developed across time? (2) What is the role of peer review today? The analysis has shown that peer review is a concept in continuous development. Where peer review evaluation formerly was carried through in the context of, to a large extent, a self-managing community of scholars, it is today carried through in a context of managerialism. Peer review evaluation has become a mediation tool between society and universities as well as a management tool within universities.

Peer review today has many faces. Classical peer review is still important first and foremost in relation to decision-making in the publication system and in relation to evaluation of dissertations. In other areas, for example, in relation to recruitment and promotion, peer review has become more standardized as sets of evaluation criteria and metrics increasingly have come into use. Also peer review practices in relation to applications of research grants in research councils seem to have become more standardized. This type of peer review was termed informed and standards-based peer review.

In recruitment and promotion contexts, managerialism encircles the peer evaluation process. On the one side, this may strengthen peer review evaluation by reducing biases and securing fairness, but on the other side, it restricts peer discretion and may be experienced by peers as not legitimate managerial intrusion. Managerialism also embraces peer review when peer review is used to evaluate the quality of research at organizational levels such as research groups, departments and universities. This type of peer review was termed modified peer review.

But managerialism not only, to an increasing extent, encircles peer review, but has also been brought into focus for peer review. This is seen in what here has been termed extended peer review, where peer panels are asked to scrutiny the organization, management and strategy in research groups, departments, faculties and universities. Further, it is seen in quality assurance practices related to the educational side of institutions. In some contexts, it is the institutional top management which uses peer review to scrutiny lower-level management. In other contexts, it is external authorities, which, as a result of political agendas, have been given the job to have a keen eye on institutional management.

While it is meaningful conceptually to distinguish between modified and extended peer review, the examples looked into show that these types may be mixed in practice. The examples of summative modified peer review from the University of Copenhagen and Uppsala University (KoF07 and KoF11) included formative advice on how to organize and thus included an element of extended peer review. And in the example of formative extended peer review from Uppsala University (KoF17), some peers reported that they found it hard to evaluate organization and management without assessing the quality of research. Peers thus seems to be more comfortable in the summative approach, with a focus on research quality, and less comfortable in the formative, with a focus on whether organizational and management practices support the development of research quality.

Peer review obviously plays different roles in different contexts, and the different types of peer review are expected to play different roles. Classical and informed, standards-based peer review play important roles in decision-making. Modified peer review plays a role as a rewarding as well as a naming and shaming technology in some situations, followed by change and improvement initiatives, and extended peer review plays a role as a learning process also in some situations, followed by change and improvement initiatives.

However, there is more to it. Modified and extended peer review processes as well as peer review related to educational evaluation constitute arenas for struggles about autonomy and legitimacy. If universities are able to be accountable and maintain order in their own house, they may keep the authorities at a distance, thereby protecting institutional autonomy. University-initiated modified and extended peer review processes thus aim at securing that the authorities experience the activities as legitimate, and authority-initiated modified and extended peer review processes aim at securing that the political leadership and citizens experience the activities as legitimate. In this way peer review processes link to policy processes and questions about how to distribute resources in university systems.

Still, one can wonder what the balance is between costs and values both in modified and in extended peer review. As the examples have shown, these processes often involve many peers as well as a considerable amount of university staff, and thus occupy many working hours, which could have been used for doing research. Likewise, quality assurance in education often demands considerable paperwork, occupying considerable administrative resources and probably building up a new administrative layer specialized in evaluation. Future studies should look into this. Further, future studies should pay attention to the linkages between managerialism and peer review practices, and put focus on the consequences of managerialism encircling peer review practices.

References

Benos, D. J., Bashari, E., Chaves, J. M., Gaggar, A., Kapoor, N., LaFrance, M., Mans, R., Mayhew, D., McGowan, S., Polter, A., Qadri, Y., Sarfare, S., Schultz, K., Splittgerber, R., Stephensom, J., Tower, C., Walton, R. G., & Zotov, A. (2007). The ups and downs of peer review. *Advances in Physiology Education, 31,* 145–152.

Bornmann, L. (2011). Scientific peer review. *Information, Science and Technology, 45*(1), 199–225.

Borum, F., & Hansen, H. F. (2000). The local construction and enactment of standards for research evaluation: The case of the Copenhagen Business School. *Evaluation, 6*(3), 281–299.

Brunsson, N., & Sahlin-Andersson, K. (2000). Constructing organizations: The example of public sector reform. *Organizations Studies, 21*(4), 721–746.

Dahler-Larsen, P. (2012). *The evaluation society.* Stanford University Press.

De Boer, H. F., Enders, J., & Schimank, U. (2007). On the way towards new public management? The governance of university systems in England, the Netherlands, Austria and Germany. In D. Jansen (Ed.), *New forms of governance in research organizations: Disciplinary approaches, interfaces and integration* (pp. 137–152). Springer.

Geschwind, L., Hansen, H. F., Pinheiro, R., & Pulkkinen, K. (2019). Governing performance in the Nordic universities: Where are we heading and what have we learned? In R. Pinheiro et al. (Eds.), *Reforms, organizational change and performance in higher education.* Palgrave Macmillan/Springer Nature. https://doi.org/10.1007/978-3-030-11738-2_9

Gröjer, A. (2004). *Den utvärdera(n)de staten. Utvärderingens institutionalisering på den högre utbildningens område.* Statsvetenskapliga Institutionen, Stockholms Universitet.

Hamann, J., & Beljean, S. (2017). Academic evaluation in higher education. In J. C. Shin & P. Teixeira (Eds.), *Encyclopedia of international higher education systems and institutions.* Springer Science+Business Media.

Hansen, H. F. (2005). Choosing evaluation models. *Evaluation, 11*, 447–462.

Hansen, H. F. (2009a). Educational evaluation in Scandinavian countries: Converging or diverging practices? *Scandinavian Journal of Educational Research, 53*(1), 71–87.

Hansen, H. F. (2009b). *Research evaluation: Methods, practice, and experience.* Danish Agency for Science Technology and Innovation.

Hansen, H. F. (2014). 'Quality agencies': The development of regulating and mediating organizations in Scandinavian higher education. In M. Chou & Å. Gornitzka (Eds.), *Building the knowledge economy in Europe. New constellations in European research and higher education governance.* Edward Elagar Publishing Limited.

Hansen, H. F., Aarrevaara, T., Geschwind, L., & Stensaker, B. (2019b). Evaluation practices and impact: Overload? In R. Pinheiro et al. (Eds.), *Reforms, organizational change and performance in higher education.* Palgrave Macmillan/Springer Nature. https://doi.org/10.1007/978-3-030-11738-2_8

Hansen, H. F., & Borum, F. (1999). The construction and standardization of evaluation: The case of the Danish university sector. *Evaluation, 5*(3), 303–329.

Hansen, H. F., Geschwind, L., Kivisto, J., Pekkola, E., Pinheiro, R., & Pulkkinen, K. (2019a). Balancing accountability and trust: University reforms in the Nordic countries. *Higher Education, 78*(3), 557–573.

Hicks, D., Wouters, P., Waltman, L., de Rijcke, S., & Rafols, I. (2015). The Leiden Manifesto for research metrics. *Nature, 520*(22), 429–431.

Langfeldt, L., & Kyvik, S. (2011). Researchers as evaluators: Tasks, tension and politics. *Higher Education, 62*, 199–212.

Langfeldt, L., & Kyvik, S. (2015). Inre spänningar och framtida utmaningar för *peer review*. In *RJ:s årsbok 2015/2016* (pp. 153–169). Riksbankens Jubileumsfond & Makadam Förlag.

Neave, G. (1998). The evaluative state reconsidered. *European Journal of Education, 33*(3), 265–284.

Olsen, J. P. (2007). The institutional dynamics of the European university. In P. Maassen & J. P. Olsen (Eds.), *University dynamics and European integration* (pp. 25–54). Springer.

Scriven, M. (1991). *Evaluation thesaurus*. Sage.

Seeber, M., Lepori, B., Montauti, M., Enders, J., de Boer, H., Weyer, E., Bleiklie, I., et al. (2015). European Universities as Complete Organizations? Understanding identity, hierarchy and rationality in public organizations. *Public Management Review, 17*(10), 1444–1474.

Stensaker, B., Langfeldt, L., Harvey, L., Huisman, J., & Westerheijden, D. (2011). An in-depth study on the impact of external quality assurance. *Assessment & Evaluation in Higher Education, 36*(4), 465–478.

Vedung, E. (1997). *Public policy and program evaluation*. Transaction Publishers.

6

Your Brother's Gatekeeper: How Effects of Evaluation Machineries in Research Are Sometimes Enhanced

Peter Dahler-Larsen

Introduction

The influence of evaluation machineries (such as bibliometric indicators) upon researcher practices is a much debated issue. One of the key points is that no indicators give a full picture. Some indicators do not show sufficient attention to publications in other languages than English (Archambault et al., 2006; Dahler-Larsen, 2018); to differences across fields and subfields in science, social science and humanities (Leydesdorff & Bornmann, 2016); to the consequences of research evaluation machineries upon the choice of research themes among researchers (López Piñeiro & Hicks, 2015); to the citations of publications rather than just publications themselves (Harzing & Mijnhardt, 2015); and, if social impact of research is taken into account, to the variations in definitions of such impact (Penfield et al., 2014), including the voices of various stakeholders throughout the research process.

P. Dahler-Larsen (✉)
University of Copenhagen, Copenhagen, Denmark
e-mail: pdl@ifs.ku.dk

© The Author(s) 2022
E. Forsberg et al. (eds.), *Peer review in an Era of Evaluation*,
https://doi.org/10.1007/978-3-030-75263-7_6

It seems that there is no indicator which finally closes the "evaluation gap," meaning a distance between what is measured and the values associated with research (Wouters, 2017). As a consequence, many have articulated advice about how to curb the influence of such machineries and secure they are used responsibly (Hicks et al., 2015). Advice include: "never use one indicator alone," "be specific about which indicators are used for which purposes" and "always use indicators as support, not as a substitute for human judgment." In a broader perspective, these warnings resonate with warnings against "automation bias," a cognitive failure where the human mind places too much trust in technological algorithms providing information, a phenomenon also leading to "death by GPS," where people drive into lakes if their GPS tells them to (Bridle, 2018).

Back to research evaluation. The underlying assumption is that somehow human judgment can and should function as a bulwark against unintended and constitutive effects of evaluation machineries. While this assumption may hold in some situations, it fails to consider the intimate interaction between human and non-human elements in research evaluation. "Peer review may already be 'informed' by metrics, albeit perhaps not in the systematic and expert led way the proponents of informed peer review would have wished for," says Wouters (2017, p. 110).

Therefore, there is good reason to study the formal and informal practices of gatekeeping which take place in this interactive space where minds and machineries are woven together and associations are made (Latour, 2005). I suggest there are many of these practices. They do not only include official decisions about publications or promotions, but a range of daily-life activities where people engage in conversations about what is and what is not likely to pass in the light of various kinds of metrics, and, importantly, in the light of not what these metrics are, but what they are becoming. If the future is uncertain, one must act with caution. For those who (claim to) know the future, it can be in one's interest to use this knowledge to position oneself and to make others act accordingly.

Empirical observations reported in this chapter come from a case study of the Danish Bibliometric Research Indicator (BRI; an indicator built upon an earlier Norwegian version of a similar nature) (Schneider, 2009). These observations reveal that quite a lot of social action takes place around this indicator based on interpretations and imaginaries.

As people take into account how they imagine future metrics, and act upon these imaginaries now, and bring others into action, too, it becomes possible to understand why human judgment in some situation helps make evaluation machineries such as bibliometrics even *more* influential than they would otherwise be.

By developing grounded hypothesis about how and why this multiplication of effects happens under some circumstances, this chapter contributes to an understanding of how the interactions between "human" and "machine-like" forms of evaluation contribute to constitutive effects of evaluation systems in research.

A key ingredient in these situations is anticipation—and co-construction—of a not-yet-constructed reality. Gatekeepers who "know the future" or anticipate a coming future play a key role, of course in combination with a range of situational factors. There is no guarantee that these acts of performativity are always successful (Butler, 2010). In fact, one of the reasons why quite a lot of "fuzz" around the BRI is found in the case study is that perhaps there is no easy, direct and linear way to a predictable form of use of the indicator. So many attempts are made by different actors with different perspectives and purposes (de Lancer Julnes, 2011, p. 67). In turn, this makes the social life of the BRI a dynamic one, but also one that remains ambiguous.

The chapter proceeds in the following way. First, a theoretical argument is provided for ambiguity and interpretability as key concepts in recent studies of the use of metrics. Secondly, the BRI and the case study of it are introduced. Then follows a number of incidents presented as vignettes illustrating interpretations and actions in relation to the BRI among Danish researchers and institutions. Finally, a short conclusion.

Metrics and Ambiguity

The concept of ambiguity refers to situations where a phenomenon can meaningfully be interpreted in multiple ways. Theoretically, the concept plays different, but not irreconcilable roles in various frameworks. In relation to the social construction of reality, ambiguity is an indication of some degree of "opening" of the future (Best, 2008). In practice studies, ambiguity is a sign of "multiple orders of worth" which actors handle in concrete situations by developing a variety of strategies (Stark, 2009).

In recent years, we have seen a number of empirical studies of metrics which resonate with these and similar theoretical orientations. The production of numbers is itself a complicated and demanding social accomplishment (Porter, 1995; Desrosières, 2011). Dambrin and Robson (2011) have shown that perfect validity of an indicator is not a requirement for practical use. Instead, ambivalence and opacity can be conducive to the implementation of otherwise flawed measures.

There is sometimes uneven implementation of the same indicator system across institutions in the same country (Hammarfelt et al., 2016; Lind, 2019; Mouritzen et al., 2018). Institutions that integrate the system most directly into their management systems are not necessarily those institutions where professional values are most consistent with the spirit of the indicator system (Lind, 2019; Mouritzen et al., 2018). The opposite may be the case if managers use the new system to induce change.

Even when a system is adopted, struggles over documentation practices sometimes remain unsettled or ambiguous over fairly long periods (Mouritzen et al., 2018; Kaltenbrunner & de Rijcke, 2017). Sometimes, people under evaluation are in position to influence the design of evaluation systems (Pollock et al., 2018), but since such influence is highly ambivalent, participatory processes run far from smoothly and may be perceived with suspicion and mixed feelings (Jensen, 2011). These observations may be particularly pertinent among researchers who usually cherish autonomy and peer review as sacred professional principles.

Once an evaluation system is in place, it may suffer from "mission drift," so that it over time serves other purposes than its original ones (Kristiansen et al., 2017). For example, when an evaluation system is connected with money streams, its main function may change from provision of information to resource allocation. Some even suggest the existence of a "runaway effect" (Shore & Wright, 2015). The potential runaway or "mission drift" may sometimes be paradoxical, however, since such effect, as we shall see, does not always hinge on financial and material implications alone. The imaginary aspect is also important.

Interpretations and behaviors among researchers themselves may in fact set in motion a "runaway effect." Managers may promise that they do not intend to use particular indicators at the level of individual research-ers, but only at collective levels such as departments or research groups. This practice of "buffering" may be seen as a good ethical practice. However, to the extent that individual scores are publicly available or can be computed by individual researchers themselves, they may use their own scores for promotion or marketing purposes (Fochler & de Rijcke, 2017). The original promise of buffering is broken, but it is researchers themselves who break it.

Researchers can also take precautionary actions against potential future use of the evaluation system, thereby setting in motion a new set of effects. Given differences between original purposes and emerging pur-poses of evaluation systems, and the differences between the design of such systems and imaginations of their functions, it becomes clear that the official promises and declarations about the purposes of evaluation systems cannot be trusted to predict the future use of such systems, even if such declarations are honestly made (Dahler-Larsen, 2013).

Evaluation systems such as bibliometrics and rankings can have dra-matic consequences for institutions, especially if strong alliances around these institutions exert pressure on managers to act upon the scores (Espeland & Sauder, 2007). In other situations, managers can intention-ally pursue a definition of reality that is constituted only by what is made visible by particular evaluative machineries, thereby reducing the com-plexity they are dealing with while leaving the difficult interface between evaluation and reality to others in the organization (Roberts, 2017).

As research evaluation based on quantification of publications and citations apparently increase in importance, researchers will potentially change their practices accordingly. Some of these practices may be unfor-tunate (such as producing more publications of lower quality, focusing on safe but trivial research questions, and slicing projects into several publications) (Osterloh & Frey, 2010). Other practices, however, may be even more problematic, unethical or illegal, including misrepresenta-tion and misconduct (Biagioli et al., 2019). For this reason, institutions see an increasing need to sharpen their regulations of ethical research conduct, documentation practices and more. One of the side effects of these endeavors may be to cast a shadow of suspicion on normal prac-tices which merely happen to not present themselves neatly in relation to the new regulations and detailed documentation guidelines (Dahler-Larsen, 2017).

In other words, while evaluation systems produce some forms of clarity and transparency, they also produce their own ambiguities. Furthermore, these studies show that the function of evaluation systems is not a physical property inherent in such systems. A more productive focus is on the activities of people in and around these systems (Becker, 1998). If the evaluative systems have "functions," they are produced by these activities. However, and that is the point, all these activities may be based on neither clear nor consensual understandings of the metrics and their meanings. If ambiguity is ever-present, and the social construction of metrics is unfinished, we can expect quite a lot of interpretive activity directed toward guessing what the metrics will bring in the future. This activity may be a constructive factor itself.

A Case Study of the Danish Bibliometric Indicator

The Danish Bibliometric Indicator (BRI) was politically decided in 2009 and came into effect in 2010. The alleged purpose was defined in terms of a "healthy competition" about resources for research (Mouritzen et al., 2018, p. 17). The BRI is basically a mechanism for distribution of bibliometric points.

Appointed groups of researchers in all disciplines and subdisciplines divide all publication outlets into two levels, while only 20% of the world production is allowed to be placed at level 2, the finest level. Level 1 is intendedly more inclusive, although only peer-reviewed publications count. All forms of publications such as articles, monographs and book chapters at levels 1 and 2, respectively, are then given a particular number of bibliometric points. Finally, a proportion of all state funding of research is reallocated across research institutions depending on how many points they scored. In Denmark, the redistribution takes place only across institutions, not across fields. Over the years, depending on a change in complicated mathematical formulae, the financial value of a BRI point has increased.

In the following case study, daily-life incidents in research institutions in which the BRI played a role will be reported in the form of short narrative vignettes. The vignettes are based on personal observations of the author (although "a research institution" is not necessarily the author's present employer).

One problem with the methodology used here is subjectivity regarding the selection and reporting of the incidents. On the other hand, however, a strength of the same methodology is its ability to capture incidents, arguments and interactions as they unfold in daily life. Furthermore, the specific purpose of this chapter is to highlight the interpretable and interpreted nature of metrics. The methodology is therefore consistent with the aim of this chapter, and it resonates with the orientation toward social practice theory, which characterizes several studies cited above.

Vignettes

Incident 1. Soon after the introduction of the BRI, there is a research symposium where a small group of senior and junior researchers at a department discuss research papers. One of the seniors claim that the level of expectations in international publication has increased in recent years. In discussion of a particular methodology commonly used, "you will not get published in the good journals," he says, unless particular requirements regarding that methodology are met. He explains what these requirements are. Presumably, junior researchers must follow his advice if they wish to hope for a future in academia.

Although his statement is not causally linked to BRI as such, it helps create a context of rising expectations. It is not specified what "good journals" means more specifically. Nevertheless, the incident exemplifies how local gatekeepers and "wise men" can use the broader context of publication pressure to channel the energy and focus of younger researchers into particular directions. Although nobody can exactly *know* what the future brings and nobody has mapped all the criteria used in all editing decisions and future promotion decisions, there are "wise" men who offer an "authoritative" view of "what is required." Ambiguity is transformed into advice about choices of paradigm and methodology. Again, although the

BRI is not referred to directly, its very existence suggests that from now on it may have more serious consequences if you do not do "what is required" since the BRI is part of university management.

Incident 2. Soon after the introduction of the BRI, managers at one university in Denmark decided to use the principles of the BRI in their internal allocation of resources across departments, in other words to reinforce the internal pecuniary repercussions of the BRI. At the same time, they defined a minimum threshold of BRI points expected from each researcher over a period of time. They declared that this was a way to prepare their university for the future. Over the years it turned out that this particular university gained from the inter-institutional reallocation of BRI funds as compared to the situation before BRI was introduced. Several other institutions did not draw any implications of the BRI for their internal allocation of resources. Some research groups ignored the BRI because they thought that the existing academic reputations of various journals provided a more serious and nuanced assessment of their value than the simple two-tiered BRI system. Perhaps they assumed that their own prestige would be strong enough to withstand any pressure from the BRI system, and they assumed that the BRI would not survive in the long run.

Incident 3. On a normal day at the department, a professor with a very good reputation talks about the qualities of a recently hired PhD student. Already, the student has got an article accepted at level 2, it is said. The example shows that although researchers often refuse to accept the BRI as a reflection of true academic value, they nevertheless use BRI terminology in some of their descriptions of great achievements.

Incident 4. One department established a system according to which all publications at level 2 release a financial reward to the authoring researcher. Over the years, this system is believed to have contributed to a significant rise in the quality and quantity of publications. The BRI system has contributed to this order of things by providing an externally defined list of journals which relieved the researchers of the otherwise painful task to internally agree on a list of what the "best" journals would be.

In a recent external evaluation, the evaluation committee recommends the elimination of the reward system in the department. The argument is that now that the system has helped raise the level of achievement, it can now be taken for granted that everybody knows the importance of high-level publications. The recommendation is implemented, although one might ask: If it is acknowledged that a given financial incentive has worked, will its disappearance not make a difference? But the evaluators and the managers assume they know the future.

Incident 5. One researcher at one institution is invited to contribute a chapter to a book edited by a researcher at another university. The editor supplies the invitation with a remark saying that of course, all authors will be given BRI points for their contribution. The invited researcher accepts the invitation.

Incident 6. In order to compensate for the somewhat rough distribution of all publications into two levels in the BRI system, and in order to sharpen the focus on the very best journals, it is suggested to introduce a level 3 in the BRI system. A consultation process is designed. A research committee discusses the proposal. (The research committee is a departmental committee responsible for strategic and practical issues related to research. It consists of leaders of all research groups and research centers in a department). The field consists of different subdisciplines, and since only a small fraction of journals (5%) can be placed in level 3, not all subdisciplines are likely to be represented there. This will create tensions between the subgroups. It is also foreseen that when it comes to the exact identification of journals to be placed in level 3, there will be intense discussions and maybe conflicts. When a committee member proposes to base the decision on an objective criterion such as journal impact factor, another member answers that based on the literature on research evaluation, the journal impact factor is not regarded as an unproblematic criterion. It also cannot be used without normalization across subfields. After lengthy discussions, the issue is brought back to the national BRI committee. It turns out that there will be no level 3, because the other research groups in the same field in the country are against the idea.

Incident 7. A well-respected researcher returns to Denmark after having worked abroad for some years. The researcher is astonished about the fuzz related to the Danish BRI system. The researcher believes that it is silly that Danes establish their own system which does not reflect the exact status that various publications have in the international world.

Another researcher argues that funds paid out through the BRI system comes from Danish taxpayers, so Danes have the right to decide whatever principles they are pleased with, as long as they are financing the consequences of their decisions. Furthermore, perhaps on a more serious note, it is suggested that there is no such thing as a single, uniform, authoritative and undebatable determination of the international reputation of all publication outlets. In addition, geography actually plays a role. For example, in Denmark it may be legitimate to prioritize Scandinavian studies or EU studies higher than they would be, for example, in the US.

Incident 8. A researcher claims to have identified a number of phony publication outlets in the BRI system. All researchers are therefore asked to participate in an official cleaning process. In the local research committee, everybody supports the removal of phony publications, but then the consensus ends. One researcher argues that the committee should not spend much energy on BRI. Intense discussions are only likely to lead to spur dissensus, but not support to major changes in the design of the BRI. In addition, the financial impact of changes in the BRI are likely to be minimal and not implemented at the department level.

Other members of the committee argue that despite the limitations of the BRI, it is likely to remain an important factor in research evaluation. For example, one never knows whether research committees in the future are likely to look at BRI scores for individual researchers in promotion and hiring situations. Even if committee members themselves would not attach much weight to BRI scores themselves, they can be instructed to do so through the terms of references given to them by university managers. For this reason, it is argued, the importance of the BRI may increase in the future, so it is important not to ignore it.

One view is that the included journals should reflect "the core of our field." Another view is that the original idea in the BRI is to be inclusive, at least at level 1, thereby stimulating plurality and diversity. It is felt that the BRI is again used to enhance a definition of the field, which is in fact not everybody's definition. The discussion is inconclusive, but the committee decides to return to the issue in future meetings.

Incident 9. As part of the clean-up process mentioned above, the Ministry initiates a review of a large selection of registrations made in the BRI system. To that purpose, it uses a new set of regulations hitherto unknown among researchers. These regulations clarify that only research publications are allowed to count. This is more complicated than it appears to be, because, for example, in social science, some books have overlapping functions, such as being a research publication and a book used in teaching, or a research publication that is also used to stimulate public debate. But researchers are insistently asked to make sure that they are clear about the primary purpose and primary audience of all their registered publications. It is also reiterated that only publications subject to peer review can be given BRI points. As a result of this clean-up process, the author contributing a chapter to the book mentioned in *incident 5,* is contacted and asked to change the registration of the book chapter. The argument is that the book is "perhaps not a research book" as it might be used in teaching. Although the official message is that within the framework of the new regulation, the responsibility to determine the type of each of his/her publications ultimately rests with the author, the researcher decides in this particular case to change the registration of the book chapter into "teaching." Paradoxically, however, the researcher thinks it deserves to be mentioned that at the institution where the editor and other colleagues work, their contributions to the same book remain registered as "research." The researcher therefore continues to believe that some element of ambiguity remains inherent in the very practices of documentation and registration which are crucial to the credibility of the BRI.

Incident 10. The Ministry of Research and Innovation finds it is time to evaluate the BRI. In order to start thinking about relevant issues and evaluation questions, the Ministry invites key academics and evaluators to a meeting. Many issues are discussed, among others whether the present version of the BRI prioritizes quantity over quality. It is also discussed whether the fuzz among researchers over the BRI is paradoxical given the fairly limited redistribution of funds caused by the BRI. However, as a counterargument, it is mentioned that once the BRI is in operation, it is easy to increase its financial impact by a simple change of an algorithm in a spreadsheet in the ministry. A few months later, it is announced that the Ministry does not wish to move forward with an evaluation of the BRI.

Incident 11. A researcher publishes a book with results of a longitudinal research project on how researchers have responded to the BRI

(Mouritzen et al., 2018). Survey data in several rounds are supplemented with interview data and bibliometric data. It is shown that the BRI has an effect on publication patterns, but only in particular fields and only in particular universities. It is also suggested that there is a correlation between use of the BRI and stress among some researchers at those institutions where it is implemented zealously.

At other institutions, researchers know very little about the BRI, the monetary consequences of BRI points, and who gets the money. For these reasons, the BRI presumably plays are very limited role in their daily life. The book also raises a number of issues about documentation practices. When the book is debated in the central committee responsible for the BRI system, its methodology is criticized and the minutes state that "the book cannot stand alone" as an assessment of the BRI system. A member of the committee disagrees with this view, arguing that the methodological weaknesses in the book are not extraordinary, and that the book contributes to the generation of relevant knowledge about the BRI and its effects.

In an interview in public media, the author of the book claims that perhaps the days of the BRI system are numbered.

Incident 12. As a new head of department is hired, a new departmental strategy is developed. The strategy includes a couple of new focus areas. One ambition is to strengthen the social impact of research. Another has to do with finding new types of funding due to financial challenges. These two areas are not totally in line with a focus on BRI points. For the most part social impact of research in Denmark is facilitated through communication channels in Danish. Research demonstrates a fairly clear trade-off between publishing in Danish and having many citations (Dahler-Larsen, 2018). Publications with many citations are often international ones. And publications on level 2 in the BRI are almost exclusively international ones. Trade-offs and compromises can be made but scoring BRI points at level 2 and having social impact are clearly different things.

Next, regarding finances, what are the realities which researchers at the department face? How can they best help the department out of its financial predicament? How much is gained from the BRI system as compared to, for example, externally funded research projects? One researcher asked himself, for example, if he wanted to earn as much money to his university through the BRI system as he got from his most recent externally funded project, for how many years would he have to publish one single single-authored monograph with a good international publisher more than he would otherwise publish? (And the question is asked under the strict assumptions that researchers at other universities do not increase their production more than they would otherwise do and that all the money gained from his publication activity goes directly to his department.) The answer is 120 years. In other words, there is no realistic way in the world that a reasonable extra individual effort via the BRI system can have any substantial effect on the financial situation at the department compared to other much more effective forms of funding.

Incident 13. The Ministry of Innovation and Research announces that time has come for a political revision of the overall financial model for publicly funded research in Denmark. A committee with international and national experts is put together with the task of describing various models and giving political recommendations. The Ministry argues that present models, including the BRI, put too much emphasis on the quantity of the research production at the expense of quality. As a consequence, the life of the BRI may take another turn.

At a meeting in the expert committee it is debated whether the BRI could be revised so that only a given number of publications every year for each researcher (say three or four) could release BRI points. In this way, researchers would focus on their best pieces of work, not on massive production. Rumors say, however, that the BRI has gotten many enemies among institutions who are not benefitting from the redistribution of funds, which follows from it. Others argue that the BRI is basically flawed and alternatives are needed. For example, some argue that peer review should play a stronger role in the research system, rather than "automatic" bibliometrics such as the BRI. In turn, some say that if peer review takes the form of expert panels visiting each research milieu at regular intervals, this model will be expensive and bureaucratic. Furthermore, in a small country like Denmark, it is not possible to recruit a sufficient number of independent experts to panels evaluating research published in the national language, which remains important at least in social science and the humanities. Another suggestion is to channel more funding through research councils, so that funding would depend more on competition among proposals. Rumors say that not only do the experts disagree about what characterizes the best model, there is also fundamentally different institutional interests at stake, because no matter which model is chosen, the choice itself has financial implications for them. Others predict that even if the spirit of the times may not be favorable to the BRI system, the Ministry itself may have an interest in maintaining it, because a lot of resources was spent constructing and maintain the whole institutional and technical apparatus that makes the BRI possible.

Incident 14. The research committee mentioned above again discusses its standpoint regarding publications and BRI. One of its members say that rumors claim that as a result of the deliberations in the national expert committee, maybe the whole BRI will not survive. The research committee decides to debate the issue again at a later point in time.

Incident 15. At the department, it is recommended that all researchers carry out a Personal Research Review at regular intervals. This is basically a consultation where a researcher discusses his or her publication strategy with a respected colleague. Before the consultation, the individual researcher prepares a document describing his or her existing production and as well as ideas for the future. One researcher explains that the normal plan, all other things being equal, is to publish articles at level 2 in the BRI whenever possible. But the main part of the consultation focuses on what can be done to increase the number of Google Scholar citations (as a sort of proxy for impact). Nevertheless, the incident shows that the BRI continues to play a role, albeit perhaps not a dominant role, in the considerations and practices among researchers in their daily life.

Discussion and Conclusion

A lot of activity is going on with regard to understanding, interpreting and sometimes influencing the BRI. There are many people playing the role of interpreters. As attempts to subject the BRI and its trajectory to sensemaking are socially distributed (Weick et al., 2005), so are the resulting gatekeeping functions.

It can be legitimately argued that the 15 incidents reported above are not only subjectively reported, they are also poorly connected as a narrative. Even rumors are reported. However, precisely the *lack* of tight connection between the incidents, and their lack of foundation in objective truth, is an important part of the story, since it opens wide spaces for interpretation. Perhaps so much activity is going on in the reported incidents exactly *because* even fairly insightful people are not in position to predict what the BRI will be like, which documentation practices will count, and which implications the BRI will have on managerial and professional practices.

Many of the incidents reported seem to include activities that have no particular finality. As stated by Butler (2010), social construction is not always successful. There seems to be much wasted energy around the BRI. Nevertheless, it is too early to tell which activities are successful given the instability of the system as such. Stories with finality in them can only be told in retrospect, when we see what the "outcome" of social processes was (Castoriadis, 1987).

Logically, because of the open-endedness of the BRI story as it goes on, there is therefore also a lot of activity that has to do with bringing oneself into a position where the BRI has at least been sufficiently taken into account so that it does not become a total surprise. We can call this *precautionary* or *preemptive use* of the BRI. Especially under complex and dynamic circumstances, precautionary or preemptive use of evaluation may be an important mechanism contributing to constitutive effects of evaluation machineries. When people take action based on what they perceive might be a reality in the future, they in fact help create a particular kind of social order (Hanson, 2000). It may contribute to this mechanism that researchers are sensitive to factors influencing their reputation (Hicks et al., 2015). So, if some see BRI points as a source of reputation-building, it is important to watch out because reputation is a positional good, and researchers are not only colleagues, but also competitors. In this context, it is an important observation that researchers are often actively using bibliometric measures even if they are also critical about the validity of such measures (Fochler & de Rijcke, 2017).

Perhaps the most critical and theoretically interesting point about the preemptive or precautionary use of the BRI system is that people who have interpretations of what the BRI will mean in the future implicate the actions of others as a consequence of these imaginations. The (imaginary construction of the) BRI leads to imperatives such as: "You have to write in this way." "We have to define the core of our discipline." "We have to establish amongst ourselves a common understanding of the hierarchy of publications in our fields." "We have to discuss publication strategies." "We have to take precautions regarding the role of the BRI scores in assessment work." "We have to collectively take a standpoint regarding registration practices." In these many ways, researchers implicate each other both as competitors, as gatekeepers and as colleagues. The fact that the social relations in which the BRI is implicated are sometimes competitive, sometimes controlling and sometimes cooperative in no way reduces the total constitutive effects of the BRI.

In all these ways, and sometimes paradoxically in the midst of confusion and fuzz, the BRI is used both as an implicit sign and as an explicit argument to incite particular understandings of research and collective action based on these understandings. Thus, a particular contribution of this chapter has been to explain that sometimes imaginations of the future of BRI constitute a key ingredient in the social construction of the effects of the BRI. We know they are imaginations because our observations show that different groups hold different views about the meaningfulness, use and future of the BRI.

These imaginaries bring in a number of other agendas with them, such as how to promote particular methodologies or particular definitions of what constitutes the "core" of particular disciplines or the "quality" of research, not to mention what constitutes the very definition of research.

This mechanism may help explain why an evaluation system such as the BRI which has fairly limited financial effects nevertheless has an effect upon minds, mentalities, debates and practices. Just because the BRI is not financially critical for individuals at the present moment in time, it does not mean that it cannot be more financially critical in the future. The precautionary or preemptive logic here contributes to understanding why a system with limited financial effects in the present can still create much fuzz. The point is not that materiality does not count. The point is rather that imagined materiality does count, and may be theoretically and practically very important. Given the strange ramifications of the imaginaries illustrated in this chapter, it has been shown that "human judgment" in all generality is not enough as a vaccine against constitutive effects of evaluation machineries.

Further research might contribute to understanding the role of imagination in the social construction of effects of evaluation systems. In the meantime, at the practical level, practitioners and gatekeepers of many kinds should perhaps reflect upon their own role in sometimes enhancing and multiplying the effects of evaluation systems. It is not enough here to merely enhance "human judgment" in contradistinction to evaluation machines, because it is "human judgment" which produces the imaginations of the evaluation systems as described in this chapter. It has got to be a more complex and well-reflected kind of human judgment. Or perhaps it is just an idea to stop and breathe for a while in the recognition that effects of evaluation systems are sometimes produced by gatekeepers who imagine they are conquering the future.

The observations presented in this chapter offer a perspective on how to meaningfully respond to the pressures from evaluation machines. Just like most fake news are spread on the internet by individuals who pass them on, the effects of evaluation machines hinge on thousands of small individual actions, which form a large network of social consequences.

As this chapter has shown, individual reactions to evaluation machines sometimes enhance the social implications of these machineries. In research you are not only your brother's keeper. You are in fact your brother's gatekeeper. You may want to take this into account when you deal with evaluation machineries.

One positive implication can be negatively articulated: Do not inadvertently use your human judgment and imagination to multiply and increase the effects of evaluation machineries. People's reactions to performance measurement are part of the construction of the political effects of performance measurement (Johnsen, 2008). However, if open protest may not be fruitful, as people may think that protesters speak up because they are themselves not able to produce good metric scores, then tacit inertness may be a meaningful practical strategy. Perhaps there is wisdom in not elevating "human judgment" to the point where one knows what the future brings and what must therefore be done. Perhaps it is better to enjoy the relative freedom in the present, the freedom that comes with ambiguity. Perhaps it is better to imagine what deserves to be published rather than what deserves to be counted.

References

Archambault, É., Vignola-Gagné, É., Côté, G., Larivière, V., & Gingrasb, Y. (2006). Benchmarking scientific output in the social sciences and humanities: The limits of existing databases. *Scientometrics, 68*(3), 329–342.

Becker, H. S. (1998). *Tricks of the trade, how to think about your research while you're doing it.* The University of Chicago Press.

Best, J. (2008). Ambiguity, uncertainty, and risk: Rethinking indeterminacy. *International Political Sociology, 2,* 355–374.

Biagioli, M., Kenney, M., Martin, B. R., & Walsh, J. P. (2019). Academic misconduct, misrepresentation and gaming: A reassessment. *Research Policy, 48*(2), 401–413.

Bridle, J. (2018). *New dark age: Technology and the end of the future.* Verso Book.

Butler, J. (2010). Performative agency. *Journal of Cultural Economy, 3*(2), 147–161.

Castoriadis, C. (1987). *The imaginary: Creation in the Social-historical Domain.* Stanford University Press.

Dahler-Larsen, P. (2013). Constitutive effects of performance indicators—Getting beyond unintended consequences. *Public Management Review, 16*(7), 969–986.

Dahler-Larsen, P. (2017). *The new configuration of metrics, rules and guidelines creates a disturbing ambiguity in academia.* LSE Impact Blog.

Dahler-Larsen, P. (2018). Making citations of publications in languages other than English visible: On the feasibility of a PLOTE-index. *Research Evaluation, 27*(1), 212–221.

Dambrin, C., & Robson, K. (2011). Tracing performance in the pharmaceutical industry: Ambivalence, opacity and the performativity of flawed measures. *Accounting, Organizations and Society, 36,* 428–455.

Desrosières, A. (2011). How real are statistics? Four possible attitudes. *Social Research, 68*(2), 339–355.

Espeland, W., & Sauder, M. (2007). Rankings and reactivity: How public measures recreate social worlds. *American Journal of Sociology, 113*(1), 1–40.

Fochler, M., & de Rijcke, S. (2017). Implicated in the indicator game? An experimental debate. *Engaging Science, Technology, and Society, 3,* 21–40.

Hammarfelt, B., Nelhans, G., Eklund, P., & Åström, F. (2016). The heterogeneous landscape of bibliometric indicators: Evaluating models for allocating resources at Swedish universities. *Research Evaluation, 25*(3), 292–305.

Hanson, F. A. (2000). How tests create what they are intended to measure. In A. Filer (Ed.), *Assessment: Social practice and social product* (pp. 67–81). Routledge Falmer.

Harzing, A.-W., & Mijnhardt, W. (2015). Proof over promise: Towards a more inclusive ranking of Dutch academics in economics & business. *Scientometrics, 102*(1), 727–749.

Hicks, D., Wouters, P., Waltman, L., de Rijcke, S., & Rafols, I. (2015). Bibliometrics: The Leiden Manifesto for research metrics. *Nature, 520*(7548), 429–431.

Jensen, C. B. (2011). Making lists, Enlisting scientists: The bibliometric indicator, uncertainty and emergent agency. *Science Studies, 24*(2), 64–84.

Johnsen, Å. (2008). Performance information and educational policy making. In W. Van Dooren & S. Van de Walle (Eds.), *Performance information in the public sector. Governance and public management series* (pp. 157–173). Palgrave Macmillan.

Kaltenbrunner, W., & de Rijcke, S. (2017). Quantifying 'output' for evaluation: Administrative knowledge politics and changing epistemic cultures in Dutch Law faculties. *Science and Public Policy, 44*(2), 284–293.

Kristiansen, M. B., Dahler-Larsen, P., & Ghin, E. M. (2017). On the dynamic nature of performance management regimes. *Administration & Society, 00*(0), 1–23.

de Lancer Julnes, P. (2011). Performance measurement beyond instrumental use. In W. van Dooren & V. de Walle (Eds.), *Performance information in the public sector. How It Is Used*. Palgrave Macmillan.

Latour, B. (2005). *Reassembling the social. An introduction to actor-network-theory*. Oxford University Press.

Leydesdorff, L., & Bornmann, L. (2016). The operationalization of "fields" as WoS subject categories (WCs) in evaluative bibliometrics: The cases of "library and information science" and "science & technology studies". *Journal of the Association for Information Science and Technology, 67*(3), 707–714.

Lind, J. K. (2019). The missing link: How university managers mediate the impact of a performance-based research funding system. *Research Evaluation, 28*(1), 84–93.

López Piñiero, C., & Hicks, D. (2015). Reception of Spanish sociology by domestic and foreign audiences differs and has consequences for evaluation. *Research Evaluation, 24*(1), ·78–89.

Mouritzen, P. E., Opstrup, N., & Pedersen, P. B. (2018). *En fremmed kommer til byen, ti år med den bibliometriske forskningsindikator*. Syddansk Universitetsforlag.

Osterloh, M., & Frey, B. S. (2010). *Academic Rankings between the "Republic of Science" and "New Public Management"*. Working Paper. Zurich: CREMA—Center for Research in Management, Economics and the Arts.

Penfield, T., Baker, M. J., Scoble, R., & Wykes, M. C. (2014). Assessment, evaluations, and definitions of research impact: A review. *Research Evaluation, 23*(1), 21–32.

Pollock, N., D'Adderio, L., Williams, R., & Leforestier, L. (2018). Conforming or transforming? How organizations respond to multiple rankings. *Accounting, Organizations and Society, 64*, 55–68.

Porter, T. M. (1995). *Trust in numbers: The Pursuit of objectivity in science and public life*. Princeton University Press.

Roberts, J. (2017). Managing only with transparency: The strategic functions of ignorance. *Critical Perspectives on Accounting, 55*, 53–60.

Schneider, J. W. (2009). An Outline of the bibliometric indicator used for performance-based funding of research institutions in Norway. *European Political Science, 8*(3), 364–378.

Shore, C., & Wright, S. (2015). Audit culture revisited: Rankings, rating and reassembling of society. *Current Anthropology, 56*(3), 421–444.

Stark, D. (2009). *The sense of dissonance. Accounts of worth in economic life*. Princeton University Press.

Weick, K. E., Sutcliffe, K. M., & Obstfeld, D. (2005). Organizing and the process of sensemaking. *Organization Science, 16*(4), 409–421.

Wouters, P. (2017). Bridging the evaluation gap. *Engaging Science, Technology, and Society, 3*, 108–118.

7

Peer Review in Public Administration: The Case of the Swedish Higher Education Authority

Agnes Ers and Kristina Tegler Jerselius

The National System for Quality Assurance of Higher Education in Sweden

Since the middle of the 1990s a national system for quality assurance of higher education has been in place in Sweden. Although both the content and the focus of the system have shifted over time, the method for quality assurance has remained remarkably unchanged. At present, the system comprises four different components: appraisal of applications for degree-awarding powers, institutional reviews of the higher education institutions' (HEIs) quality assurance processes, programme evaluations and thematic evaluations.

The method used in all four components is based on self-evaluations written by the HEI under review. The self-evaluation is assessed by a panel of independent peers, so-called peer review. The peers conduct interviews at the evaluated HEI and write a report stating their findings.

A. Ers (✉) • K. Tegler Jerselius
Swedish Higher Education Authority, Stockholm, Sweden
e-mail: kristina.tegler.jerselius@uka.se

© The Author(s) 2022
E. Forsberg et al. (eds.), *Peer review in an Era of Evaluation*,
https://doi.org/10.1007/978-3-030-75263-7_7

Decision, based on the peer report, on the quality at the HEI is taken by the accountable governmental authority, at present the Swedish Higher Education Authority (hereafter referred to as the Authority). This is a well-recognised and internationally accepted method for quality assurance of higher education following the European Standards and Guidelines (ESG).

The national quality assurance system has been described as a "peer review method of evaluation," peer review thus seen as a core element in the assessment processes. Hypothetically, the use of peers and peer review has also been important to the identity of the administrators working at the quality assurance Authority. In other words, peer review is at the heart of the Authority's operations and therefore deserves further attention.

The History

To fully grasp the changes that have taken place within the national system for quality assurance in Sweden from the late 1990s up until today, it is important to look more closely at the historical developments over time. Already in 1992, "the Umeå model" for assessing quality was developed by researchers/evaluation experts at Umeå University. The model consisted of three main activities: self-evaluation, peer review and follow-up. The evaluation was carried out by groups of peers, and assessments were based on general aspects and criteria within a known framework of standards and guidelines. The aspects and criteria were aimed at guiding the institutions and the peers in the assessment processes. To a large extent, there were possibilities for the peers to interpret these quality aspects and criteria freely (Franke & Nitzler, 2008).

The model, although modified over time, has been used by the Authority for quality assurance from 1993 and onwards. Initially the Authority carried out institutional reviews assessing HEIs' quality assurance processes and appraisal of applications for degree-awarding powers. However, the focus has shifted over time. In the period 2001–2006, programme evaluations were conducted at a large scale. Basically all programmes and courses leading up to a degree were evaluated until 2006. The assignment from the government implied that the evaluations were to be made systematically and fully during a long period of time. The framework for the reviews consisted again of general quality aspects, and it was stressed that they were not to be seen as "final or exhaustive." They were designed to catch the aim of the reviews: to assess and present the quality in higher education and represent a stimulus for development and renewal of the programmes. During this time sanctions were also introduced into the system, meaning that a negative outcome of a review could result in a revocation of the right to award a degree (Franke & Nitzler, 2008).

Over time the Authority grew, and with it, its department of evaluation. The number of staff is still today approximately 35. During this period, a strong consensus was established in the sector of higher education that it was necessary to have a full national system for quality assurance of higher education. The number of peers involved in the reviews grew steadily. The system was successful, but by the end of this period the debate about how this system should be designed became more and more intense. At the heart of the debate—even though not always pronounced—was the method of peer review.

Critique

The method for assessing quality was put into question. A lot of the critics focused on the fact that the system was *arbitrary*, or rather that the results of the reviews and evaluations were arbitrary, that is, that assessments differed too much between the different evaluations and reviews, and that it seemed as if the result of the reviews depended too much on the individual views of the peers. There was said to be a *lack of transparency* in the system: it was not clear how and on what grounds peers choose what aspects, and according to which criteria they were examining the different programmes and courses (Franke & Nitzler, 2008).

The critics also pointed out a *lack of comparability*; they thought that similar programmes were reviewed differently and according to different standards. In short, the results differed too much between similar programmes and universities. Consequently, the system was said to suffer from a *lack of legal certainty*. This became a pressing issue due to the fact that a negative outcome of a review could lead to a revocation to award a degree.

One crucial point of critique, finally, was the assessments' focus on processes and conditions of programmes and courses, instead of looking at students' results. The argument went that if the students' results are of high quality, then in what ways the institutions ensure quality in programmes and courses is of no importance. In addition, in the aftermath of an autonomy reform in the higher education sector initiated just by that time by the government, the argument that the state (in this case represented by the Authority) should keep away from reviewing higher education institutions' design of programmes and courses grew strong (Segerholm et al., 2019).

2011: A New System Is Launched

After much debate, in 2011 a new national system for quality assurance was launched. This new system entailed a substantial shift in focus in both content and form. Now, the reviews should first and foremost focus on results: results of programmes, courses and students. As opposed to earlier systems, the students' final theses were an important basis for the assessments of peers. The shift can be seen in the guidelines from the Authority as well as in the reports from this period. To a large extent, it can be seen as a result of the radical change in how the mission given to the Authority by the government was phrased. Before that, the Authority's mission had been formulated in a broad way, giving the Authority a clear mandate to organise its operations, including its reviews, independently of the government in a way it saw fit. However, the new assignment was detailed to an extent previously unprecedented.

Transparency and predictability were now pinpointed as primary and essential conditions for the reviews. Decisions by the Authority based on the reviews had to be made on the basis of principles and criteria well known by the assessed institutions and programmes beforehand. The predictability was crucial. This can be seen as a result of the critique of the earlier system for being arbitrary and legally insecure. Also, in order to increase the equivalency in reviews, the Authority developed a new web-based tool for peers and HEIs for handing in data and material for the assessment.

Aside from the Authority developing new tools for carrying out the evaluations, other changes took place that affected the quality assurance system, and with that, the way peer review was carried out. During the evaluation cycle of 2011–2014 a new element of monetary reward was introduced as a result of a governmental reform. This was a new feature in the landscape of national quality assurance of higher education in Sweden, although it followed international trends. Thus, in 2011 performance-based funding was introduced, which meant that a positive result in reviews performed by the Authority rendered more money for the HEI in question. Connecting the results of the reviews to the redistribution of funding for higher education affected the conditions of peer review by strengthening the demands for legal security and transparency as overarching principles in the assessments.

The new system was heavily criticised by parts of the higher education sector, especially so in the beginning of the evaluation cycle. The critique predominantly focused on one thing: that the assessment model focused too strongly on students' results (achieved learning outcomes), and that results tended to be measured by the "quality of students' theses."

However, the method of peer review was not in itself questioned, as far as we can find proof of in the studied material. Despite the complicated and somehow contradictory relationship between, on the one hand, ideals of peer review stressing constructive dialogue and exchange of ideas, and, on the other, demands for transparency and legal certainty, peers continued to be used in the programme evaluations, but the recruitment of peers was said to be in need of more transparency. As a consequence, the Authority produced routines regarding bias and how to handle biased peers within the framework of the evaluations. In addition, in order to ascertain the comparability and legal security of the evaluations, the information and, above all, the introduction of the method and the assessment criteria to the peers were to be much more extensive.

In fact, we have found only one example of critique, or worry rather: that if the assessment criteria were too fixed—in the name of predictability—it would undermine the method of peer review] for clarity. A few years before the new system was launched in 2011, in a report from 2009, peers were invited to reflect upon the method of peer review (HSV 2009: 8 R, p. 22.). The text, written in Norwegian, gives the impression of being unedited by the Authority, thus enabling the peers to voice their opinions outside the set framework of the assessment criteria. In the report, a fear is voiced that the assessment criteria (which the panel were involved in creating) would make assessments "mechanical," encouraging a "box-ticking mentality" among both HEIs and reviewers. According to the panel this would not enhance quality (HSV 2009: 8 R, pp. 23–24):

> The panel members represent expertise within specific fields on which they base their assessments, and they neither can nor should be set within an assessment matrix as if the evaluation were absolute science. It is important to stress that the assessments are made by "peer review" by groups that are to some extent heterogeneous, and that the final comments, assessments and recommendations are the result of a democratic discussion within the panel. As a result, the final assessments are a synthesis of diverse impressions and discussions. In addition it should be noted that the quality work at a HEI is complex matter, and that assessments thereof cannot be restrained by a simple templet. (HSV 2009: 8 R, p. 24)

The quote pinpoints the difficulty in guarding the integrity of peer review without making concessions, while at the same time attempting to make assessments more alike by using templates and assessment matrixes. Noteworthy, thus, that this chapter was written a couple of years before the launching of the new system of quality assurance that would standardise assessments further. However, this innate tension between peer review and governance, openness and restrain, seems to end up in the shadow of the discussion of results and performance-based funding.

Peer Review

As a starting point to understand how peer review processes are used under different circumstances and in different settings, we have identified four ideal types of peer review in the literature.

The core idea behind what is sometimes referred to as the *classical peer review* is that only individual experts with a research field closely matching those they are assessing are able to comprehend research output and therefore pass judgment on scientific quality. This type of peer review focuses on the performance of the individual, and might be distinguished from assessments focusing on group performance. For the latter type of peer review the term *modified peer review* is sometimes used (see, e.g., Langfeldt, 2002; Sandström & Harding, 2002).

A third model, like the second model mainly used for assessing group performance, is the performance *indicator model*, which is based on bibliometric indicators and economic input-output models. One advantage of this model, which is often emphasised, is that it can be conducted remotely and therefore is of low cost. However, some researchers argue that the indicator model is unable to provide the opinion and broader insight generated by a more qualitative approach. Thus, a fourth model has been developed, sometimes called the *informed peer review model*. This is a mixture of modified peer review and the performance indicator standard within which panels of experts are asked to review groups using both quantitative indicators and more qualitative assessment material such as self-evaluations or interviews. The key benefit of this approach, often stressed, is that it combines "hard" data with "soft" opinions, resulting in a more comprehensive picture of the assessed unit (Hammarfelt et al., 2016).

As types, these models rarely occur in their ideal form but rather in variations. We would argue that the peer review model used by the Authority is a mixture of all four models, but mainly resembles the modified peer review and informed peer review models.

As discussed in earlier research, the method of peer review carries both positive features and possible pitfalls. One feature about peer review, which is often emphasised as something positive by the academic community, is that the individual assessor can keep in mind several variables at the same time and see complex relationships that cannot always be quantified. In addition, the peer can see potential in a way that quantitative data cannot capture. Moreover, the peer review process provides the opportunity for idea flow and feedback between the reviewer and the reviewed. However, there are also potential weaknesses in peer assessments such as the risk of protectionism, which is often stressed by critics in the academic community, along with the risk that peers might believe that what they like is also the best (see, e.g., Carlsson et al., 2014).

Peer review is also closely linked to what Sahlin and Eriksson-Zetterquist (2016) refer to as collegiality, which is a form of decision-making and governance. According to Sahlin and Eriksson-Zetterquist, collegiality, when properly designed, puts knowledge and search for knowledge at the centre. In addition, collegiality as a form of governance permits autonomy and creativity, and gives space to independent action and thinking. It builds on, and supports, the idea that organisations should be built bottom-up or, more precisely, starting in activities and competences in interaction between free and reflective individuals and groups. Collegial processes and structures should, according to Sahlin and Eriksson-Zetterquist (2016), give space for discussions, criticism and trial of arguments.

Within the system of national quality assurance of higher education in Sweden, peer review has been used as a way of assessing quality, and collegiality has played a prominent role in decision-making and governance. As shown by Sahlin and Eriksson-Zetterquist, evaluation of research and higher education are often performed by mixing collegial governance with bureaucratic and result-based assessments and decision-making (2016, s. 70). We are interested in exploring how these different forms of management are mixed, within the system for quality assurance of higher education in Sweden and how this has changed over time. Not least are we interested in exploring the possible tensions this has created and how they have been discussed by the Authority and by the peers involved in the assessments.

Sources

To get a comprehensive picture of how peer review has been used within the national system of quality assurance of higher education in Sweden, we have analysed a large variety of documents published by the Swedish Authority of Higher Education and its predecessors (referred to as the Authority) during the period 1995–2017. These include the following categories:

- Descriptions of quality assurance systems
- Guidelines for peers

- Guidelines for HEIs
- Panel reports
- Reports covering specific government assignments to the Authority regarding quality assurance of higher education
- Other reports, for example, reports with methodological discussions and meta-analysis of evaluations and reviews performed under the supervision of the Authority

Going through the sources, we have looked for descriptions of the peer review process and instructions to peers on how to perform the evaluations. We have looked for statements of how the peers themselves describe the peer review process of which they have been part. In addition, we have analysed the content of the reports regarding the assessments to get information on what level of freedom was given to the peers within the framework, as opposed to the peer assessments being fitted inside pre-defined templates. Reports covering government assignments have given us a more comprehensive picture of how the method of peer review has been discussed and presented in different contexts. Other reports have been studied in order to see if the method of peer review has been analysed and problematised from a broader comparative perspective.

We have also been interested in quality aspects and criteria used in the evaluations and how they have been formulated and what level of freedom of individual interpretation and variation has been permitted. In addition, we have been interested in the relationship between the quality criteria and the assessments in the panel reports; that is, how the criteria have been used and to what extent they have functioned as a way of structuring the content in the assessments in the reports. Furthermore, we have looked for more detailed descriptions of how the peer review process has been carried out both according to instructions from the Authority and in accounts from the peers describing the work they have done. We have used this information to draw conclusions about the extent of freedom in the peer review process; *if* and, if so, *how* this level of freedom has changed over time within the framework of national quality assurance of higher education in Sweden. In the analyses below, we have selected examples from two reviews, 1997–1998 and 2016–2017, that serve to illustrate what we have detected in the larger material.

The Role of the Peers and the Method of Peer Review

We start our analyses by looking more closely at the ways in which the role of the peers and the method of peer review has been presented in official descriptions, policies and guidelines regarding the Swedish national quality assurance system(-s) for higher education during the period 1995–2017. Two examples have been chosen to illustrate what we have seen in the larger material.

Early Assessments 1997–1998

In the Authority's guidelines for peers from 1997 the starting points for the reviews and assessments of HEIs' quality assurance work are described. The peers are instructed on what should be assessed and how the assessment should be carried out. The guidelines emphasise that

> the assessment is not about judging right or wrong. Rather, the [adequate] metaphor would be the public defence of a doctoral thesis and its exchange of opinions and viewpoints, but [it] clearly also [contains] elements of judgment and control of applied methods and their reliability. (HSV 1997: 33R, p. 10)

Thus, the assessment is compared to an academic seminar, or even the act of public defence of a doctoral thesis. As such, it contains both an exchange of opinions and "obviously" an act of control (HSV 1997: 33R, p. 10). The guidelines stress that the external review functions as a way for the peers to, in collaboration with the HEI, reach insights into, and draw conclusions regarding, the quality assurance work of the HEI. This is described in the following manner:

> The main task of the assessment panel is to initiate discussions, create reflection and contribute with material for problem solving. Such an open approach to the assessment means that the consultative role of the assessment panel is accentuated and that the self-evaluation of the HEI is given a prominent role in the assessment. (HSV 1997: 33R, p. 19)

Hence, when the role of the assessment panel is described, the open attitude is stressed, and the role of the peers is mainly characterised as "consultative" by "initiating discussions," creating reflection and contributing with support in order to solve problems.

This description and understanding of the role of the peer clearly resembles the *classical peer review* described above. Ideally, the basis of classical peer review is that decisions should be built on knowledge as well as on critical assessments of such knowledge. It should be based on an ongoing trying and critical conversation in which there is, at the same time, a review and development of knowledge claims. Ideally, the best argument wins regardless of whether your own group gains or loses from it. Peer review in this sense is often compared to the seminar or round where the colleague or peer is someone you listen to, receive criticism from and give criticism to. When seeking a solution to a common problem, one talks and listens to each other's arguments; one is also prepared to change one's opinion if the opposing party has stronger arguments. The individual participant in a peer review process represents their competence, not an interest or a group (Sahlin & Eriksson-Zetterquist, 2016, p. 34f). This concept of giving and taking, of receiving criticism and negotiating knowledge claims, resembles the way decision-making is described in the guidelines. In the report discussed above, to accentuate the consultative role not only of the peers but also of the Authority, it is stressed that the final decision on quality at the HEI is a result of negotiations between the Authority, the HEI and the external peers (HSV 1997: 33R, p. 19).

Later Assessments 2016–2017

In the guidelines for peers published by the Authority almost 20 years later, in 2016, the method of peer is not discussed but rather taken for granted, as an underlying assumption. The same can be said about the role of the peers, which is not deliberated upon or explained. Instead, it is established that the review is made by an independent panel consisting of experts from the discipline that is being assessed (earlier referred to as peers) and representatives of working life and students. The main point made is that the panel should be unbiased and that all panel members should participate in the evaluation on equal terms (*Vägledning för granskning av lärosätenas kvalitetssäkringsarbete, pilotstudie*, UKÄ 2016). This use of the term "equal" does not refer to the content of the assessment, that it should be fare and well grounded. Instead, it refers to the way in which the assessment panel should work during the peer review process.

Regarding the focal point of the assignment for the peers it is stated that the main objective of the review is the assessment of

> the result of the quality work, that is, that it assures the quality and develops the programme in a systematic and effective way. (Vägledning för granskning av lärosätenas kvalitetssäkringsarbete, pilotstudie, UKÄ 2016, p. 13)

The way of describing the assessment has changed—words like "result," "assures the quality," "systematic" and "effective" are crucial in this quote. Here is one example of the vocabulary used in the guidelines and in the assessments described:

> The outcome of the assessment panel's judgment of the HEI's fulfilment of the bases of assessment for the reviewed aspect areas and perspectives is stated in a report which functions as the basis for UKÄ's decision. All aspect areas and perspectives have to be deemed satisfactory for the overall judgment to be positive. (Vägledning för granskning av lärosätenas kvalitetssäkringsarbete, pilotstudie, UKÄ 2016)

In this quote the role of the peers is pinpointed: it is to assess whether the HEI fulfils the "areas of focus" and meets the criteria for the "aspects" and "perspectives" decided by the Authority. In order for the HEI to receive the judgement "approved," *all* the aspect areas and perspectives must be judged as meeting the criteria. Compared to the example from the end of the 1990s, the HEI is no longer part of the decision-making, which is no longer described as a negotiation between Authority, the HEI and the peers. The seminar-like meetings, and the discussion between colleagues, have been replaced by stricter governance, and the ambition to measure the HEI's fulfilment of pre-defined quality criteria is in focus.

Grasping the Concept of Result

As the examples above have shown, the role of peers and the descriptions of the peer review process have undergone radical changes over the last 20 years. Closely connected to this is the concept of result. We are interested in how it has been defined and used within the national framework of quality assurance of higher education, and whether it has undergone similar changes over time. Therefore, we continue our analysis by exploring how the concept of result has been described and realised by the Authority and by the peers. Again, we start with an analysis of the early assessments in the late 1990s and move forward to the later assessments within the current national system of quality assurance, letting examples illustrate our findings within the larger material.

The Early Assessments 1997–1998

To fully grasp how the peer review process has changed over time we have also looked at peer reports, focusing on the concept of result, in terms of peer judgement. In a report from 1998, the peers who had been assessing the quality assurance work at comprehensive university put it this way:

> The additional material we were given access to over time in addition to the many discussions during the site visit gave us a new and more comprehensive picture of the rootedness of the quality assurance work and its strategies, as well as the various approaches to the distribution of roles at a university, depending on where in the organisation a person is located. (HSV 1998: 38 R, p. 34).

The text concerns the main *strategies* for quality assurance at a comprehensive university at the time. It seems like the peers were working "organically" and had time to collect complementary material in order to "see the whole picture." Even though peers stated earlier in the report that they had followed the guidelines from the Authority in assessing the "quality work" at the university, this is not completely clear in the text. In the report peers do not explicitly refer to the aspects and guidelines. It seems instead as if the author of the text has been free to dispose it in her own manner.

When reading the report, the impression given is that the panel to a large extent was able to make independent decisions on working methods, and that these decisions were made in dialogue with the university. This includes, for example, how and when assessment material should be handed in and how site visits should be organised. In addition, it is clear from the report that there was room for the panel chair to have an independent view of the purpose of review, which might or might not correspond to the aims of the Authority. The peers were also at liberty to make independent decisions on what the review should focus on:

> In our review we have chosen to focus our attention on some overarching characteristics at [...] University. Some of them have an indirect connection to the quality work, others a more direct. (HSV 1998: 38 R, p. 13)

These characteristics included "the HEI's self-image," "the idea of the existence of a set of shared values," "views on identity," "boundaries," "student participation" and the "somewhat mythical concept of the local university spirit," which is explained as the "prevailing notion of the informal and close contacts between students, teachers, researchers, administration and management on campus" (HSV 1998: 38 R, p. 29ff).

As stated above, what ideally characterises classical peer review is the ongoing discussion between peers in which knowledge and quality is (re-)defined. New knowledge is created in a dialogue in which the strongest argument wins. The way peers describe their assessment process, and coming to conclusions in this early assessment, does lead thoughts to the ideal of classical peer review. An example is the following quote from the panel's report:

> Another conclusion is therefore that the panel's first assumption that the quality work at the university was characterized by "top-down" principles must be abandoned. There is today a clear impact of a bottom-up [approach]. (HSV 1998: 38 R, p. 34)

New material and interaction with faculty result in change of perception and the report gives a clear account of how this change has taken place. The peers are present in the text as individuals, not instruments. They even describe an emotional reaction, "surprise," when confronted with colleagues at the assessed HEI:

> When we have discussed this with representatives of the HEI our observations have not always been met with an understanding attitude, something which has surprised us. (HSV 1998: 38 R, p. 34)

The discussion in the report covers a large spectrum of topics, some of which might seem far from the actual aspects the panel was set to assess. One example of this is the topic of co-creation of knowledge, loosely connected to discussions on student participation and student influence. The way in which these issues are covered is very different from today's reports. In addition, the panel expands on broader issues such as universities' role in society, and mass education and its ramifications (HSV 1998: 38 R, p. 36ff). This way of contextualising the results is prominent in the reports from the late 1990s (see, e.g., HSV 1996: 6 R; HSV 1997: 1 R; 1997: 38 R; 1998: 27 R). Also more academic texts are mentioned in the list of references at the end of the report, thus making it clear that the panel has drawn its conclusions within a larger scientific setting, relating its findings to research on topics connected to the review. This broad leeway for peers to relate the outcomes and results of the reviews to a larger societal setting disappears somewhere between the mid-1990s and today.

Later Assessments 2016–2017

In contrast to the panel report from the 1990s, in an example from an institutional review of the quality assurance procedures at the same university in 2016, the text in the report strictly follows the structure decided on beforehand by the Authority. The text is written in a fixed form, a template, and the peers account for their assessments in the separate aspect areas. The peers put it this way in the report:

The system for quality assurance of higher education at first and second cycle is seen as a system with elements at all levels of the operations [at the HEI]. The eleven items create a whole with a clear sequential structure. (UKÄ 2019, reg.no 411-00483-16, pp. 9–10)

Words like "system," "element" and "structure" can be seen as typical in the kind of language that is used in order to fulfil the task, that is, to review the different "aspect areas." Of course, it is difficult to fully grasp what this really means taken out of context, but nonetheless it is a typical way of expression in reports today (see, e.g., UKÄ 2019, reg.no 411-00488-17; UKÄ 2019, reg.no 411-00483-17; UKÄ 2019, reg.no 411-00486-17).

More general observations, and historical or societal contextualisation, which held a prominent role in the report from the late 1990s discussed above, are absent in today's reports. No references are made to research related to topics assessed during the review. There is no detailed account of how the review has been carried out. The panel and its views are presented in a formal and bureaucratic way, strictly following the quality criteria set up for the assessment:

The assessment panel finds that the quality assurance system reflects and is constructed in a way that makes systematic and proactive quality work possible at xx university. (UKÄ 2019, reg.no 411-00483-16, p. 8)

Again the systematic way of working with quality is stressed, thus following the criteria set up by the Authority. The panel continues:

A quality indicator to assure high quality in doctoral theses and the public defence is that the examination is done in a legally secure way, according to rules and regulations. (UKÄ 2019, reg.no 411-00483-16, p. 9)

Quality, in this quote, is linked to following rules and regulations and working in a way that is "legally secure." There is little room for discussions between the panel and the university, and no room for emotions such as surprise. The discussions that take place follow a structured form and are conducted as interviews, which function as a way to corroborate the statements in the self-evaluation (or not).

Thus, the panel report from 2016–2017 clearly shows that today's quality assessments are made in order to deliver sharp and clear judgements of the quality (or quality assurance work) in higher education. It is also clear that the national quality assurance system has been made more efficient in order for the Authority to be able to fulfil its assignment from the government. If the 1990s panel report mentioned above is more similar to a research report, with almost tentative reflective analysis, texts in panel reports from the latest panel reports focus directly on quality assessments and the impression is almost that it can be assessments of the quality of any product, not specifically higher education. How does this affect the peer review method?

Collegiality as a Form of Governance

Peer review, as it has been practised within the national system for quality assurance of higher education in Sweden, contains several elements. Most notably, and especially in the beginning of the period in the 1990s, it contains the element of discussion between peers, the trial of arguments and counterarguments resulting in a shared view of quality. This is maybe most clearly expressed in the co-making of decisions about the outcome of institutional reviews in the 1990s in which not only the peers and the Authority, but also the reviewed HEI, took part. However, this co-creation of knowledge claims about quality was gradually replaced by other ways of understanding the peer review process and its outcomes. These new ways entailed linking the assessments to pre-defined quality criteria and ideals of transparency and legal security. One way of understanding how and why these changes took place is by connecting them to the new ways of public monitoring and control that were gradually introduced during this time. In the 1990s the term "new public management" was launched, referring to new types of governance and control. The term is related to what Michael Power calls the "audit society." Power shows how the amount of auditing activities exploded in the United Kingdom and in North America from the 1980s and onwards (Power, 1997). The creation of a national, full system of quality assurance of higher education in Sweden might be seen in itself an expression of the audit society and the emergence of new public management. A tentative analysis is that the development of peer review within the Swedish system of quality assurance of higher education from the beginning of the 1990s until today can be seen as an example of how result-based management, which characterises new public management, is gradually strengthened within the governmental sector in Sweden during this period.

Looking at the national system for quality assurance of higher education, the institutional reviews in the 1990s were based on ideals of collegiality. As shown by Sahlin and Eriksson-Zetterquist, in this form of governance, knowledge is built through argumentation and the negotiation of truth claims is seen as a core ideal. As a consequence, this way of decision-making is often both complicated and time-consuming. In contrast, the system of quality assurance at work since 2011 shows different characteristics. Quality is still a core value, but from 2011 and onwards, the concept of quality is linked to concepts such as results, control and efficiency. The reviews, which previously took place within a loose and bendable framework, from 2011 and onwards take place within a set system. Any discussions, criticism and trial of arguments between individuals and groups are bound to the system and its pre-defined criteria for quality. Although decisions are based on peer review, the reports, including the findings, are calibrated by the Authority beforehand to assure equivalency. Moreover, reports from this time show that the Authority acknowledged that time was a problem.

Hypothetically, we suggest that this creates tensions within the framework of national quality assurance between conflicting ideals of openness and trial of arguments, on the one hand, and predictability and legal security, creating needs for calibration, on the other hand. Although the assessments in the end were made by the peers and not by the Authority, the strict framework of quality criteria can be seen as inhibiting the more open-ended, explorative side of the peer review process. In other words, contrary to ideals of collegial decision-making, the decisions within the national system for quality assurance of higher education became increasingly the result of a top-down process.

Although collegiality can be seen as an ideal closely corresponding to the ideals of peer review, as pointed out by Sahlin and Eriksson-Zetterquist (2016), in reality most organisations are governed by a mixed form of governance. Thus, collegiality is often blended with more bureaucratic forms of governance. This mix of forms can be a good thing, when the different forms complement each other. But the mix might just as well lead to continual compromises, troublesome and unclear structures for governance, and consequently "institutional unclearness" (Sahlin, 2014). As a result, one has to ask oneself: When does the interplay between the different forms of governance lead to undermine individual forms of governance—in this case collegiality? When do compromises and the transformation of a form of governance go too far—when does it become "perverted" (Hernes, 1978), or lose its purpose and become empty words?

Peer Review in the National Quality Assurance System

Well known from previous research (e.g. Sahlin & Eriksson-Zetterquist, 2016), different forms of governance are often interacting and exist in organisations side by side or, rather, intertwined—hopefully complementing and strengthening each other and promoting better governance overall. In the quality assurance processes at the Authority, the collegial form with peer review at its core, has been an important identity factor for project managers—they themselves educated at universities and quite often with a PhD degree. Somehow, it seems logical using peer review in a system assessing and reviewing the quality of higher education. The Authority "borrows" an academic touch—or more than that, an academic method and orientation—to its bureaucratic governance. As discussed by Franke and Nitzler (2008, p. 114), one reason for the use of peer review within quality assurance of higher education is to gain acceptance for the reviews from those affected by it. In other words, an important objective for using the method of peer review might well be that the peers involved in the exercise render the evaluation legitimacy.

All along, since 1993, collegiality has been mixed with bureaucratic decisions and increasingly along the way with (new public) management, that is, performance management or result-based management. In this case, performance-based funding might be the most obvious expression. It was abandoned in 2015, but it remains crucial to look at performance or results in the quality assurance system today. How this mix and interaction has expressed itself more in detail still remains to be examined more closely. But it is clear that the changes that the national system of quality assurance in higher education underwent in 2011, and which are still working to a great extent, implicate challenges for the peer review method. Moreover, these challenges have not really been addressed in a thorough way by the Authority.

Claims for transparency, predictability and equivalency are difficult to combine with ideals of collegiality, as they might undermine the authority of the peers. The claims for transparency, predictability, equivalence and the focus on quality defined as "good student performances," that is, to what extent students achieved the intended learning outcomes within the national qualifications framework, coincided with the introduction of performance-based financing in 2011. Introducing economic benefits as a result of the assessments carried out by the Authority radically changed the conditions for quality assurance. Demands for transparency and legal certainty overruled ideals of creativity and independence in search of new knowledge. Peers, instead of being consultative discussion partners, became judges, making verdict and exercising control.

As shown above, we have found that the national system for quality assurance has been surprisingly consistent, and that it has undergone remarkably few changes from 1995 up until 2011. We argue that the new quality assurance system launched in 2011 amounted to a turning point, during which the method of peer review changed concordant with result-based management and the emergence of the audit society (Power, 1997). However, although we have identified 2011 as a turning point, the changes that took place then were part of trends that started almost 20 years earlier in 1993, when the first national system for quality assurance of higher education in Sweden was launched. Already in the early 1990s, and as a result of the reforms in higher education that took place at that time, the state had identified needs to install a stronger and more systematic way of exercising control. In 2011, this element of control, already introduced, was enhanced due to a change of government, opening up new possibilities of governance along the lines of new public management and introducing monetary rewards into the national quality assurance system, thereby strengthening demands for predictability, transparency and legal security, thus changing the preconditions for peer review.

The process of change in 1993 today remains to be studied more in detail; in this chapter, we have only been able to highlight a few examples to make our point. The system launched in 2011 came to an end in 2017, when yet another system of quality assurance was put in place. Result-based management and new public management have been vividly criticised in the latest years and seem to be gradually replaced by trust-based public management, which has already affected the peer review processes performed by the Authority. In what way, and with what result, is for the future to tell.

Sources

Directions on Assessments

Maj 2011. Reg.no 12-4013-10. *Riktlinjer till bedömargrupper för viktning av underlagen samt för framtagande av bedömargruppens förslag till samlat omdöme.*

Reports Describing Whole Quality Assurance Systems

Högskoleverkets rapportserie 2001a: 2 R. *Nationella ämnes- och programutvärderingar.*
Högskoleverkets rapportserie 2006: 57 R. *Nationellt kvalitetssäkringssystem för perioden 2007–2012.*
Högskoleverkets rapportserie 2007a: 59 R. *Nationellt kvalitetssäkringssystem för perioden 2007–2012. Reviderad 2007-12-11.*
Högskoleverkets rapportserie 2009a: 25 R. *Kvalitetsutvärdering för lärande. Högskoleverkets förslag till nya kvalitetsutvärderingar för högskoleutbildningar.*
Högskoleverkets rapportserie 2010a: 22 R. *Högskoleverkets system för kvalitetsutvärdering 2011–2014.*

Högskoleverkets rapportserie 2012a: 4 R. *Högskoleverkets system för kvalitetsutvärdering 2011–2014. Examina på grundnivå och avancerad nivå. Fastställd 21 December 2010. Reviderad 3 April 2012.*

Högskoleverkets rapportserie 2012b: 15 R. *Högskoleverkets system för kvalitetsutvärdering 2011–2014. Examina på grundnivå och avancerad nivå. Fastställd 21 December 2010. Reviderad 19 June 2012.*

Guidelines for HEIs and Peers

Högskoleverkets rapportserie 1997a: 33 R. Granskning och bedömning av kvalitetsarbete vid universitet och högskolor. *Utgångspunkter samt angrepps- och tillvägagångssätt för Högskoleverkets bedömningsarbete.*

Högskoleverkets rapportserie 1997b: 33 R. *Granskning och bedömning av kvalitetsarbete vid universitet och högskolor. Andra reviderade upplagan.* September 1997.

Högskoleverket 1997a. *Vägledning för lärosäten vid bedömning av kvalitetsarbete. Bilaga 1 till Granskning och bedömning av kvalitetsarbete vid universitet och högskolor.*

Högskoleverket 1997b. *Handledning för bedömare av kvalitetsarbete vid universitet och högskolor. Bilaga 2 till Granskning och bedömning av kvalitetsarbete vid universitet och högskolor.*

Högskoleverkets rapportserie 1998a: 21 R. *Fortsatt granskning och bedömning av kvalitetsarbetet vid universitet och högskolor. Utgångspunkter samt angrepps- och tillvägagångssätt för Högskoleverkets bedömningsarbete.*

Högskoleverkets rapportserie 2001b: 4 R. *Examensrättsprövning. Utgångspunkter och tillvägagångssätt för Högskoleverkets examensrättsprövning.*

Högskoleverket maj 2003, Utvärderingsavdelningen. *Nationella ämnes- och programutvärderingar. Anvisningar och underlag för självvärdering. Reviderad May 2003.*

Högskoleverket oktober 2007. *Granskning av lärosätenas kvalitetsarbete. Anvisningar för självvärdering.*

Högskoleverket 2009. Reg.no 641-3701-09. *Högskoleverkets prövning av ansökan om rätten att utfärda examen på forskarnivå inom ett område.* (Vägledning till högskolor)

Högskoleverket mars 2010. *Högskoleverkets prövning av högskolors ansökan om tillstånd att utfärda examen på forskarnivå inom ett område. Informationskompendium till sakkunniga mars 2010.*

Högskoleverkets rapportserie 2011a: 4 R. *Generell vägledning för självvärdering i Högskoleverkets system för kvalitetsutvärdering 2011–2014.*

Reports on Reviews Conducted Within the Framework of National Systems for Quality Assurance of Higher Education

Högskoleverkets rapportserie 1996a: 6 R. *Kvalitetsarbete vid universitet och högskolor 1994/95.*

Högskoleverkets rapportserie 1997c: 1 R. *Granskning och bedömning av kvalitetsarbete vid fem lärosäten—en sammanfattning.*

Högskoleverkets rapportserie 1997d: 38 R. *Magisterexamensprövning vid elva högskolor—Examensrättsprövning.*

Högskoleverkets rapportserie 1998b: 27 R. *Vetenskapsområden. Bedömning av tre högskolor.*

Högskoleverkets skriftserie 1998: 38 R. *Granskning och bedömning av kvalitetsarbetet vid Umeå universitet.*

Högskoleverkets rapportserie 2007b: 46 R. *Prövning av masterexamensrätt 2007.*

Högskoleverkets rapportserie 2008: 21 R. *Högskoleverkets prövningar av masterexamensrätt 2008.*

Högskoleverkets rapportserie 2009b: 8 R. *Granskning av kvalitetsarbetet vid nio lärosäten 2008.*

Högskoleverkets rapportserie 2009c: 30 R. *Granskning av kvalitetsarbetet vid sex universitet 2009.*

Högskoleverkets rapportserie 2010b: 2 R. *Granskning av kvalitetsarbetet vid åtta högskolor 2009.*

Högskoleverkets rapportserie 2011b: 9 R. *Högskoleverkets prövningar av tillstånd att utfärda masterexamen 2011.*

Accounts of Government Assignments

Högskoleverkets rapportserie 1996b: 12 R. *Kriterier för benämningen universitet—En utredning.*

Högskoleverkets rapportserie 1996c: 24 S. *Rätt att inrätta professurer. Högskoleverkets prövning av Högskolan i Kalmar, Karlstad, Växjö, Örebro samt Mitthögskolan och Mälardalens högskola.*

Högskoleverkets rapportserie 1997e: 37 R. *Rätt att inrätta professurer—Högskoleverkets prövning av Högskolan i Halmstad, Högskolan i Karlskrona/Ronneby, Högskolan i Örebro, Idrottshögskolan samt Mitthögskolan.*

Högskoleverkets rapportserie 2007c: 43 R. *Utvärdering av arbetet med breddad rekrytering till universitet och högskolor. En samlad bild.*

UKÄ: 2016, *Pilotgranskning av kvalitetssäkringsarbetet vid Umeå Universitet,* reg. nr 411-00483-16

UKÄ: 2019a, *Granskning av lärosätets kvalitetsarbete, Bedömargruppens yttrande över kvalitetssäkringsarbetet vid Mälardalens högskola,* reg.no 411-00488-17

UKÄ: 2019b, *Granskning av lärosätets kvalitetsarbete, Bedömargruppens yttrande över kvalitetssäkringsarbetet vid Högskolan i Borås,* reg.no 411-00483-17

UKÄ: 2019c, *Granskning av lärosätets kvalitetsarbete, Bedömargruppens yttrande över kvalitetssäkringsarbetet vid Stiftelsen Högskolan i Jönköping,* reg.no 411-00486-17

Meta-analyses of the Authority's/the Authority's Own Reviews

Högskoleverkets rapportserie 1997f: 41 R. *Kvalitetsarbete—ett sätt att förbättra verksamhetens kvalitet vid universitet och högskolor? Halvtidsrapport för granskningen av kvalitetsarbetet vid universitet och högskolor.*

Högskoleverkets rapportserie 1998c: 1 R. *Hur står det till med kvaliteten i högskolan?*

Högskoleverkets rapportserie 2002: 20 R. *Metautvärdering av Högskoleverkets modell för kvalitetsbedömning av högre utbildning. Hur har lärosäten och bedömare uppfattat modellen?*

Högskoleverkets rapportserie 2007d: 31 R. *Hur har det gått? En slutrapport om Högskoleverkets kvalitetsgranskningar åren 2001–2006.*

Högskoleverkets rapportserie 2012c: 21 R. *Från granskning och bedömning av kvalitetsarbete till utvärdering av utbildningsresultat—ger utvärderingen en bild av kvaliteten på utbildningen vid universitet och högskolor?*

Other Reports

Högskoleverkets rapportserie 2001c: 21 R. *Akademisk frihet—en rent akademisk fråga?*

Högskoleverkets rapportserie 2003: 17 R. *Enklare och nyttigare? Om metodiken för ämnes- och programutvärderingar. (Kvalitetsgranskning).* Författare: Karl-Axel Nilsson.

Other

Odat. Dokumentet har namnet gemensam-plattform.pdf, rubriken inne i dokumentet är Ett kvalitetsdrivande resurstilldelningssystem. Förmodligen är dokumentet från 2009 och utgör anteckningar från en informell arbetsgrupp som tillsattes när Björklund ville införa ett system med fokus på resultat.

Promemoria *Kvalitetssäkring av högre utbildning*. U2016/1626/UH. Reg nr 131-176-15.

References

Carlsson, H., Kettis, Å., & Söderholm, A. (2014). *Research quality and the role of the university leadership*. Sveriges universitets- och högskoleförbund, SUHF.

Franke, S., & Nitzler, R. (2008). *Att kvalitetssäkra högre utbildning: en utvecklande resa från Umeå till Bologna*. Studentlitteratur.

Hammarfelt, B., Nelhans, G., Eklund, P., & Åström, F. (2016). The heterogeneous landscape of bibliometric indicators: Evaluating models for allocating resources at Swedish universities. *Research Evaluation, 25*(3), 292–305.

Hernes, G. (Ed.). (1978). *Forhandlingsøkonomi og blandingsadministrasjon*. Universitetsforl. cop.

Langfeldt, L. (2002). Decision-making in expert panels evaluating research constraints, processes and bias, Norsk institutt for studier av forskning og utdanning.

Power, M. (1997). *The audit society: Rituals of verification*. Oxford University Press.

Sahlin, K. (2014). Global themes and institutional ambiguity in the university field: Rankings and management models on the move. In G. S. Drori, M. Höllerer, & P. Walgenbach (Eds.), *Global themes and local variations in organisation and management: Percpectives on globalisation* (pp. 52–64). Routledge.

Sahlin, K., & Eriksson-Zetterquist, U. (2016). *Kollegialitet: En modern styrform*. Studentlitteratur.

Sandström U. and Harding T. (2002). Tvärvetenskap och forskningspolitik, Spänningsfält: politiken - tekniken – framtiden, (red.) L. Sturesson. Stockholm: Carlssons förlag.

Segerholm, C., Hult, A., Lindgren, J., & Rönnberg, L. (2019). *The governing—evaluation—knowledge nexus*. Springer International Publishing.

8

Performance-Based Evaluation Metrics: Influence at the Macro, Meso, and Micro Level

Gustaf Nelhans

Introduction

Performance-based research evaluation using quantitative indicators has many purposes. But regardless of their use for ranking purposes, quality evaluation, or the distribution of funding, from a research policy perspective, quantitative indicators are thought of as tools to evaluate research without steering it directly. Often, these metrics (such as citation counts) are interpreted as indicators of "quality." In bibliometric research, the sociological basis of their use is founded on the Mertonian so-called CUDOS norms. This acronym stands for "Commun[al]ism, Universalism, Disinterestedness, and Organized Skepticism" and is often described as a role model for how research should be conducted. But norms do not determine actual practice, and even though quantitative indicators are often claimed represent notions of quality, it could be argued that several

G. Nelhans (✉)
University of Borås, Borås, Sweden
e-mail: gustaf.nelhans@hb.se

© The Author(s) 2022
E. Forsberg et al. (eds.), *Peer review in an Era of Evaluation*,
https://doi.org/10.1007/978-3-030-75263-7_8

prerequisites have to be met for the fulfillment of such claims. For example, it is expected that the indicators chosen should be distinct objective measures, and that data is unobtrusively collected so that those who are evaluated are not influenced or affected directly by the measurement. In this view, the use of quantitative indicators makes it (relatively) easy to operationalize performance goals based on bibliometric indicators.

In this chapter the following lines of thought are pursued: Firstly, it is argued that the so-called representational model of bibliometric indicators as described above is questionable in practice because goal displacement over time will alter which representation should be chosen, but also that in the light of future developments, representations tend to lose their stability and become contingent on external factors. And secondly, that the uncertainty in relevant choices is not merely a technical problem that is solved by larger samples, better accuracy, or more sophisticated statistics, but that it is inherent in the kind of linear model that is used as the basis for measurement. It is therefore argued that a performative notion on scientometric indicators needs to be developed that takes account of the variability and uncertainty of the aspects of research that is to be evaluated.

This performativity will be investigated using empirical examples at three levels of scale from the perspective of the Swedish research policy. At the macro level, the Swedish performance-based funding system (PRFS) for reallocating parts of the national funding to universities using citation-based bibliometric indicators will be discussed against the background of other available PRFS at the time of its inception in 2009, predominantly, the Norwegian point-based system.

While controversial already at its inception, with suggestions that it should be evaluated after an initial period of use posed both by government officials themselves and by actors across the university sector (Nelhans, 2013), the Swedish system has never been subjected to a formal review (Kesselberg, 2015). Instead, this PRFS has been used relatively untroubled as an established part of research policy (except for the years it has not been used; see below). At the meso level, there are both self-initialized evaluations within universities as well as their internal funding systems, where higher education institutions (HEIs) risk losing self-government due to the establishment of standardized performance-based indicators, leading to a "hands-tied" situation for vice-chancellors when steering has to be negotiated in light of results. Finally, there is the micro level of the individual researcher, who in daily practice must navigate between different sets of norms and directives coming from the other levels as well as discipline-related notions and specific knowledge demands coming from the actual research field at hand.

Algorithmic Historiography and the Birth of Scientometrics

Evaluation of academic research has been with us for a long time. For a century-based timeline, it is enough to go back to the early notions of evaluating *American Men of Science* (first printed in 1906, later renamed *American Men and Women of Science*) (Cattell, 1921). This volume, containing biographical sketches for thousands of American scientists, also had a ranking system, whereby an asterisk was affixed to about a thousand entries of scientists that were "supposed to be most important," "by order of merit" (ibid.). The exact method of calculation was not disclosed, but it is stated that it involved the ranking of subjects within 1 of 12 different natural or exact sciences, ranging from chemistry physics and astronomy, past the biological and earth sciences, mathematics, psychology, and medical sciences, to anthropology. The ranking was performed by ten "leading students of the science" (ibid.). Thereafter, statistical methods of ordering the names were used to finalize the ranking. The darker background for this exercise was to determine a group of leading "American men of science" for scientific study "[t]o secure data for a statistical study of the conditions, performance, traits, etc., of a large group of men of science." While not clearly stated, it is could be implied that there was a nationalistic air to the endeavor that built on studies by one of his intellectual forefathers, Francis Galton, who published *English Men of Science*, with the subtitle "Their Nature and Nurture" (Galton, 1876).

Another work prompting for evaluation and distinction of scientific work was made by the first science policy adviser to President Franklin

D. Roosevelt, Vannevar Bush, who famously argued for increased support for basic research in public and private colleges in his recommendation for the instigation of a government agency called the National Research Foundation (NRF). Here, Bush argued for a balance between the Foundation's adherence to the "complete independence and freedom for the nature, scope, and methodology of research carried on in the institutions receiving public funds"; at the same time he argued for the Foundation "retaining discretion in the allocation of funds among such institutions" (Bush, 1945, p. 27). Consequently, Bush suggested the creation of a permanent Science Advisory board composed of disinterested scientists without the intervention of either the legislative or the executive branch of the government. To a large extent, these views still permeate the academic research system and are at heart in the argument for "academic freedom."

In a parallel historical setting, 60 years ago, the citation index was founded as a means of analyzing the history of science by quantitative methods—for algorithmic historiography (Garfield, 1955). Its use was intended for information retrieval, although some form of evaluation was implied even from the start: "[A] bibliographic system for science literature that can eliminate the uncritical citation of fraudulent, incomplete, or obsolete data by making it possible for the conscientious scholar to be aware of criticisms of earlier papers" (Garfield, 1955).

Although the incentive to measure scientific publications quantitatively was at least 30 years older (Lotka, 1926), it was not until a citation theory of sorts was formed that interest in the citation as a measure of scientific quality or the merits of research was introduced. Compare, for example, Urquhart's ranking lists of loans of scientific periodicals (Urquhart, 1959) and Derek de Solla Price's first notions of "Quantitative Measures of the Development of Science," first published as early as in 1951 but more generally known from his monographs *Science Since Babylon* and *Little Science, Big Science* (Price, 1951, 1961, 1963). In the sixties, together with the development of Sociology of Science and the studies of the institutional structure of science, with Cole and Cole, Zuckerman, and others as members (Cole & Cole, 1967; Zuckerman, 1967), building on the works of Merton (1973a), a new view of the scientific publication and the reference was formed that paved the way for equating quantitative measures of references indexed as citations in the citation index (Kaplan, 1965; Price, 1963, pp. 78–79).

Soon the citation and the metrics that were derived from the practices of publishing within journals had come to use in very different settings. On the one hand, a new field, bibliometrics (Pritchard, 1969), or sciento-metrics (from Naukometrija; Nalimov & Mul'chenko, 1969), had evolved from the older notion of "statistical bibliography" (Hulme, 1923). To a large degree, this followed the notions from the earlier use of quantitative measures in the library field, both for classification purposes and for journal selection criteria.

Early bibliometrics focused on the network model of the structure of publications (Price, 1965) and the development of bibliographic cou-pling and co-citation analysis at the article level for identifying "scientific specialties" (Griffith et al., 1974; Kessler, 1963; Small & Griffith, 1974). Co-citation analysis was subsequently developed into focusing on co-citation at the authorship level (McCain, 1986; White & Griffith, 1981), and later at the journal level (McCain, 1991a, 1991b). Notions of co-citations as a means of illustrating the intellectual base of a research area and bibliographic coupling for identifying research fronts have been sug-gested (Persson, 1994).

On the other hand, the sociological interest in the bibliometric tools led to a closer relationship between bibliometrics and research policy studies. As noted above, the view of the citation as an indicator of quality stems very much from the Mertonian *norm system of science.* It is com-monly owned; universal in terms of being valid everywhere and for any-one; that scientists should be disinterested, meaning that they should not have personal bonds toward the research that they pursue; and they should strive for originality in their research, while at the same time they should hold a skeptical attitude toward all claims, both their own and those of fellow scientists (Merton, 1973b [1942]).

All this lays the ground for the peer review system, by which research is refereed before getting published, but it stands also as a guarantee for us, referring to previous research in a true and timely manner—"standing on the shoulders of giants," meaning that science is a cumulative knowledge-making process and that we should always acknowledge our predecessors.

Against this, there is another line of argument, stemming from the more critical stance of Science and Technology Studies, stating that, although the norms of doing research are important as the goal, for various reasons they are impossible to follow to the letter in practice.

If we take the Mertonian norms for "how science should be done," one could talk about a set of counter-norms (Mitroff, 1974), which substitute actual practices of research that many would attribute to problematic issues which limit academic freedom such as external influence or steering of research or more broadly: poor practice. It is not implied that this is a binary distinction, but rather that there is a continuum between the two end-points. These "counter-norms of scientific practice" have been spelled out by the British theorist of science John Ziman as *PLACE*: *P*roprietary, *L*ocal, *A*uthoritarian, *C*ommissioned, and *E*xpert, as opposites to *CUDOS*. Ziman showed that in practice, for every virtue of the Mertonian "ideal scientist," there is a contextual counterforce intertwined, which is hardly possible to break free from. Here, the elevated norms of science meet practice and we get to the first clash between how science "ought to be done" and the practical implication of research being performed in practice.

Bibliometrics for the Evaluation of Research

The Journal Impact Factor (JIF) has become a testament to this duality. It was developed with a certain set of journals in focus, empirically tuned to the publication patterns of the coverage of Science Citation Index (SCI) in the latter half of the 1960s, where the bulk of citations for a paper were found to have been received within a two-year window from its publication (Garfield, 1972). Even though it was not created as a tool for evaluating research(ers), but for calculating the inclusion in SCI, it was almost immediately used in that way upon the publication of the first citation index. This led Eugene Garfield to write an early commentary about the sociological use of his invention, stating: "One purpose of this communication is to record my forewarning concerning the possible *promiscuous* and *careless use* of quantitative citation data for sociological evaluations, including personnel and fellowship selection." Furthermore, he stated quite unequivocally that "*[i]mpact* is not the same as *importance* or *significance*" (Garfield, 1963).

Other bibliometricians even suggested that every bibliometric study should be accompanied by a warning:

> The warning reads: "CAUTION! Any attempt to equate high frequency of citation with worth or excellence will end in disaster; nor can we say that low frequency of citation indicates lack of worth." (Kessler & Heart, 1962)

But while there was stark criticism against this use, by the mid-1970s, there was already an established textbook on "the use of publication and citation analysis in the evaluation of scientific activity" (Narin, 1976).

The Citation as Mediator: The Performativity of "Being Cited"

Here, I would like to briefly discuss the key arguments for and against using citations in evaluation by mentioning the two positions in the controversy regarding indicator use in evaluation. On the one hand, there is the notion that citations indicate the actual use and influence of previous research, and that citation could be seen as a reward (Cole & Cole, 1967) or currency in a scholarly "quasi-economy." On the other hand, there is the view that researchers cite persuasively, and that citation could be viewed as a rhetorical device (Cozzens, 1989; Gilbert, 1977; Gilbert & Woolgar, 1974). The implications of this perspective in the citation system are important. On the one hand, it could be described as, although researchers cite the sources that have influenced them, giving credit where credit is due, on the other, there are other motivations for citing a reference. It could be done to note that the cited author is wrong (negative citations) or to cite authorities, for example.

Borrowing a notion from the Actor-Network Theory, a bibliometric indicator such as the citation can be described as either an intermediary or a mediator. In the first case, according to Michel Callon (Callon, 1986; Callon et al., 1991), an (1) intermediary only transmits the information from one point to the other without transforming it, while Bruno Latour (2005) has noticed the role of (2) mediators as entities that actually transform the meaning and thus need to be explained in terms of "other activities" (such as the social realm), since these entities not only transfer meaning but translate it. We will not dig deeper into this, but to note that when Derek de Solla Price noted that citations were a viable way of measuring impact in the 1960s, he regarded citations as "unobtrusive" indicators (*intermediaries*) of scholarly activity, something that could be studied without exercising an influence on those who were to be measured. In this view, then, to theorize about the citation and its role as a mediator of scientific work would be *not to view it as a representation* of the research that is studied, *but rather as a performative agent.* The citation as *mediator* implies the notion of the citation being performative rather than *representative* in practice (let us leave nature out here).

So, to what consequences does this lead? Well, for one thing, it renders the citation into an object of sociological study and opens up an interesting venue to act upon. Of course, it also brings into question that the quantity of citations implies quality. If we return to the classic debate, as noted above, traditionally, citations are given to research as a reward. This means that citation performance that is observed in the citation index could be viewed as a true representation of "true impact" or quality. From the perspective sketched here, it is more relevant to talk about the performativity of "being cited," and that we as researchers act both in accordance to this dynamic and in a reflexive mode. In this view the different ways that research is published or the different practices of publishing and citing one's results are not static, but instead are co-produced by both internal demands from research and the social and, in this internet-based world we live in, technical demands or affordances that are provided by such systems as the citation databases in Web of Science (WoS) or Google Scholar (GS).

Consequences for Research Policy

Moving the discussion to research policy, we can make a similar trajectory from a traditional "linear perspective" of the relationship between science and politics to one which includes multiple feedback loops. For this, we need to ask ourselves: Why would we evaluate research and allocate funding resources based on indicator models?

First of all, research policy needs tools to allocate funds without steering research *directly*. Secondly, there is also the idea that indicators would mean that evaluation would be based on notions of "quality." According to the position that was sketched above, these would be the Mertonian *CUDOS* norms, which would ensure that this is the fact. Of course, this would build on the prerequisites that citation indicators are objective measures and that they are unobtrusive in their actions on researchers (Price, 1963). Lastly, quantitative models are (quite) easy to operationalize, meaning that they can easily be separated from the object under study. Still, they are not easy to interpret and play different roles in different contexts.

In the following, I will make a historical account of research policy and paraphrase an argument made by science policy scholar and Emeritus Professor of Theory of Science Aant Elzinga (personal communication). After World War II three research policy "regimes" can be identified (Table 8.1). There is a move from a so-called linear model where, given enough funding and not disturbing researchers too much, the resulting knowledge could be directly implemented in technology. Next, by the 1960s–1970s there was a notion that by focusing on funding in a specific strain, according to the needs of society, it would be possible to increase the output of useful knowledge. Then, at present, we are in a more heterogeneous constellation where science, technology, and society are much more interlinked and driven by the economic promises of future application of research.

At the same time, there is a general sense in academia that its relation to society has shifted from a social contract of trust to mistrust with regard to control mechanisms being instigated. This is not least driven by governing phenomena such as New Public Management (NPM)—better called "outcome-oriented public management," empowered by new bureaucratic layers, such as branding and high-ranking list scores—as well as digitalized audit society (Power, 1999), seeking to foster cultures of bibliometric compliance in academia.

Table 8.1 Three "regimes" of research policy after WWII

Politics for science (basic research)	–	Science for politics (sectorialization)	–	Science & Technology for economy (commercialization)

The concept of "epistemic drift" (Elzinga, 1984, 1997, 2010), first noted in connection with sectionalization policy, is relevant here. It states that politically driven agendas can crowd out internal quality control criteria in favor of external relevance. With "economization" pressures nowadays, we find the same risk. With regard to scientometric indicators, one could add the notion of "bibliometric creep," where, in practice, bibliometric measures are constantly tuned to external needs not linked to internal research values or needs.

This situation could be explicated with the notion of the "co-production" of science and society. Scientometric indicators play an increasing role in research, both by its design as a reflection of the act of citing scientific references and as a result of being constructed—and used—in valuation practices of perceived scientific quality. Therefore, it is important to study what is here labeled "the performative nature of citations," either critically (1) from the outside, or as a more (2) reflexive endeavor within the metric community, taking into account that practitioners in the scientometric community create or employ indicators in different ways that have an impact on those who are measured. Among other things, it is often argued that single indicators of research (such as the citation) could not be regarded as the actual representation of how research "is done." Instead, it is important to establish their origin and subsequent development to get an understanding of how they work in practice.

Co-production of Science and Society

From a constructivist position within science and technology studies (STS)—which generally has been highly critical to the use of quantitative indicators to represent research—what scientific research is and what it should be are empirical questions that are context dependent. This means that cognitive and social factors cannot be separated from each other. In this chapter, this critique is taken at face value in that the technical conditions, society's demands to measure research performance and researchers' pursuit of knowledge, are treated as expressions of what has been described as a co-production (Jasanoff, 2004) between science, technology, and society, wherein there is mutual interdependency between science/technology and society (Felt et al., 2017).

But there is yet another level that co-production works on that is relevant to the topic of this chapter: *At the policy level, governments and other policy-setting organizations introduce new indicators to* evaluate research based on how it performs in the societal realm as "societal impact" or as "collaboration with society." Today, it is even stated that *"[s]ocietal impact should increase"* (Prop. 2016/17: 1, 2016, p. 20). In later budgets, economic incentives for collaboration with societal actors have been introduced in the national funding of higher education institutions in Sweden (Prop. 2019/20: 1, 2019).

Performance-Based Research Allocation Models at Three Levels

In the following, I will argue that the way indicators are implemented in different contexts means that the actors in the academic system are torn between different ways of evaluating the academic impact of research, which risks making it problematic for the individual researcher to navigate the evaluational landscape. By taking the Swedish PRFS as an example, we will pick three different levels of bibliometric evaluation for the allocation of funding: the macro level, the national renegotional model for funding HEIs; the meso level, within universities; and lastly, at the micro level, using the individual researcher as the object of study. It should also be noted that "evaluation" will be used as a term to describe the respective performance-based funding systems, since, in their presentation, they are described as quality-based models (Prop. 2008/09: 50, 2008, p. 55) (Universitets- og høgskolerådet, 2004, p. 35).

To contextualize the development of the Swedish PRFS, this section outlines the immediate research policy background against which it was developed. In this new era of epistemic drift, as proposed above, at the turn of the century, one could identify three means of using publication-based evaluation indicators apart from a pure peer-review-based model (Hicks, 2012). Such a system had been used in the United Kingdom since

1986, using a purely qualitative evaluation system with peer review panels to evaluate the research within universities. The Research Assessment Exercise (RAE), generally performed every six years (since 2014, redeveloped into the Research Evaluation Framework) has, in review, been found to be resource-intensive and expensive (RAE, 2009). In the mid-1990s, Australia developed a publication-based indicator that evaluated HEIs based on the count of Web of Science (WoS)-indexed publications and award funding accordingly. It was not met favorably by experts. For example, one influential study showed that while the numbers of publications rose in the Australian higher education system during the time, researchers seemed to publish their studies in lesser ranked journals, as measured with JIF (Butler, 2003). Notions such as *salami slice publishing* and *Least Publishable Unit* (LPU) became household terms during this time.

When it was time for the Nordic countries to select a model for evaluating research, both the British RAE model and the Australian straight counting were dismissed. As will be shown below, different choices were made. On the one hand, in Norway, an evaluation model using the perceived quality of the publication channel was introduced, while in Sweden, the perceived quality of the actual specimen using citation counts as an indicator was introduced. Arguably, it is only here that we can speak of bibliometric systems that involve calculations which attempt to compare the results between research fields. The next section presents the main features of the Norwegian and Swedish PRFS before we compare some of their main characteristics.

PRFS at the Macro Level: A Comparison Between the Norwegian and Swedish Systems

As noted above, in Norway, the funding model was developed in light of the Australian model and its perceived shortcomings regarding concerns of mass publishing of journal articles (Universitets- og høgskolerådet, 2004). Therefore, it was found necessary to introduce a quality-based notion into the evaluation indicator. In its simplest form, it can be described as measuring impact based on publication "channel" and "quality level," where normal publication channels are evaluated at the basic level (1), and high-quality publication channels are evaluated at level 2.

For journals, in disciplines with a tradition of publishing in (double-blind) peer-reviewed journals found in citation indices such as WoS, these levels were based on JIF, and for other disciplines and publication channels such as book publishers, they were based on the degree of internationalization. Panels representing scholars from different fields were then in charge of evaluating the publication channels and suggest substitutions of sources at different levels. In Sweden, then, the performance-based allocation model for the bibliometric indicator used field normalized citations and Waring distributions of publications were used to evaluate the performance of universities. This model was inspired by a system used in the Flanders region at the same time. Here, only research publications indexed in WoS were included (Sandström & Sandström, 2009; Vetenskapsrådet, 2009). Additional features are that a four-year moving average is used, together with author fractionalization, and lastly, an additional (arbitrary) weighting that awards different research areas differently so that Medicine and Technology are multiplied with 1.0, the Sciences with 1.5, the Social Sciences and the Humanities with 2.0, and other areas with 1.1. The source of this weighting scheme has not been recovered, but seems to be old-established in Swedish research policy (Nelhans, 2013; Prop. 2008/09: 50, 2008; Prop. 2012/13: 30, 2012)

If we compare the Swedish and Norwegian models, we find several contrasting features (Table 8.2).

Table 8.2 Comparison between the Swedish and the Norwegian PRFS

	Swedish model: *Field-normalized citation of specimen*	Norwegian model: *Publication channel (impact)*
Selection	Only published material that is indexed in WoS ISI	Broader publication channels (monographs, conf. proc, journal articles)
Source of data:	Already available data (WoS ISI)	An authorization index must be created (Cristin, NSB) and publication lists must be updated.
Measure of quality	Citation measures, field normalized	"Secondary peer review"
Transparency	Variables in the calculated model are relative to the performance of each publication	Pre-determined "point system"

First of all, the Norwegian model distributes 2% of the total funding (Hicks, 2012, p. 257), while the Swedish model first used 5% and later 10% of the renegotiated funding. While both models are designed to measure the performance of the unit under study, the Swedish model evaluates performance based on received citations for each specimen in four years, the Norwegian uses an impact-factor-styled model, which evaluates the judged channel of the publication, rather than the actual impact in itself. These features result in differences concerning the selection, sources of data, measures of quality, and transparency of the evaluation exercises. First of all, the underlying coverage of the Norwegian model consists of all publication channels, regardless the form of publications that are found to have a scientific, peer-reviewed status. Journals, monographs, and edited books can be eligible. An authoritative list of all channels is maintained and all publications that are reported by Norwegian researchers are matched against this list. The Swedish model, in turn, uses an external source, the WoS journal indices, which means that only publications in journals indexed by this database are covered. Concerning the actual quality indicator, the Swedish model uses field-normalized citations to the (likewise) field-normalized publications from researchers at the respective HEIs. In Norway, as noted above, the evaluated level of the publication impact, instead of the actual impact, is noted. Lastly, with regard to the transparency of each model, the Norwegian uses a pre-determined system of points (0.7, 1, 5 at level 1 and 1, 3, 8 points respectively at level 2 for chapters in edited books, journal articles, and monographs). In the Swedish model, the indicators are not transparent but relate not only to the number of citations but also to the calculated weight factors for both citations and publications, as well as the total number of publications and citations in the whole model. Instead, the actual "worth" of one point in the Norwegian model could be established quite easily and was calculated to the sum of about 40,030 NOK in the year 2007 (SOU, 2007, p. 81, 2007, p. 385). By dividing the total funding each year in Norway with the accrued number of points performed in the Norwegian system, it can be found that there is a significant devaluation of the performance of Norwegian researchers, as valued in funding per point. As shown in Fig. 8.1, for each year, a point is worth less, and in 2019, it was calculated to 23,572 NOK, based on data from

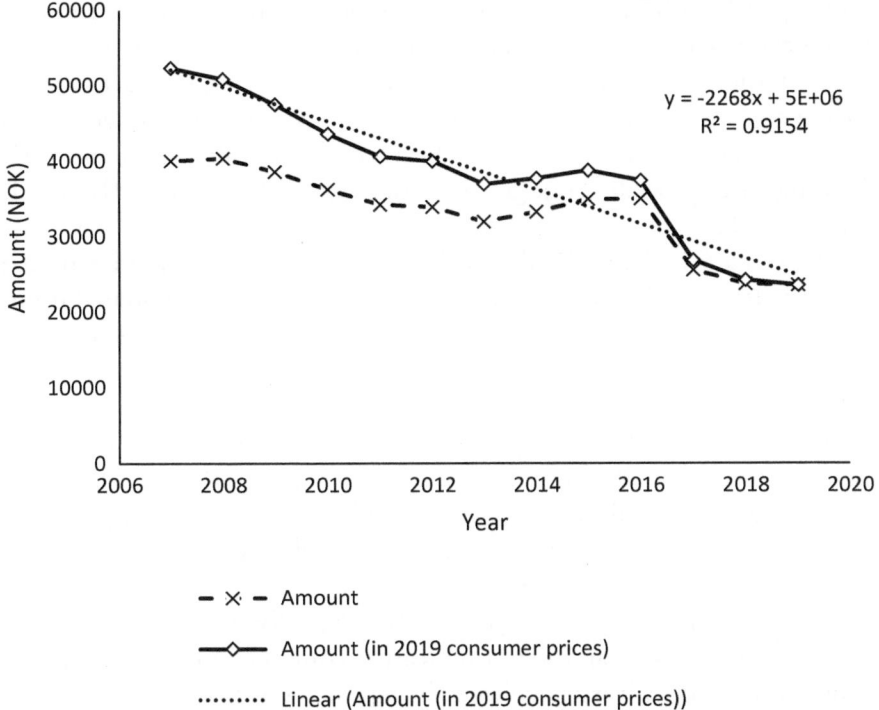

$$y = -2268x + 5E+06$$
$$R^2 = 0.9154$$

Fig. 8.1 Yearly amount in NOK for each Norwegian point. (Data from the Norwegian state budget (e.g. Kunnskapsdepartementet, 2018). The data calculated in 2019 consumer prices were found at Statistics Norway webpage (Consumer Price Index, 2020))

Kunnskapsdepartementet (Kunnskapsdepartementet, 2018). If consumer price index is taken into account, a Norwegian publishing point is worth roughly 45% of the buying power of a Norwegian point in 2007.

On the other hand, while the Swedish model seems to be straightforward to calculate, in practice, there have been several issues that were reported at the outset, and that have never been corrected. Most importantly, the Swedish Research Council, which has been tasked with performing the calculations, has reported every year that they are not able to calculate the Waring distribution for field-normalizing the publication counts (Swedish Research Council, 2009; Vetenskapsrådet, 2011). Instead, pre-set values as described already in 2007 have been used (SOU,

2007, p. 81, 2007). Any new publications that have been introduced have been given the same reference value as the field that is most prominent in the new sources' reference lists (e.g. Vetenskapsrådet, 2012). While not properly documented, by inspecting each yearly budget, it can be found that the model has not been used for reallocating the basic funding in the Swedish budget since the year 2017 (Prop. 2016/17: 1, 2016, p. 208).

The outcome of the Swedish PRFS analysis shows some unexpected systemic features that are relevant to note. Below, the renegotiated funding for four HEI types that are found in Sweden was calculated. For administrative uses, the Swedish HEI system is divided into comprehensive universities (e.g. Uppsala and Lund), special universities, (e.g. Karolinska Institutet, KI, and Chalmers University of Technology), newly formed (~2005) universities (e.g. Karlstad and Linnaeus Universities), and university colleges, roughly correlating with polytechnics in the United Kingdom, which only have the right to award doctorates in select subjects (Hansson et al., 2019).

As noted in Fig. 8.2, for the years 2010–2016, for which the renegotiation model was used, an interesting feature could be noted. Only special universities and university colleges have a net positive performance in bibliometric performance as calculated in SEK, based on government data. These results are somewhat unexpected, given that basic funding is heavily weighted toward universities, with very small parts distributed to university colleges. That these latter perform so much better than expected is an open question that would need further attention. For instance, a 1 MSEK in extra funding based on performance is a quite large sum for a university college with basic funding amounting to less than 100 MSEK on average, while it is a rather small sum for one of the comprehensive universities like Uppsala University with basic funding for research at about 2000 MSEK (e.g. Prop. 2016/17: 1, 2016).

A last feature of the Swedish performance-based renegotiation model needs to be described. By way of how it is presented in the annual budgets, this model has never been recorded to render any significant negative economic impact for universities, regardless of their performance, a seemingly magical feat. This is due to how the results are presented, combined with new funding added to the university sector each year. In all

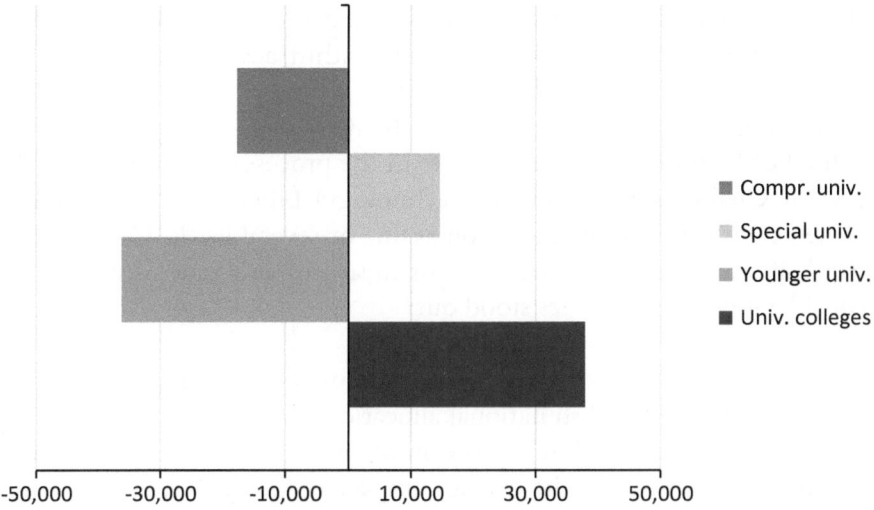

- ■ Compr. univ.
- ▨ Special univ.
- ▨ Younger univ.
- ■ Univ. colleges

Fig. 8.2 Cumulative renegotiated funding by HEI type according to the state budget for the years 2010–2016. (Data graciously provided by Lars Olof Mikaelsson at the Swedish Government Offices)

Table 8.3 Combined "new" funding and results of renegotiation for the University of Kristianstad in the budget for 2012. (Adapted from Prop. 2011/12: 1, p. 169)

HEI (numbers in thousands of SEK)	Strengthening of basic funding	Renegotiation	Total
University of Kristianstad	636	–691	–55

the years the model has been used for renegotiation, the combined funds have shown a negative result just once, as shown in Table 8.3, where the university college in Kristianstad received net negative funding of 55,000 SEK, about €5000 (Nelhans, 2015; Prop. 2011/12: 1, 2011, p 169).

At the Meso Level, Within Universities

In two studies, we set out to map and describe different bibliometric models and indicators that are used in the allocation of funds within Swedish HEIs and the collective aim was to invite a critical discussion about the advantages and disadvantages and the relative value of using such indicators for allocation of funds within academia (Hammarfelt et al., 2016; Nelhans & Eklund, 2015).

We found that at the time, all HEIs in Sweden—except Stockholm School of Economics—used (or were in the process of start using, in the case of Chalmers University of Technology) bibliometric measures to some extent for resource allocation at one or several levels. On the other hand, it was found that the types of measures and models used differ considerably, but two types stood out:

- "Actual impact" calculated as the share of publications and citations (as used in the Swedish national allocation model
- "Point-based evaluation" of the number of publications, combined with an appraisal of the average impact of the publication channel (similar to the Norwegian/Danish/Finnish model).

Two specialized universities, Karolinska Institutet (KI) and Royal Institute of Technology in Stockholm (KTH), used state-of-the-art models, including field-normalization of citations at the aggregate level, for the performance-based model at department or school level. Additionally, comprehensive universities with a broad range of disciplines often use a range of measurements depending on faculty type.

At the Micro Level, Individual Researchers

In our studies, we also found that, sometimes, performance-based funding models were used also at the micro level. Here we found several distinct ways in which bibliometric indicators were used for funding individual researchers. Three of these are mentioned here.

Even before the Swedish national PRFS was introduced, a performance-based model including a bibliometric component has been used since 2008 at the humanities faculty at Umeå University. Here, researchers and teachers apply for funds in competition, and publication measures are an important part of the application process. In an evaluation performed already in 2011, it was found that "[m]any believe that the system has a negative impact on the work climate," and that "many cite that they experience individual stress and press." (Sjögren, 2011).

At Linnaeus University, 2.5% of the allotted research funding was distributed at the individual level using a "field-normalized publication point model" where publication points were translated to actual currency. Between 8000 and 150,000 SEK were distributed directly to the researcher based on publication performance. At the same time, publication points <8000 SEK were distributed to the department instead, probably so that the model would not stigmatize researchers who do not perform well. Additionally, an excellence share was distributed as a "bonus"; 20% of researchers with the highest share of publication points receive an additional 15,000 SEK per individual.

At Luleå Technical University, an "economic publishing support" was awarded at the departmental level, but directly based on a price tag per publication with an ISSN. At the Norwegian level 1, 35,000 SEK was awarded and double that amount for a publication at level 2 or indexed in WoS. An interesting feature of the model was that since there were set price tags, in 2016, researchers "broke the bank" when they performed better than the amount of funding allowed. The result was that for conference papers with an ISSN at level 2, the lower amount was awarded, while at level 1 the support was removed (Luleå Technical University, 2016).

Conclusions: The Performativity of Performance-Based Research Funding on Different Levels

Here I would like to reflect on how researchers are affected by the above-mentioned examples and how the culture of bibliometric compliance is performed in practice, and what role does peer review play in this equation.

As noted above, peer review is at the center of the publication tradition mentioned here. In general, only peer-reviewed publications are used in the bibliometric evaluation and as a basis for PRFS. But that is not to say that peer review in itself has issues that may affect how it impacts bibliometric compliance. While peer review in general is seen as a guarantee against unsubstantiated claims and subjective arguments, in practice, reviewers are also actors in the performative setting that involves the publication of research. Bibliometric data is seldom used without an evaluator that is using them, and, as discussed at several instances here, decisions are partly based on the data, but often in light of expected outcomes and normative views about how it "should be." For example, it is expected that bibliometric models are constructed so that all disciplines can perform "on par"; otherwise, there would be claims that the model is biased, even though we know that researchers in different disciplines and subspecialties publish in different ways. Some may output loads of short papers in collaboration, while others would call these working papers or even small studies for a manuscript in a book-length format that is published by a respected publication house. Still, as in the Norwegian model, there is an inherent claim that you can compare between different kinds of research. And in Sweden, the citation-based system normalizes both the production and citation impacts with European averages at the discipline level to make the numbers comparable. And when the numbers cannot be calculated, or when the government decides to add a weighting factor based on established ratings, an arbitrary factor is included, thus yielding the bibliometric system less effective.

While it is hard to directly pinpoint the effect of a single incentive on researchers' publication practice, there is an extensive literature that discusses the issues (e.g. de Rijcke et al., 2016; Hammarfelt & de Rijcke, 2015; Hicks et al., 2015; Wilsdon et al., 2015). These authors argue that there are visible effects on researchers' publishing practice and that these could be related to the introduction of PRFS at different levels. Rather than reiterating their arguments, I will exemplify these issues with several practices that I have documented at different levels in academia.

First, at the individual level, let us start by making a Google search on the terms "Curriculum vitae" AND "H-index." H-index is a measure of the rate of articles having a certain number of citations, which has become an increasingly used shorthand for scholarly excellence (Bar-Ilan, 2008). We find that researchers seemingly have gotten the incentive to add the H-index value to their CVs, sometimes even at the title level of the document, such that it states: "Curriculum vitae for [name], [title], H-index =

[h]." Of course, this does not show a direct correspondence between the funding model and scholarly practice, but it does show that researchers are catching on when new indicators are constructed.

Many other activities are used to "game the system." With regard to citation data, there is self- (or colleague-) citing of references to one's own work or editor coercion, where editors or reviewers suggest references to articles published in the same journal or to themselves, for example, "Manuscript should refer to at least one article published in '*** Journal of *** Sciences' [the title of the same journal]" (Nelhans, 2013). According to the ethics guidance for peer reviewers published by the Committee on Publication Ethics (COPE Council, 2017), reviewers should "refrain from suggesting that authors include citations to your (or an associate's) work merely to increase citation counts or to enhance the visibility of your or your associate's work; suggestions must be based on valid academic or technological reasons."

Citation cartels, where editors join up to cite each other's journals to artificially raise each journal's JIF, have been identified (Davis, 2012) and, in some cases, this has led to the removal of the journals from citation indices.

At the meso level, the will to perform well in university rankings could lead to the following suggestions by administrators. "[A]nother way of advancing on the list would be to appoint highly cited researchers since they 'bring with them' their earlier citations"(Gunnarsson, 2013). This claim is not even true, since citations are linked to the affiliation at the time of publication and will not follow the researcher to a new institution. Other suggestions have been to publish review articles or methods papers, especially if these are cited extensively outside the specialty.

Finally, I would like to reflect on the consequences of using bibliometric indicators for evaluating and funding research. As already noted above, using indicators for evaluating research builds on the notion that (aspects of) quality can be measured. The goal of this study has not been to show that indicators themselves are bad, and should be abolished altogether, but instead to critically engage with their performative nature and to explore ways in which they could influence scholarly practice at many levels.

For instance, at the *research policy level*, the occurrence of bibliometric models has been regarded as a supposedly objective and unobtrusive tool to "tap" the research system for information about its intrinsic qualities, but without influencing the research analyzed. But here, unobtrusiveness has been questioned, at the individual level, not least in how new practices of searching for one's publications in online search indices to find out how many references one's works have received or to calculate one's H-index.

For *research* in itself, it can be seen that different bibliometric models, such as publication-point-based models or citation-based models, suit some disciplines better, while others fare worse. Additionally, the introduction of performance-based models creates incentives for researchers to publish according to the yardstick used, rather than to further knowledge as such. This creates false competition.

For *researchers*, performance-based evaluation incentives are forced on them in a top-down direction, as performance-based models have trickled down at all levels in the research practice. To conclude, the impact on individual researchers is discussed as they grapple with adapting their performance to different and sometimes contradictory quantitative benchmarks.

Indicators have been shown to work not only as intermediaries between research and research policy but, at the same time, as mediators, which add performative aspects in terms of incentives, shifted policy goals, and evaluation practices that act to change how research gets done.

To end on a somewhat more positive note, rather than arguing for the abolishment of metric indicators in evaluation and funding allocation altogether, I would like to argue for some sort of constructive pragmatism (echoing a call for methodological pragmatism by (Sivertsen, 2016)) in their use. The proposed program will build on what I label "meta-metric evaluation," which adapts its use of indicators for evaluation in a specific situation or for a specific subject area from within a pool of possible measures that are combinable in creative ways. This would mean that the universalistic claim of having an indicator to evaluate them all would be lost in favor of lining up indicator-based measurements with skilled expertise that is trained to evaluate research using both qualitative and quantitative data. Instead, transparency and openness would be the guiding principles, while at the same time, a communal agreement of what works best for the discipline or subspecialty at hand would be needed. While the latter is a setback for the prognosis of future metric performance, at the same time, it counteracts the use of gaming metrics or adapting to the scale, and instead helps researchers to focus on the "best possible research" regardless of whether its metric footprint will be the largest according to some external measure of past performance.

Still, since at the aggregate level, there is a quite good correlation between many scientometric indicators and peers' evaluation of the underlying research, it would be counterintuitive to not use them at all. The mission instead is to curb the scientometric indicators to tailor them into measuring what we *intend* to measure instead of what we *can* measure.

References

Bar-Ilan, J. (2008). Informetrics at the beginning of the 21st century, A review. *Journal of Informetrics, 2*(1), 1–52. https://doi.org/10.1016/j.joi.2007.11.001

Bush, V. (1945). *Science: The Endless Frontier a Report to the President by Vannevar Bush, Director of the Office of Scientific Research and Development, July 1945 (Web version)*. United States Government Printing Office. http://www.nsf.gov/od/lpa/nsf50/vbush1945.htm

Butler, L. (2003). Explaining Australia's increased share of ISI publications—the effects of a funding formula based on publication counts. *Research Policy, 32*(1), 143–155. https://doi.org/10.1016/S0048-7333(02)00007-0. http://www.sciencedirect.com/science/article/B6V77-455VNDT-1/2/ebf2bd1294bd7f0aef6e2631791d2bec

Callon, M. (1986). Some elements of a sociology of translation: Domestication of the scallops and the fishermen of St Brieuc Bay. In J. Law (Ed.), *Power, action and belief: A new sociology of knowledge?* (pp. 196–233). Routledge & Kegan Paul.

Callon, M., Courtial, J. P., & Laville, F. (1991). Co-word analysis as a tool for describing the network of interactions between basic and technological research: The case of polymer chemistry. *Scientometrics, 22*(1), 155–205. https://doi.org/10.1007/BF02019280

Cattell, J. M. (Ed.). (1921). *American men of science* (3rd ed.). Jacques Cattell Press, The Science Press.

Cole, S., & Cole, J. R. (1967). Scientific output and recognition: A study in the operation of the reward system in science. *American Sociological Review, 32*(3), 377–390. http://www.jstor.org/stable/2091085

Consumer Price Index. (2020). Statistics Norway. https://www.ssb.no/en/kpi

COPE Council. (2017). *Ethical guidelines for peer reviewers.*

Cozzens, S. E. (1989). What do citations count? The rhetoric-first model. *Scientometrics, 15*(5), 437–447. https://doi.org/10.1007/BF02017064

Davis, P. M. (2012). The emergence of a citation cartel. *Scholarly Kitchen.* https://scholarlykitchen.sspnet.org/2012/04/10/emergence-of-a-citation-cartel/

de Rijcke, S., Wouters, P. F., Rushforth, A. D., Franssen, T. P., & Hammarfelt, B. (2016). Evaluation practices and effects of indicator use—a literature review. *Research Evaluation, 25*(2), 161–169.

Elzinga, A. (1984). Research, bureaucracy and the drift of epistemic criteria. In B. Wittrock & A. Elzinga (Eds.), *The University research system. The public policies of the home of scientist* (pp. 191–220). Almqvist & Wiksell International.

Elzinga, A. (1997). The science-society contract in historical transformation: With special reference to "epistemic drift". *Social Science Information, 36*(3), 411–445. https://doi.org/10.1177/053901897036003002

Elzinga, A. (2010). Globalisation, new public management and traditional university values. *1st Workshop of the Nordic Network for International Research Policy Analysis (NIRPA) 7–8 April 2010*, Swedish Royal Academy of Engineering Sciences (IVA), Stockholm.

Felt, U., Fouché, R., Miller, C. A., & Smith-Doerr, L. (Eds.). (2017). *The handbook of science and technology studies* (4th ed.). MIT Press.

Galton, F. (1876). *English men of science: Their nature and nurture.* Macmillan & Co..

Garfield, E. (1955, July 15). Citation indexes for science: A New dimension in documentation through association of ideas. *Science, 122*(3159), 108–111.

Garfield, E. (1963). Citation Indexes in Sociological and Historical Research. American Documentation, 14, 289–291.

Garfield, E. (1972). Citation analysis as a tool in journal evaluation. *Science, 178*, 471–479. (Essays of an Information Scientist, Vol1, pp. 527–544, 1962–73)

Gilbert, G. N. (1977). Referencing as persuasion. *Social Studies of Science, 7*(1), 113–122. http://www.jstor.org/stable/284636

Gilbert, G. N., & Woolgar, S. (1974). The quantitative study of science: An examination of the literature. *Science Studies, 4*(3), 279–294.

Griffith, B. C., Small, H. G., Stonehill, J. A., & Dey, S. (1974). The structure of scientific literatures II: Toward a macrostructure and microstructure for science [Article]. *Science Studies, 4*(4), 339–365. https://doi.org/10.1177/030631277400400402

Gunnarsson, M. (2013). *SHANGHAIRANKINGEN 2013: En analys av resultatet för Göteborgs universitet. PM 2013: 07* (Dnr: V 2013/621).

Hammarfelt, B., & de Rijcke, S. (2015). Accountability in context: Effects of research evaluation systems on publication practices, disciplinary norms, and individual working routines in the faculty of Arts at Uppsala University [Article]. *Research Evaluation, 24*(1), 63–77. https://doi.org/10.1093/reseval/rvu029

Hammarfelt, B., Nelhans, G., Eklund, P., & Åström, F. (2016). The heterogeneous landscape of bibliometric indicators: Evaluating models for allocating resources at Swedish universities. *Research Evaluation, 25*(3), 292–305. https://doi.org/10.1093/reseval/rvv040

Hansson, G., Barriere, S. G., Gurell, J., Lindholm, M., Lundin, P., & Wikgren, M. (2019). *The Swedish research barometer 2019: The Swedish research system in international comparison.* Swedish Research Council.

Hicks, D. (2012). Performance-based university research funding systems. *Research Policy, 41*(2), 251–261. https://doi.org/10.1016/j.respol.2011.09.007

Hicks, D., Wouters, P., Waltman, L., Rijcke, S. d., & Rafols, I. (2015). The Leiden Manifesto for research metrics. *Nature, 520*, 429–431. https://doi.org/10.1038/520429a

Hulme, E. W. (1923). *Statistical bibliography.* Butler & Tanner; Grafton & Co.

Jasanoff, S. (2004). *States of knowledge: The co-production of science and the social order.* Routledge.

Kaplan, N. (1965). The norms of citation behavior: Prolegomena to the footnote. *American Documentation, 16*(3), 179–184. https://doi.org/10.1002/asi.5090160305

Kesselberg, M. (2015). *Forskningsresurser baserade på prestation.* Universitetskanslersämbetet.

Kessler, M. M. (1963). Bibliographic coupling between scientific papers [Article]. *American Documentation, 14*(1), 10–25. https://doi.org/10.1002/asi.5090140103

Kessler, M. M., & Heart, F. E. (1962). *Concerning the probability that a given paper will be cited* (pp. 19). Cambridge: Massachusetts Institute of Technology.

Kunnskapsdepartementet. (2018). *Orientering om forslag til statsbudsjettet 2019 for universitet og høgskolar.* https://www.regjeringen.no/contentassets/31af8e 2c3a224ac2829e48cc91d89083/orientering-om-statsbudsjettet-2019-for-universiteter-og-hogskolar-100119.pdf

Latour, B. (2005). *Reassembling the social: An introduction to actor-network-theory.* Oxford University Press.

Lotka, A. J. (1926). The frequency distribution of scientific productivity. *Journal of the Washington Academy of Sciences, 16,* 317–324.

Luleå Technical University. (2016). *Ekonomiskt publiceringsstöd 2015–2016.* https://www.ltu.se/cms_fs/1.79817!/file/Ekonomiskt%20publiceringsst% C3%B6d%202015-2016.docx

McCain, K. W. (1986). Cocited author mapping as a valid representation of intellectual structure. *Journal of the American Society for Information Science, 37*(3), 111–122. https://doi.org/10.1002/(sici)1097-4571(198605)37:3 <111::aid-asi2>3.0.co;2-d

McCain, K. W. (1991a). Core journal networks and cocitation maps—new bibliometric tools for serials research and management [Article]. *Library Quarterly, 61*(3), 311–336.

McCain, K. W. (1991b). Mapping economics through the journal literature—an experiment in journal cocitation analysis [Article]. *Journal of the American Society for Information Science, 42*(4), 290–296. https://doi.org/10.1002/(sic i)1097-4571(199105)42:4<290::aid-asi5>3.0.co;2-9

Merton, R. K. (1973a). *The sociology of science: Theoretical and empirical investigations.* The University of Chicago Press.

Merton, R. K. (1973b [1942]). The normative structure of science. In *The sociology of science: Theoretical and empirical investigations* (pp. 267–278). The University of Chicago Press.

Mitroff, I. I. (1974). Norms and counter-norms in a select group of the Apollo moon scientists: A case study of the ambivalence of scientists. *American Sociological Review, 39*(4), 579–595. http://www.jstor.org/stable/2094423

Nalimov, V. V., & Mul'chenko, Z. M. (1969). *Naukometriya. lzucheniye Razvitiya Nauki kak Informatsionnogo Protsessa [Translated title: Measurement of Science. Study of the Development of Science as an Information Process. Washington, DC: Foreign Technology Division, U.S. Air Force Systems Command, 13 October 1971. 196 p.].* http://www.garfield.library.upenn.edu/ nalimov/nalimovmeasurementofscience/book.pdf

Narin, F. (1976). *Evaluative bibliometrics: The use of publication and citation analysis in the evaluation of scientific activity*. Report. (pp. 459): Cherry Hill, New Jersey: Computer Horizons, Inc.

Nelhans, G. (2013). *Citeringens praktiker. Det vetenskapliga publicerandet som teori, metod och forskningspolitik. (The practices of the citation: Scientific publication as theory, method and research policy)* University of Gothenburg]. Gothenburg. http://hdl.handle.net/2077/33516

Nelhans, G. (2015). Meaningful citation analysis? In *20th Nordic Workshop on Bibliometric and Research Policy*, Oslo.

Nelhans, G., & Eklund, P. (2015). *Resursfördelningsmodeller på bibliometrisk grund vid ett urval svenska lärosäten*. http://urn.kb.se/resolve?urn=urn:nbn:se:hb:diva-21

Persson, O. (1994). The intellectual base and research fronts of JASIS 1986–1990. *Journal of the American Society for Information Science, 45*(1), 31–38. https://doi.org/10.1002/(SICI)1097-4571(199401)45:1<31::AID-ASI4>3.0.CO;2-G

Power, M. (1999). *The audit society: Rituals of verification*. Oxford University Press.

Price, D. J. d. S. (1951). *Quantitative measures of the development of science* VI Congrès International des Histoire des Sciences, août 1950, Amsterdam.

Price, D. J. d. S. (1961). *Science since Babylon* ([1975] ed.). Yale University Press.

Price, D. J. d. S. (1963). *Little science, big science*. Columbia University Press.

Price, D. J. d. S. (1965). Networks of scientific papers. *Science, 149*(3683), 510–515.

Pritchard, A. (1969). Statistical bibliography or bibliometrics? *Journal of Documentation, 25*(4), 348–349.

Prop. 2008/09:50. (2008). *Ett lyft för forskning och innovation* [Proposition]. Ministry of Education.

Prop. 2011/12: 1. (2011). *Förslag till statsbudget för 2012: Utgiftsområde 16* (Vol. 2011/12: 1) [Proposition]. Ministry of Education.

Prop. 2012/13: 30. (2012). *Forskning och innovation*. Ministry of Education.

Prop. 2016/17: 1. (2016). *Utgiftsområde 16, Government budget*. Ministry of Finance.

Prop. 2019/20: 1. (2019). *Utgiftsområde 16, Government budget*.

RAE. (2009). *RAE 2008 accountability review*. PA consulting services limited. http://www.hefce.ac.uk/pubs/rdreports/2009/rd08_09/rd08_09.pdf

Sandström, U., & Sandström, E. (2009). The field factor: Towards a metric for academic institutions. *Research Evaluation, 18*(3), 243–250.

Sivertsen, G. (2016). A welcome to methodological pragmatism. *Journal of Informetrics, 10*(2), 664–666.

Sjögren, D. (2011). *Utvärdering av kvalitetsbaserat resurstilldelningssystem på humanistiska fakulteten.* Rapport till fakultetsnämnden 16/9 2011. Umeå: Institutionen för idé- och samhällsstudier, Umeå universitet. https://www.aurora.umu.se/globalassets/dokument/enheter/humfak/for-vara-anstallda/forskningsfragor/utvarderingav-kvalitetsbaserad-resurstilldelning-.docx

Small, H. G., & Griffith, B. C. (1974). The structure of scientific literatures I: Identifying and graphing specialties. *Science Studies, 4*(1), 17–40.

SOU. (2007). Resurser för kvalitet: Slutbetänkande av Resursutredningen. SOU 2007:81. Utbildningsdepartementet. Stockholm: Fritzes.

Swedish Research Council. (2009). *Bibliometrisk indikator som underlag för medelsfördelning. Svar på uppdrag enligt regeringsbeslut U2009/322/F (2009-01-29) till Vetenskapsrådet.*

Universitets- og høgskolerådet. (2004). *Vekt på forskning: Nytt system for dokumentasjon av vitenskapelig publisering.* http://www.uhr.no/documents/Vekt_p__forskning__sluttrapport.pdf

Urquhart, D. J. (1959). Use of scientific periodicals. In *Proceedings of the International Conference on Scientific Information,* Nov.16-21 1958, Washington.

Vetenskapsrådet. (2009). *Bibliometrisk indikator som underlag för medelsfördelning. Svar på uppdrag enligt regeringsbeslut U2009/322/F (2009-01-29) till Vetenskapsrådet.*

Vetenskapsrådet. (2011). *Svar på regeringsuppdrag U2011/1203/F: Uppdrag till Vetenskapsrådet att redovisa underlag för indikatorn vetenskaplig produktion och citering.* http://www.bibl.liu.se/bibliometri/litteratur/1.275135/VR-svar_110607_Regeringsuppdrag_U2011-1203-F.pdf

Vetenskapsrådet. (2012). *Svar på regeringsuppdrag U2011/1203/F: Uppdrag till Vetenskapsrådet att redovisa underlag för indikatorn vetenskaplig produktion och citering.* http://www.bibl.liu.se/bibliometri/litteratur/1.354375/VR-svarpregeringsuppdragU2011-1203-F_Underlagfrindikatornvetenskapligproduktionocitering.pdf

White, H. D., & Griffith, B. C. (1981). Author cocitation—a literature measure of intellectual structure [Article]. *Journal of the American Society for Information Science, 32*(3), 163–171. https://doi.org/10.1002/asi.4630320302

Wilsdon, J., Allen, L., Belfiore, E., Campbell, P., Curry, S., Hill, S. A., Jones, R., Kain, R., Kerridge, S., Thelwall, M., Tinkler, J., Viney, I., Wouters, P., Hill, J., & Johnson, B. (2015). The metric tide: Report of the independent review of the role of metrics in research assessment and management. https://doi.org/10.13140/RG.2.1.4929.1363

Zuckerman, H. (1967). Nobel laureates in science: Patterns of productivity, collaboration, and authorship. *American Sociological Review, 32*, 391–403.

9

Peer Advocacy: Expressions of Loyalty in Peer Review

Lars Geschwind and Kristina Edström

Introduction

Evaluations and assessments are always guided by particular aims and methods, but the evaluators also influence how evaluations are conducted. Evaluators bring to their task different methodological competences and levels of knowledge about the object of evaluation. In addition, the identities, values, beliefs, interests, preferences, personalities, and idiosyncrasies of evaluators may influence their assessments. This is also the case in peer review, the dominant and most legitimate form of evaluation in higher education.

In recent years, the use of peer review has proliferated well beyond the assessment of publications and grant proposals. Policy-makers and university managers also use evaluations based on peer review to inform decision-making, quality assessment, and allocations of funds (Hansen

L. Geschwind (✉) • K. Edström
Department of Learning in Engineering Sciences, KTH Royal Institute of Technology, Stockholm, Sweden
e-mail: larsges@kth.se

© The Author(s) 2022
E. Forsberg et al. (eds.), *Peer review in an Era of Evaluation*,
https://doi.org/10.1007/978-3-030-75263-7_9

et al., 2019). Organisations appoint peer reviewers and assign them the task of assessing the preconditions, processes, or outcomes of their peers within a given field (Westerheijden et al., 2007).

Earlier research has shown that academic staff identify to a high degree with and are loyal to their discipline or scientific field (Henkel, 2005). This chapter considers how this sense of loyalty may be reflected in evaluation practices when peer reviewers act as advocates for the peers whom they evaluate (the evaluands). We will use case studies to identify peer advocacy in various forms, whether for the object they are asked to evaluate or for the stakeholders involved. This issue sheds light on and problematizes evaluation practices and evaluation roles, with a particular focus on advocacy in the higher education sector (Morris, 2011).

Aims and Uses of Peer Review

Evaluations serve different purposes. The main ones are summative, formative, and informative. Summative evaluations are made to *control* institutions and actors or their activities. High-stakes evaluations can have important consequences for the evaluands and the stakeholders, and some are performed by authorities who control sanctions and resources. Other evaluations are more formative or process-oriented, aiming to support the *enhancement* of activities. A third common aim is to provide information, often destined to feed into a decision-making process. In higher education and research sectors, the balance between control and enhancement has frequently been discussed in relation to national quality systems (see Ers and Tegler Jerselius' chapter on the Swedish system in this volume). Various national systems have been criticised for placing too much emphasis on control and "sticks" rather than "carrots" (Laughton, 2003), for the unreflective implementation of evaluation machineries (Dahler-Larsen, 2011), and for allowing evaluations to proliferate too much in numerous areas (Hansen et al., 2019). There is also extensive research that focuses on the *uses* of evaluations, which are related to but not synonymous with their aims. Vedung describes how the intended and unintended uses of evaluations may include not only control and enhancement, but also routines, rituals, and window-dressing, with the aim of addressing internal and external stakeholders well beyond the evaluands (Vedung, 2017; see also Karlsson, 2016).

As we have learned from this volume, peer review was first developed as an academic practice. Its roots are found in academic publishing; the practice later spread to other academic decision-making processes such as appointments. It is regarded as a cornerstone of life and culture in academia, and as one of the main ingredients of collegial responsibility in the academic profession (Sahlin & Eriksson-Zetterquist, 2016). Subjecting one's work to the scrutiny of peers and taking on the task of evaluating the work of others are obligations crucial to academic work.

The legitimacy of peer review is based on the idea that only peers can provide a sufficiently high level of expertise to make judgements in a particular area. At the same time, a certain distance is implied between the peer and what is being evaluated, because someone external and without personal interest in the matter might be less prone to favouritism. External reviewers are also more at liberty to give an honest appraisal, especially when blind (anonymous) review is practised. Hence, there is simultaneously an expectation of necessary *affinity*, as the reviewers are assumed to have a similarity in competence for understanding the object of evaluation, and of necessary *distance*, since they are external and disinterested (Merton, 1968).

Peer review as a form of evaluation is intrinsic to academic organisations and academic work. It is routinely used in many decisions and processes in education, research, and in the institutional structures of higher education. However, the boundaries of peer review and the proliferation of peer review as a tool for exercising authority and for management have also been discussed. For instance, peer review panels are increasingly used in areas that were not previously assessed in this way. Their function can differ slightly depending on whether the aim is *ranking* (candidates in appointment cases), *grading* (research grant applications, doctoral theses), judging against *standards or threshold levels* (accreditation of educational programmes, promotion cases), or *informing decisions* together with formative comments (journal manuscripts, projects), to mention some of the most common aims.

It is well known, and documented in previous research, that peer review does not necessarily produce a non-biased or "neutral" opinion (Lee et al., 2013). Systematic disadvantages for certain groups have been identified in peer review, contributing to inequalities in the higher education sector (Wennerås & Wold, 1997; van den Brink & Benschop, 2012). No matter how careful the selection of peer reviewers might be, they have ideas and perceptions that may be reinforced or changed during the peer review process. In the case of peer review panels, the inherent group dynamics add complexity (Lamont, 2009). In national systems for evaluating research, there has been much discussion of the methodological question: should the focus be on peer review or on metrics? And if both are used to evaluate research, one may still discuss what should be the relation between these methods (see Butler & McAllister, 2009). There are instances when reviewers interact with stakeholders during review processes to gather information, often in interviews. In more "democratic" approaches to evaluation, intense stakeholder involvement is a key component (Fagrell et al., 2020).

It is natural that stakeholder representatives seek opportunities to further their own interests and perspectives in the process of evaluation. They frequently come to interviews with a clear agenda, presenting arguments and analyses of their own. It can be hard for the evaluators to sift through wise analyses made with the expertise and insights of the stakeholders, and to identify self-interested appraisals that may be detrimental to other stakeholders. Therefore, it is important for peer reviewers to be able to withstand persuasion and charm from stakeholders who try to engage them as advocates.

Advocacy in Academia

In this chapter, we seek to shed light on one type of peer reviewer behaviour: namely, when reviewers act as advocates and promoters of the evaluand or any of the stakeholders involved. Our aim is to increase awareness of this phenomenon and to discuss the risks it poses for the legitimacy of peer review. Advocacy in evaluation is not a new topic in evaluation research. Ultimately, it relates to our views on knowledge production and the role played by the evaluator/researcher. Going back to Guba and Lincoln (1989), different generations or, in Vedung's phrase, (2010) "waves" of evaluations and evaluators have embraced different epistemic and ontological paradigms. The positivist view of evaluations saw impartial measurement from a distance as the preferred (or only) way to produce valid and reliable knowledge. Other evaluation models have emphasised a more constructivist and relativist view of evaluation practices. Some of these models have implied a closer, even intertwined relationship between the evaluator and the evaluand. Green, for instance, has suggested that advocacy is inevitable, an unavoidable part of evaluation inquiry. She also admits that it has been, and still is, deeply controversial in evaluation communities:

> *the very notion of evaluation as advocacy invokes shudders of distaste and horror among most members of today's evaluation community, theorists and practitioners alike… Advocacy is the antithesis of fair evaluation, according to these founding visions and ideals. To advocate is to espouse and promote a partisan belief or stance, to embrace and advance a cause. To evaluate is, according to tradition, to judge fairly the quality, merit, and worth of a program based on impartial, scientifically gathered information.* (Greene, 1997, p. 26)

For Green and others, this is not only a question of scientific standpoint. It is also about power and values: whose interests should evaluation advance? Which values should it represent? Some scholars argue that sympathies with one or several stakeholder groups or preferences, or for or against a particular programme, should not only be brought out in the open and disclosed; they should even be emphasised as part of a democratic and pluralistic agenda (House, 1990). However, serious issues arise when discussing evaluation advocacy, including integrity, fairness, and long-term credibility, and this chapter seeks to address them. Our chief contribution is to discuss advocacy specifically in a peer review context, with the higher education sector as the particular arena. The role of a peer, as mentioned earlier, implies a specific relation to the evaluand, although the notion of peership has changed over time (see Chap. 1 in this volume for a discussion).

Our aim is not to paint peer advocacy as an obvious threat to the integrity of evaluation processes and results. Instead, we approach the phenomena studied here with great curiosity and openness. After all, it is an intrinsic feature of peer review that judgements are made by human beings who are capable of empathy for other human beings and their work (Stake, 2004). As some evaluations are formative, with enhancement as their main purpose, there are situations when it is not necessarily problematic for peers to help the evaluand. The question, however, is when and to what extent this kind of help is appropriate, and when there is risk of tilting the balance between stakeholders or compromising the integrity of the evaluation. This makes peer advocacy a potentially sensitive subject. To raise the stakes further, we have chosen to study cases in which we ourselves were involved. Taking Sweden as the empirical arena, we will discuss and exemplify four different experiences of peer advocacy.

Cases of Advocacy

We will analyse four cases from a Swedish higher education context. We selected the cases as personal experiences where we felt that something was at play that we have come to call peer advocacy. The authors were either involved as evaluators or had insight into the process for other reasons.

- The first case concerns a national review of subjects and programmes. The first author was the panel secretary.
- The second case comes from a Swedish university college that initiated an evaluation of its unit for teaching and learning support. Two evaluators were commissioned and instructed to make separate evaluations; the second author was one of them.
- The third case is a peer review launched by a Swedish university college as part of the application process for becoming a fully-fledged university. The first author was at the time employed at the state agency responsible for evaluating applications for university status.
- The fourth case is an assessment exercise of administrative processes initiated at a Swedish university. Both authors were employed at that university at the time and involved in appointment and promotion processes there.

The materials used are evaluation reports supplemented with reflections from the authors who were involved in the cases. We explored the cases "from the inside" both through self-reflection and by interviewing each other. The experiences are lie the past, up to 20 years ago. The temporal distance has the advantage of encouraging honesty, notwithstanding potential embarrassments. In this regard, it helps that the authors have gained many new experiences since these particular cases; they have also grown older. The disadvantage is that their memory of details of the cases and their contemporary contexts may have faded somewhat.

In the following, the cases are described without the distraction of too many details. All quotes are translated from Swedish; some are lightly edited for brevity.

Case 1: National Reviews of Subjects and Programmes

The first case is taken from national evaluations of educational programmes and subjects leading to general degrees. The aim was to check that subjects and programmes were of sufficient quality, with the possible sanction of a warning that the degree-awarding rights of that particular higher education institution might be revoked if it fell short of standards. The chosen evaluation model focused on institutional preconditions, processes, and results (Franke, 2002). Panels consisted of academic staff from Swedish and other Nordic institutions and representatives of students and doctoral students, aided by a secretary from the national agency. The evaluation model was based on self-evaluations by the institutions, and site visits by the external peer review panel. During the period studied here, the national assessments in Sweden resulted in 30 reports. A final report by the agency included, as a separate part, the peer review panel report (Stensaker, 2000).

We have gone through the summaries and conclusions in each report. What we found particularly interesting in relation to the topic discussed here are the frequent remarks about working conditions for teachers. This theme was present in 24 of the 30 reports. The panels comment especially on heavy faculty workloads, the limited scope for research and development, the pressure to apply for external funding, and insecure conditions of employment. These topics were also mentioned in a conference paper written by the secretary at the agency:

What we have seen in the national assessments is first of all that many teachers do an excellent job. They work hard, they are enthusiastic, and they want what is best for their students. Many think they have the best job there is. The drawbacks are long working hours, insecure employments, especially for junior academics, and limited scope for research and development. In the past decade, salaries and societal status seem to have decreased as well. (Geschwind, 2004, p. 12)

That the panels brought up such themes in the evaluations is unsurprising, as the preconditions for education were indeed included in the evaluation criteria. What is striking is the balance and proportions. The overall impression of the evaluations is that much less attention was paid to results and to education processes such as teaching methods and assessment than to the difficulties faced by teachers themselves.

This case shows that the national evaluations of education in this system included a fair amount of peer advocacy. By focusing on the preconditions for education, and more specifically on working conditions for academic staff, most evaluation reports discussed topics such as workload, time available for research, and even the Swedish funding system. Although the evaluand as such was the quality of educational programmes, colleagues from the Nordic countries and the secretary at the agency instead seemed inclined to assist their colleagues by acting as advocates who drew attention to difficult working conditions on behalf of their peers, and recommending improvements.

Perhaps as a consequence of this, the architects of the next national evaluation model faced a good deal of political pressure to focus on results only, resulting in a period of conflict in the higher education sector between the Ministry of Education and the agency (see Ers and Tegler Jerselius in this volume).

Case 2: Institutional Review of an Academic Development Unit

The next example is an evaluation of the academic development unit of a higher education institution. The unit's main purpose was to provide courses in teaching and learning in higher education for academic staff.

At the time, professors and associate professors were required to have ten weeks of such education. This requirement was mandated in the Higher Education Ordinance together with a set of nationally agreed learning objectives (Lindberg-Sand & Sonesson, 2008). The Pro-Rector who had supported the establishment of the unit commissioned the evaluation shortly before her retirement. It is likely that the evaluation was made a part of the handover in order to protect the unit from reorganisation, as this type of unit was often reorganised when managers were replaced (Gosling, 2008; Challis et al., 2009; Palmer et al., 2010).

Two evaluators, both active at similar units in Swedish universities, were instructed to work separately without communicating. They were given the same documents and made separate site visits, interviewing mainly the same stakeholder representatives, including the unit's staff, university leadership at all levels, faculty, and students. The brief asked for a thorough analysis and documentation of the unit in the light of comparable national and international work, together with recommendations for its future development.

When the evaluations were presented to the institution in an open seminar, it became clear that they were very different. One evaluator's report was twelve pages long, presenting an analysis and recommendations supported by brief descriptions. The perspective emphasised was that of faculty members who were participants in courses given by the unit. About half the report was devoted to complaints from faculty members about the courses being compulsory:

> *The problem lies in what is described as "studentification" of the teachers who participate in the courses. This is seen as mistrust in the ability of staff to take responsibility for their own pedagogical development. They also see it as an expression of low respect, or lack of trust in the participating teachers' ability to define their own needs for further pedagogical education. For instance, they mention how the courses are assessed, requiring them to cite course literature. The focus is on course participants being able to demonstrate knowledge, or to mention the regulations that steer education. The critique is that the same patterns are followed that course participants use on their students. This is seen as very problematic, not least by teachers who hold PhDs.*

Most of the recommendations made at the end of the report aimed to oblige those who criticised the unit's courses, for instance suggesting that the unit should switch to a softer approach, a "carrot" model.

The other evaluator devoted 54 pages to explaining academic development in a national and international perspective, analysing the needs and conditions at this institution in particular, and evaluating the work of the unit in some detail. The overall judgement was positive, and recommendations aimed at organisational stability and increased resources. Here, the main perspective was to further academic development at the institution, but also more widely. That the courses were required was taken as a given, as this was regulated at the national level. The critique from a small minority of faculty was depicted from the viewpoint of the academic developers leading the courses. What appeared is a mirror image of the preceding quote:

> *The pedagogical courses do not take place in a social vacuum. This activity has various identities—it is the long arm of the university leadership while also being academic or emancipatory, and this often creates dissonance... Assessment is a particularly sensitive matter, since it puts to the test both the format for assessment and the legitimacy of those who make the assessment. That the courses are de facto compulsory pushes every such issue to its limits, and course leaders encounter various forms of resistance... This friction must be seen in the light of its complex political and social context—only then can workable strategies be found. The course leaders must be able to navigate these troubled waters, through a well-developed understanding of their role.*

This evaluator was explicitly situated the academic development unit in a context of several diverse interests, positioning faculty members as just one voice in a stakeholder chorus:

> *The task of the pedagogical courses is to enable and drive the development of the educational programmes by supporting staff development. Pedagogical competence must be interpreted more widely than just the individual teacher's desires and needs; it must be seen in relation to the aims and challenges of the educational programs and the institution.*

Despite the different approaches reflected in these two evaluations, both were essentially positive and made strikingly similar recommendations regarding increased resources to the unit. But while both evaluators brought up other interests in their analyses, they clearly held one particular perspective in the foreground. If one evaluator took a stance defending the views of academic staff, the other can be seen as a peer advocate for academic development endeavours. These positions were likely influenced by their own backgrounds. One evaluator was a full professor whose part-time engagement in academic development was tied to a particular research area. The other was invested in academic development, also on national and international levels. We note that it was only the way that the evaluation was designed, with two separate and parallel evaluations instead of a single collaborative one, that revealed these different perspectives. If the evaluators had worked together, they would likely have conciliated their interests and presented a unified result.

Case 3: External Evaluators Appointed to Promote an Application for University Status

The third case is that of a university college preparing an application for full university status. The government had rejected a previous application; in fact, it was not even sent out for review. This time, as part of the preparations, the rector commissioned an evaluation that was conducted by an external peer review panel of distinguished professors. The composition of the group is worth mentioning in some detail. One panel member had previously been a driving force for establishing the university college as a new seat of learning. Another member was then Rector at an institution that had just been granted university status. Other prominent professors from major Swedish universities made up the rest of the review panel. A manager from the university college acted as secretary.

In the report, the panel declared that their experiences were important:

> *The Board of the University College decided to update the application for university status. In parallel with this update, the Rector of the University College has hired us to make a situation assessment based on our experience of university business and of assessment work.*

The evaluation used the same criteria that had previously been applied for university status. The first part of the report is based on key figures that compare the university college with the last three Swedish institutions that had been "promoted" to university status. The comparison showed that the university college was on par with, or even above, the level of these three universities. To assess educational standards, a meta-evaluation of the National Agency for Higher Education's reports was utilised. In their conclusion, the panel praised the university college and strongly recommended that it should be elevated to university status:

> *At the University College there is accumulated capital in the form of high ambition, good work morale, high-class infrastructure, and not least a corps of academic staff with high academic competence who can be expected to work well under the conditions that a university status confers.*

This case shows us how a peer review panel can be commissioned and composed, and the evaluation set up, for a specific purpose—in this case to support an application for university status. The positive outcome of the evaluation was only as expected. Nevertheless, the government resisted, and did not award university status. Instead, it communicated that this avenue was closed, and a period of mergers followed (Benner & Geschwind, 2016).

Case 4: HR Competencies in Academic Appointment Processes

The last case comes from a university that initiated a comprehensive internal evaluation of its administrative processes. The evaluation followed the logic of previous exercises in research (RAE) and education (EAE) assessment, and it was named the Administrative Assessment Exercise (AAE). The evaluation design was also similar to these other exercises, comprising a self-evaluation written by the evaluands and a panel of external experts who analysed the written material, undertook a site visit on campus, and delivered a report.

While the overall aim of the exercise was to improve the quality of administration, two other interesting aims were also identified:

The AAE was also expected to have a number of positive side effects. Amongst these was to increase the administrative staff's knowledge about evaluations and quality work, including a greater understanding of the processes that teachers and researchers continuously undergo in exercises such as RAE and EAE. Through the AAE, administrative work would also become more visible throughout the organisation. This development, in turn, would facilitate better communication between administration, faculty, students, and other stakeholders.

It was thus seen as important for professional support staff to learn more about evaluations as phenomena that have profound effects on academic life. But a further purpose of the AAE was to give administrative work more recognition within the organisation.

Unlike many other evaluations, including previous education and research assessment exercises at this university, the AAE focused on processes rather than organisational units. One of these processes was the appointment of faculty, described as "a strategic process that requires intense administration-academy interaction." A senior administrative officer chaired the panel, which also comprised experienced human resources officers, with the exception of a professor emerita. The panel report includes a number of interesting passages. Under the heading "Competence," the tension between the academics and HR officers is discussed. Unclear roles and a blurred division of labour had created, overall though not everywhere, a situation of mutual distrust. It seems that this lack of trust was rooted in traditional practice:

There is a strong tradition that has great legitimacy in academia to appoint academic staff after peer review, without the involvement of HR, which would be unthinkable in many other organisations. However, there is a trend in many universities to ask for and appreciate HR competence.

One of the suggested ways to better utilise HR competence is to focus more on the candidates' personal abilities and leadership issues. According to the panel, other assessment criteria than the current ones would also bring HR into the process of selecting candidates for academic positions, a task that was perceived at the time as being largely in the hands of external peer reviewers.

Further arguments were put forward about the importance of HR competences:

> *Also from a cost perspective, HR competence is important in the recruitment process. It contributes not only to finding the right person for a position, but also to avoiding appointing the wrong person. A failed recruitment carries enormous costs. A trained and experienced interviewer might more easily identify possible risk personalities and investigate concerns more closely providing a better basis for decisions about employment.*

Finally, in the concluding section two out of three panel recommendations for the future revolve around the roles and contributions of HR personnel. The first is to "strengthen awareness of and confidence in what different roles contribute" and the second is simply to "increase HR competence in the process."

Discussion

All these cases involve evaluations in higher education, and some show how peer review is used in new academic processes. We chose them not because they are representative of all or most cases of peer review, but we wanted to illustrate a phenomenon so common that we have seen several instances in our own experiences. In every case, we can see that the evaluators demonstrate strong loyalty towards their peers. They recognise and show understanding of their colleagues' situation and needs. It is beyond the scope of this chapter to discuss advocacy in relation to results, outcomes, or effects, but other studies have shown that peer reviewers can focus on difficult working conditions as way to explain and sometimes excuse poor performances (see Geschwind, 2016). Furthermore, it becomes clear that one is a peer *in a context*. The nature of the commission and particular evaluation models, methods, interactions, and expectations can present different opportunities for the peer role. A particular situation may offer more or less opportunity for peer advocacy, and the instinct to further the interests of one's peers may be aroused to various degrees.

Case 1 shows that while the evaluation model at the time encompassed working conditions, processes, and results, the peer role seemed to shift the balance significantly towards the former. Although the aim was to evaluate the quality of programmes, the peer reviewers also cast a strong light on other factors beyond the evaluands. In practice, the panels delivered what amounts to a wish list related to funding systems and institutional governance. The reviewers seemed to be torn between their commission to evaluate quality and their loyalty towards frustrated peers. One way to limit opportunities for peer advocacy, then, might be to use more structured evaluation models that leave reviewers less room to comment on matters that lie beyond the review's main focus. The composition of a panel may also help to keep a balance, since the presence of other stakeholders might hold in check the instinct to promote the perspective of one's peers in a one-sided way. In this case, however, the presence of other stakeholder representatives—who in all likelihood had a different set of interests and loyalties—was not sufficient to balance the academic peers of discontented faculty, who were in the majority.

Case 2 shows that it is not always obvious who the peer reviewer is a peer to, when there are several stakeholders. Here we saw conflicting peer advocacies over the same evaluand. Hence, the peer role varies with reviewers' relations to stakeholders. While a peer review panel is often composed to ensure that different stakeholder interests are represented, a sole evaluator has more room to engage in peer advocacy. In this case, the different perspectives became apparent only because there were two evaluators working separately. Had the two evaluators been asked to collaborate, they would probably have weighed various interests against each other in their deliberations and presented a unified result. Had there been just one evaluator, the result could have gone either way. But in the absence of a comparison, the peer advocacy might have gone unnoticed. Since a proliferation of biased evaluations would undermine the legitimacy of peer review, on balance the model based on separate evaluations has advantages.

Case 3 is an example of how an evaluation might be initiated with a specific aim and intended use. It seems clear that the desired result was implicit from the outset, as the evaluation was to be used for a specific purpose: namely, to strengthen the application for full university status. The evaluation was obviously aimed at mobilising a group of friendly supporters; it was wilfully rigged with peer advocacy as the sole purpose. We note, however, that the university college's application was unsuccessful. Perhaps the quality of the application made little difference in a situation where they lacked the necessary political support, and not even this group of illustrious peers could help their cause. The institution's efforts of self-promotion may also have been too conspicuous. They did succeed, however, in establishing an image of the institution as on a par with some universities. In future attempts to gain university status, they can always pull up this report from the archives and use it again. This case also raises the question of where evaluators should draw the line to preserve their own integrity and legitimacy.

Case 4 has shown how peer advocacy can also work to further the interests of university managers. In most university settings, appointment processes are based on the classic form of peer review, where colleagues from the same academic field are trusted to assess candidates' merits. In this case, the panel, consisting of HR officers and other senior administrators, stressed the need for greater clarity regarding division of labour between management and academics, and called for increased mutual respect. But they also took the opportunity provided by the evaluation process to propose fundamental changes in appointment processes. They advanced the idea that HR competences could be more extensively utilised, even if this meant changing certain criteria for appointing academic staff. It is not clear to us, however, whether these HR advocates fully understand the academic context. We wonder if the perspective of their own type of expertise makes them see some essential and even valuable characteristics of academia as flaws that ought to be corrected.

Conclusions and Recommendations

Peer review is the most legitimate way to judge quality in higher education, and often the only feasible one. It involves the judgements of experts who are also ordinary humans with a capacity for empathy towards the people involved. However, too much bias carries the risk of watering down legitimacy, thereby corrupting peer review as an institution. It is not merely the results of any specific evaluation that are at stake, but also the role of peer review as a cornerstone of academic life, culture, and autonomy. This suggests that anyone involved in peer review processes has a great responsibility to preserve the integrity of this form of evaluation.

Peer advocacy should not only be understood as emanating from peer reviewers themselves. As we have shown, it can also be enabled or limited by the way in which a review is commissioned, the specific evaluation model, the criteria and conditions that the commissioning client sets out, and by both real and perceived expectations. We recommend, firstly, that the client who commissions an evaluation should strive to be transparent and clear about their chosen evaluation model, methods, and criteria. Secondly, clients may want to take into account the interests of stakeholders when appointing peer reviewers. For an evaluation to be credible, thirdly, the evaluators must have the chance to undertake an actual investigation, and base the results on materials that they collect and document; the result should never be given *a priori*. Finally, peer reviewers too should bear all these issues in mind. There may be situations when it is appropriate to turn down an invitation, to protest during a skewed process, and to deliver results that were not what the client desired.

References

Benner, M., & Geschwind, L. (2016). Conflicting rationalities: Mergers and consolidations in Swedish higher education policy. In R. Pinheiro, L. Geschwind, & T. Aarrevaara (Eds.), *Mergers in Higher Education. The Experience from Northern Europe* (pp. 43–58). Springer.

Butler, L., & McAllister, I. (2009). Metrics or peer review? Evaluating the 2001 UK research assessment exercise in political science. *Political Studies Review, 7*(1), 3–17.

Challis, D., Holt, D., & Palmer, S. (2009). Teaching and learning centres: towards maturation. *Higher Education Research & Development, 28*(4), 371–383.

Dahler-Larsen, P. (2011). *The Evaluation Society*. Stanford University Press.

Fagrell, P., Gunnarsson, S., & Fahlgren, A. (2020). Curriculum development and quality assurance of higher education in Sweden: The external stakeholder perspective. *Journal of Praxis in Higher Education, 2*(1), 28–45.

Franke. (2002). From Audit to Assessment: A national perspective on an international issue. *Quality in Higher Education, 8*(1), 23–28.

Geschwind, L., (2004). Working Conditions for University Teachers in Sweden: Findings from the assessments of subjects and programs 2001–2003, paper presented at the 26th EAIR forum, Barcelona, 5–8 September 2004.

Geschwind, L. (2016). Academic core values and quality: the case of teaching-research links. In M. Elmgren, M. Folke-Fichtelius, S. Hallsén, H. Román, & W. Wermke (Eds.), *Att ta utbildningens komplexitet på allvar: En vänskrift till Eva Forsberg* (pp. 227–238). Acta Universitatis Upsaliensis.

Gosling, D. (2008). Educational Development in the UK, Report to the Heads of Education Development Group (HEDG), February 2008.

Greene, J. C. (1997). Evaluation as advocacy. *Evaluation practice, 18*(1), 25–35.

Guba, E. G., & Lincoln, Y. S. (1989). *Fourth generation evaluation*. Newbury Park, Calif.: Sage.

Hansen, H. F., Aarrevaara, T., Geschwind, L., & Stensaker, B. (2019). Evaluation practices and impact: Overload? In R. Pinheiro, L. Geschwind, H. F. Hansen, & K. Pulkkinen (Eds.), *Reforms, organizational change and performance in higher education: A comparative account from the Nordic countries*. Palgrave Macmillan.

Henkel, M. (2005). Academic identity and autonomy in a changing policy environment. *Higher Education, 49*(1–2), 155–176.

House, E. R. (1990). Methodology and justice. *New Directions for Program Evaluation, 1990*(45), 23–36.

Karlsson, S. (2016). The active university: Studies of contemporary Swedish higher education. (Doctoral dissertation, KTH Royal Institute of Technology).

Lamont, M. (2009). *How professors think*. Harvard University Press.

Laughton, D. (2003). Why was the QAA approach to teaching quality assessment rejected by academics in UK HE? *Assessment & Evaluation in Higher Education, 28*(3), 309–321.

Lee, C. J., Sugimoto, C. R., Zhang, G., & Cronin, B. (2013). Bias in peer review. *Journal of the American Society for Information Science and Technology, 64*(1), 2–17.

Lindberg-Sand, Å., & Sonesson, A. (2008). Compulsory Higher Education Teacher Training in Sweden: Development of a national standards framework based on the Scholarship of Teaching and Learning. *Tertiary Education and Management, 14*(2), 123–139.

Merton, R. K. (1968). *Social Theory and Social Structure, enlarged version*. The Free Press.

Morris, M. (2011). The good, the bad, and the evaluator: 25 years of AJE ethics. *American Journal of Evaluation, 32*(1), 134–151.

Palmer, S., Holt, D., & Challis, D. (2010). Australian teaching and learning centres through the eyes of their directors: characteristics, capacities and constraints. *Journal of Higher Education Policy and Management, 32*(2), 159–172.

Sahlin, K., & Eriksson-Zetterquist, U. (2016). Collegiality in modern universities–the composition of governance ideals and practices. *Nordic Journal of Studies in Educational Policy, 2016*(2–3), 33640.

Stake, B. (2004). How far dare an evaluator go toward saving the world? *American Journal of Evaluation, 25*(1), 103–107.

Stensaker, B. (2000). Quality as Discourse: An Analysis of External Audit Reports in Sweden 1995–1998. *Tertiary Education and Management, 6*(4), 305–317.

Van den Brink, M., & Benschop, Y. (2012). Gender practices in the construction of academic excellence: Sheep with five legs. *Organization, 19*(4), 507–524.

Vedung, E. (2010). Four waves of evaluation diffusion. *Evaluation, 16*(3), 263–277.

Vedung, E. (2017). *Public policy and program evaluation.* Routledge.

Wennerås, C., & Wold, A. (1997). Sexism and nepotism in peer-review. *Nature, 387*(6631), 341–343.

Westerheijden, D. F., Stensaker, B., & Rosa, M. J. (Eds.). (2007). *Quality assurance in higher education: Trends in regulation, translation and transformation* (Vol. 20). Springer Science & Business Media.

10

Is Peer Review Fit for Purpose?

Malcolm Tight

Introduction

Peer review is endemic to judgement in higher education, as well as throughout the social world. Indeed, it should not be surprising to learn that its origins—like those of higher education in general—are religious, involving the pre-publication judgement on whether a book should be permitted to be put on sale or burnt, perhaps along with its author, as heretical (Lipscombe, 2016).

In contemporary higher education, it is widely assumed that when we need to make a judgement on the quality of something—for example, a student's performance, the employment or promotion of an academic, the importance of an academic publication, whether someone should get a research grant—then we may rely on the assessment of one, two or multiple peers, typically but not always more senior academics. While available quantitative, and thus seemingly more 'objective', data also has

M. Tight (✉)
Lancaster University, Lancaster, UK
e-mail: m.tight@lancaster.ac.uk

© The Author(s) 2022
E. Forsberg et al. (eds.), *Peer review in an Era of Evaluation*,
https://doi.org/10.1007/978-3-030-75263-7_10

an increasing role to play in these assessments—for example, in the context of the four examples given, course grades, student evaluations of teaching, numbers of citations, and previous grants held, respectively—the final judgement will typically be taken by a small number of academics (and increasingly perhaps professional administrators or managers).

This chapter will illustrate and challenge the assumptions underlying peer review, and assess how 'fit for purpose' it is in twenty-first century mass higher education. The chapter will focus on different practices of peer review in the contemporary higher education system as practiced by academics (i.e. it will not consider peer review between students, an increasingly popular means of both developing student skills and reducing academics' workloads). It will question as to how well they work, how they might be improved and what the alternatives are. While the presentation will be grounded in the UK experience, and in English language publications, the discussion will spread out internationally and comparatively.

Three main examples of academic peer review will be introduced and discussed: the refereeing of academic journal articles, the assessment of doctoral degrees and the UK Research Excellence Framework (REF). These have been chosen not simply for their broad and international significance—this is self-evident in the case of journal articles and doctoral degrees, while the UK REF and its predecessor have served as models elsewhere—but because the author has a lot of experience in each of them. Thus, in the first example I have been a widely published author and, across my career, an editor of multiple journals. In the second example, I have been involved, as research supervisor or examiner, in well over 100 doctoral degree examinations; chiefly in the UK, but (mainly at a distance) in a number of other countries as well. In the last example, I served as a member of the Education Sub-Panel in the 2014 REF exercise.

Following a brief discussion of the methodology adopted in this chapter, I will proceed by considering each of my three examples in turn, before drawing some more general conclusions.

Methodology

The methodology adopted in this chapter may be characterized as being informed by systematic review and personal experience. While this may seem a somewhat odd combination, it is an approach that I have refined over the years: you focus on something of interest and explore all of the existing research that you can access and analyse.

Systematic review (Torgerson, 2003) principles have been used to identify relevant published articles on the topics discussed, using keywords and relevant online databases (chiefly Scopus and Google Scholar). However, rather than conduct a full systematic review—which a chapter of this length doesn't really allow scope for—an informed selection has then been made from among the thousands of articles identified for discussion in this chapter.

Personal experience has been used, as already indicated, both in the selection of the topics to be discussed, and in knowing where to look for useful evidence. It has also then, naturally enough, underlain the critique presented. Given the mixture of systematic review and personal experience, the discussion necessarily mixes the third and first persons.

No new empirical data is, therefore, presented or used in this chapter. Rather it rests on the accumulation and interrogation of evidence from past research into the topics of interest. It is, thus, also an example of documentary research (Tight, 2019a).

Refereed Journal Articles

Nearly 20 years ago, with many years' experience of the journal publication process—both as an author and an editor—already under my belt, I decided to carry out a personal test of the veracity of the article reviewing process (Tight, 2003). I had kept copies of all of the reviews of the articles I had submitted to journals over the previous ten years, together with the decisions taken on them by the editors concerned and the comments made by referees.

I went through each article, assessing whether the reviews were positive or negative in tone, or a mixture of the two. While this was, admittedly, a subjective assessment, it was surprisingly easy to do. I then cross-tabulated the results against the editors' verdicts, which (analogous to the PhD examination process discussed in the next section) were typically one of four decisions: accept, minor revisions, major revisions or reject.

The pattern this exercise revealed was quite striking: the relationship between referees' ratings and editorial decisions was far from clear. Highly criticized articles had sometimes been accepted with little or no amendment required, while positively reviewed articles were sometimes rejected. Where one or more referees were positive, and one or more negative, the editorial decision might, of course, go either way.

Clearly, then, the opinion of one's peers was not the only factor that mattered—other considerations were also in play. From personal editorial experience, I would say that two of these additional factors are the editor's own opinions (also a form of peer review of course) and the limitations imposed by the amount of publication space available in the journal (i.e. some editors are looking for reasons to reject articles, while other editors are looking for reasons to accept them), but there are doubtless other factors as well.

Perhaps unsurprisingly, this topic has also been the subject of more extensive, and less personal, research. It would be strange if such a central aspect of academic life had not attracted such attention:

> *Authors, manuscripts, reviewers, journals and readers have being [sic] scrupulously examined for their qualities and competencies, as well as for their "biases", faults or even unacceptable behavior. This trend has risen with the pioneering work of Peters and Ceci (1982) who resubmitted to journals articles that they had already published, simply replacing the names of the authors and their institutions with fictitious names and making minor changes to the texts. Much to their surprise, almost all of the manuscripts were rejected, and, three exceptions aside, without any accusation of plagiarism. Thirty years later, hundreds of studies on manuscript evaluation are now available. The diverse arrangements of manuscript evaluation are thus themselves systematically subjected to evaluation procedures. For example, in order to comparatively valuate single blind and double blind, studies have increasingly used randomized controlled trials, leading to opposite results and recommendations for journal editors.*
> (Pontille & Torny, 2015, p. 75)

Many contemporary studies have focused on the experience of a specific journal or nation. Thus, Hewings (2004) analysed 228 reviews submitted to the journal *English for Specific Purposes*, finding that 'reviewers take on multiple roles, at the same time discouraging the publication of work that fails to meet the required standards and offering encouragement to authors and guiding them towards publication' (p. 247). Atjonen (2018) surveyed the opinions of 121 Finnish researchers on the ethics of peer review, concluding that:

> *Out of nine ethical principles honesty, constructiveness, and impartiality were appreciated but promptness, balance, and diplomacy were criticized. According to two open questions, a third of authors praised and blamed reviewers as experts and non-experts. The accuracy of feedback was more often present in the best rather than in the worst experienced review processes. Journals' editors and their decision-making called forth more negative than positive accounts. (p. 359)*

In a third example, Falkenberg and Soranno (2018) analysed 49 reviews of 26 submissions to *Limnology and Oceanography: Letters*. They found that 'editor perception of review quality was based on review content rather than if there was agreement on the manuscript decision' (p. 1), which is somewhat reassuring.

Journal article peer review has also been the subject of large-scale research synthesis. Bornmann et al. (2010) undertook a meta-analysis of studies of the journal peer review process. This involved identifying all of the previous quantitative studies of the topic which they could, and combining their data, focusing on the inter-rater reliability of reviewers (i.e. the extent to which article reviewers make the same recommendations). They identified 70 reliability coefficients from 48 studies, which together had examined the assessment of 19,443 manuscripts. They found that the inter-rater reliability was low; that is, journal reviewers seldom agreed with each other. Meta-regression analyses found that neither discipline nor the method of blinding (i.e. anonymizing) manuscripts impacted on this result.

This might, of course, be at least partly explained by the relatively simple rating scale used by many journals; the accept/major/minor/reject scale already referred to. What constitutes major revisions to one reviewer might, for example, easily be called minor revisions by another. There are also, however, the occasional cases where one reviewer recommends that an article should be accepted without any further work, and another reviewer recommends rejection. While the obvious response of seeking a third opinion (perhaps that of the editors themselves) is pragmatic, it does ignore the underlying disparity of judgement.

These analyses suggest that both the practices of peer review of academic journal articles and the accuracy of its results may be challenged. Editors, of course, would probably not want to encourage too much of this: the editorial role is demanding enough as it is.

Another response, however, is to question just how much this matters. After all, authors receiving reviews of their work—even when it is rejected by the journal in question—are hopefully receiving at least some useful information, which they may use in revising their articles for possible publication elsewhere. There are usually many alternative journals available, with higher or lower acceptance thresholds, in which publication may be sought. Academic authors simply have to get used to the rough and tumble of the article publication process (to which they themselves contribute as reviewers) if they are to succeed.

It is also possible, in certain circumstances—including that the authors concerned have a strong sense of the worthwhileness of their work, which should come with experience, and are able to defend or respond to criticisms of it—for the authors to negotiate with journal editors, and even through the editor with their reviewers, over the treatment of their submission after a decision has been taken on it (Kumar et al., 2011). This can work to mutual benefit. Thus, in their study of selected science and engineering articles, Kumar et al report that:

> *Most types of negotiations helped authors to improve presentation of their underlying concepts, quality, clarity, readability, grammar and technical contents of the article, besides offering an opportunity to rethink about several other aspects of the article that they overlooked during the preparation of manuscript. (p. 331)*

It is, of course, unrealistic in any case to expect unanimity of judgement (the closest we might get to 'objectivity') amongst academics. Some may warm to a particular line of argument, theoretical framework and/or methodology, while others will be put off by it. The academic world, at least in research terms, is built to a large extent on competition and disagreement (a brief visit to any academic conference should confirm this). To some extent, reviewers might also be said to be acting in a 'zero-sum' game; that is, if they recommend the rejection of an article they are reviewing, there is potentially that much more space available for their own publications.

It would be hard, however, to argue that the academic journal article review process works well, for, in addition to taking up an inordinate amount of (typically unpaid) time and effort, it causes a great deal of emotional upset among those whose efforts are being judged. It may be, of course, that the growing moves towards online and freely available publication, and towards researchers self-publishing their articles on their own websites, will go some way towards resolving these issues.

The Assessment of Doctoral Degrees

Over my career I have been involved in well over 100 doctoral degree examinations: initially my own (which was a disaster), and then as internal examiner, external examiner, supervisor and independent chair, and even as a third examiner brought in to adjudicate between the first two. While the vast majority of these examinations or vivas have been in the UK, I have also participated at a distance in several doctoral examinations in Australia and Hong Kong, and witnessed them (as a member of the public) in Finland and Sweden.

My own doctoral examination was an interesting induction into this experience: my viva lasted for 20 months! I had two external examiners, rather than the more typical UK pattern of an internal examiner (from the candidate's department) and an external examiner, because my department was then trialling measures for making doctoral examinations more robust. Unfortunately, however, these two external examiners could not agree on their recommendations, and it took the department 20 months to bring them to an agreement. As the university I was studying with then had no regulations for dealing with these circumstances, as universities typically did not then (when the idea of the student as customer had yet to take hold), my viva technically just went on and on.

Since that induction, I have experienced a wide range of viva experiences: ones where the candidate cried, one where the candidate tried to hit one of the examiners, one where the candidate was invited to contribute to a forthcoming book being edited by one of the examiners, one where the candidate was failed for plagiarizing the entire thesis, one where the candidate failed to give a straight answer to any of the examiners' questions, even one where I fell asleep (I was the non-participating attendant supervisor for that one)! However, the vast majority, if sometimes somewhat underwhelming, have led—usually after major or minor revisions, but occasionally awarded straight away—to a doctoral degree.

While it seems clear that the doctoral examination process in the UK works—if it did not, there would have been increasingly strident calls for change before now—this is not to say that it works well and consistently. Even within a single country, there is considerable variation in institutional and disciplinary practice (Tinkler & Jackson, 2000), and a variety of doctoral models. In the last few decades, professional or taught doctorates have become popular, alongside the traditional format of a lengthy thesis produced by an individual after a few years of supervised research, changing the dynamic and expectations.

Examiners can vary a great deal in the attention that they give to a thesis. One examiner may produce a report of 20, closely typed, pages, while another may turn in a single paragraph. And, of course, how one examiner interprets major or minor revisions may be very different to another. A lot may actually come down to whether the examiner in question 'likes' the candidate and topic.

The scope for variation was demonstrated to me clearly when a candidate at another university appealed against the examiners' decision (which was 'major revisions'), winning their case on what might be termed a technicality. Another viva was held with two new examiners, and, while they also returned a verdict of 'major revisions', the viva was 'friendlier' and the revisions were much more achievable (as well as being different from the first set). While the notion that the 'academic judgement' of the examiners chosen, as opposed to the examination process, cannot be challenged is still upheld, it is clearly coming under some threat.

Whether the verdict of two academics on a doctoral thesis—widely seen nowadays as the entry point to an academic career—can be relied upon, therefore, is highly debatable, even if the doctorate is only seen as setting a certain minimum standard. Much clearly depends critically on which two—or which panel of—academics are picked as examiners. I now warn all of my research students that they might be lucky and get the only two academics in the world who would be prepared to pass their thesis as their examiners, or they might be unlucky and get the only two who would fail it. This may be an exaggeration, but only slightly, and it certainly does not make the task—typically born by the candidate's supervisor—of choosing examiners any easier.

Some authors have argued that it is time, at least in the UK:

for a radical review of doctoral education assessment across disciplinary boundaries to consider systematic and universally agreed criteria and scrutiny procedures to quality assure the award. For example, measures such as the convening of a public panel for the viva on the continental model are worthy of consideration, with this open forum removing the secrecy element from the process. With an increase in work-based, professional and 'taught' doctorates, a further aim of such a review would be to develop standardised procedures across disciplines and institutions to 'benchmark' standards in both the written and oral assessment components. Also, the establishment of codes of practice concerning examiner selection that, for example, might move towards the appointment of anonymous reviewers and the mandatory nomination of an independent chair for the viva, might increase confidence in the integrity of the doctoral assessment process. Whilst supervisors may perceive such policies as a threat to their academic autonomy and thus might resist their implementation, they may help to positively transform the current disparities through which inequalities and inconsistencies are maintained. (Watts, 2012, pp. 379–380)

One does, however, have to question just how realistic some of these recommendations are, and also whether Watts' understanding of 'the continental model' is as complete as it might be. To expect 'universal agreement' 'across disciplinary boundaries' is, after all, asking rather a lot of a disparate professional group for whom the favourite managerial metaphor is 'herding cats'! On the other hand, 'standardised procedures' and 'codes of practice' do already exist, at least at a disciplinary level, but the point is that they remain subject to individual interpretation.

Of course, as Watts recognizes, practices regarding the assessment of doctoral degrees also vary significantly from country to country. In some, including many European countries, it is a public event, but the result is pre-determined in private beforehand (van der Heide et al., 2016). In other countries, such as in North America, it is a committee decision and a viva may not be held. In some countries, the doctorate is even graded, such that only those who pass with a certain grade are eligible for academic appointments.

The extent to which the more labour intensive of these practices, however 'good' they might be, can survive as doctoral education—like the rest of higher education—increasingly becomes a mass market, is, though, debatable. In addition to consistency and transparency, the assessment of doctoral degrees needs to be time efficient.

Again, though, as for the refereeing of journal articles, it might be questioned how much the deficiencies identified in the doctoral examination process really matter. The vast majority of students who work hard on their research and thesis over a period of years, with their supervisor's support, are awarded a doctoral degree, usually after some more work following their viva (i.e. they are recommended to undertake minor or major revisions). The doctorate is 'a PhD, not a Nobel Prize' (Mullins & Kiley, 2002), an indication that the candidate is judged fit to undertake independent research in their own field, and to supervise others. It is merely an early staging post in the academic 'journey', not its end.

The UK Research Excellence Framework (REF)

The UK REF—and its predecessor the Research Assessment Exercise (RAE)—is a particularly high-stakes exercise, determining a large part of the research funding received by universities and their component departments. Despite calls to make use of available metrics (such as citation rates), its judgements remain based on peer review by 'expert' panellists (Koya & Chowdury, 2017; Marques et al., 2017; Mryglod et al., 2013).

From personal experience on the Education Sub-Panel in 2014, I can confirm that this involved a great deal of reading, discussion, benchmarking, cross-checking and debate. But it was not as onerous a task as some would have you believe. After all, as a supposed expert on a particular field, you were already familiar with the work of many of its researchers, and had already read—but in a different, less judgmental, context—many of the outputs that you were now required to rate.

Curiously, the details of this expensive and time-consuming exercise are not made available. All that the submitting institutions receive is a brief summary for each unit of assessment. This means that, at best, they can only 'second guess' what individual ratings led to the overall ratings given, and that they may then base their future planning on erroneous interpretations.

The relevance of the REF and the RAE to UK higher education is obvious, but it has also been influential in many other countries which have adopted their own versions. These countries include, for example, Australia, China, Estonia, Hong Kong, Ireland, Italy and Japan.

Not surprisingly, since they were designed to assess and reward the research prowess of all UK academics and their employing institutions—or, at least, those research-active academics that their universities chose to submit—the RAE and REF have been the subject of a great deal of speculative, critical and evaluative research by UK (and other) academics. Some of these studies will be quoted here to illustrate the range and depth of the critique.

Interestingly, for the purposes of the present chapter, Bence and Oppenheim (2004) drew the analogy between peer review, as practiced for journal articles, and the peer review of submitted outputs (typically published journal articles) undertaken by the subject panels of experts set up for the RAE. They argued that the secondary assessment of articles that had already passed peer review was poor practice:

> *The links between the RAE, journal peer review and quality are complex. The use of peer review for refereeing papers submitted for publication has evolved to become a self-policing mechanism for the community, by the community, which attempts to maintain quality standards and to an extent guard the reputation of journals… the academics doing the judging [in the RAE] are from other*

institutions in the same sector, essentially competing for the same resources, and yet are relying on secondary subjective judgements of earlier peer-review decisions. This would be fine if everyone trusted the outcomes of peer review; but they do not. We conclude that because of the many criticisms of peer review, it may be unwise to base funding decisions on second level peer review of articles that have already undergone initial peer review. (pp. 363–364)

Others addressed the issue of the apparent improvement in research ratings revealed by the RAE over time and attempted to explain this. For example, Sharp (2004) examined the RAE results for the three successive exercises of 1992, 1996 and 2001, focusing on differences between years and between units of assessment (i.e. subjects or disciplines):

The results show that mean ratings have improved markedly over time, particularly between 1996 and 2001, but that this upward shift is unevenly spread across Units of Assessment. In both 1996 and 2001, mean ratings varied significantly across Units of Assessment, with higher means being associated with Units in which there were fewer submissions. (p. 201)

While Sharp was very careful and measured in his comments and conclusions, it seems clear that—given that the assessment panels were recruited from the departments being assessed, and particularly when the panels were relatively small—the possibilities for some, perhaps unconscious, inflation in ratings given, so as to protect or enhance the relative standing of one's own discipline, were there.

Others could afford to be rather more overt in their critique of what became widely derided as 'game playing', but might simply represent pragmatic institutional decision-making designed to maximize their RAE results and the financial benefits that followed them. Thus, Moed (2008), based in the Netherlands, was able to persuasively chart how institutional strategies changed over time to respond to the continual changes made to the RAE's methodology:

A longitudinal analysis of UK science covering almost 20 years revealed in the years prior to a Research Assessment Exercise (RAE 1992, 1996 and 2001) three distinct bibliometric patterns, that can be interpreted in terms of scientists' responses to the principal evaluation criteria applied in a RAE. When in the

RAE 1992 total publications counts were requested, UK scientists substantially increased their article production. When a shift in evaluation criteria in the RAE 1996 was announced from 'quantity' to 'quality', UK authors gradually increased their number of papers in journals with a relatively high citation impact. And during 1997–2000, institutions raised their number of active research staff by stimulating their staff members to collaborate more intensively, or at least to co-author more intensively, although their joint paper productivity did not. This finding suggests that, along the way towards the RAE 2001, evaluated units in a sense shifted back from 'quality' to 'quantity'. (p. 153)

Alongside the pervading critique of the RAE and REF as a grotesque game, the purpose of which was to deliver the lion's share of the available research funding to the oldest and best-established universities and departments, the most prevalent critique, however, has probably been about whether peer review was the best way of undertaking the evaluation. This critique has had a number of elements. Thus, in economic terms, it has been argued that the costs of the RAE and REF, principally in terms of the hours of academic and administrative time taken in putting together departmental and institutional submissions, and then in evaluating them, were very hard to justify. After all, would not this time be better spent in actually doing some more research?

There is also a disciplinary element to this critique, however, arguing that bibliometric methods—the obvious alternative to peer review, and used by some RAE and REF panels to inform their decisions—are not appropriate to all disciplines. For example, in the case of social policy and social work, McKay (2011) argued that:

Using quantitative evidence it seems possible to base estimations of research environment on observable data, or at least to regard such data as a valuable check on the assessment. The same may not be said of the evaluation of research outputs, at least in SPA [social policy and administration] and SW [social work], although there are disciplines where journal rankings correlate very strongly with the outcome. (p. 540)

The underlying argument is that in some disciplines, notably the hard sciences—where quantitative methods dominate—the relative standing of particular journals is widely acknowledged and citation rates are substantial. In other disciplines, however, notably the arts, humanities and social sciences—where qualitative methods are much more popular—there is not such a clear 'pecking order' among journals and citation rates are generally low, making bibliometric data a much less useful guide to quality.

This argument is somewhat supported in the Italian context by Abramo et al. (2011), who examined the experience of the first Italian research assessment (the VTR) and its planned replacement (the VQR). They considered the use of bibliometric exercises as a replacement for peer review, arguing that the former, as well as being less time-consuming, would yield a better result:

> For the Italian VTR, the objective was to identify and reward excellence: in this work we have attempted to verify the achievement of the objective. To do this we compared the rankings lists from the VTR with those obtained from evaluation simulations conducted with analogous bibliometric indicators... The results justify very strong doubts about the reliability of the VTR rankings in representing the real excellence of Italian universities, and raise a consequent worry about the choice to distribute part of the ordinary funding for university function on the basis of these rankings. One detailed analysis by the authors shows that the VTR rankings cannot even be correlated with the average productivity of the universities. Everything seems to suggest a reexamination of the choices made for the first VTR and the proposals for the new VQR. The time seems ripe for adoption of a different approach than peer review, at least for the hard sciences, areas where publication in international journals represents a robust proxy of the research output, and where bibliometric techniques offer advantages that are difficult to dispute when compared to peer review. (p. 940)

Note the qualifying statement they carefully include—'at least for the hard sciences'—thus effectively supporting McKay's position.

In the Irish context, Holland et al. (2016) extend McKay's argument to the whole of the arts, humanities and social sciences, and lay the blame for what they clearly regard as an unwarranted imposition on the forces of neoliberalism (an oft-chosen target for academics and others at the present time, and one that is sufficiently nebulous not to need to fight back: see Tight, 2019b).

> The dynamic of what is being valued within research assessment exercises in higher education in Ireland and elsewhere is changing as a result of the re-

emergence of neoliberalism in the context of the global recessionary economic climate. AHSS [arts, humanities and social sciences] researchers are becoming increasingly concerned at the lack of inclusivity in what is being valued as research outputs, and in what can be counted within research assessment exercises. Evidence is emerging that quantitative metrics are more valued within neoliberal agendas, and that this is changing the behaviour of researchers towards engaging in and disseminating research that can readily contribute to such quantitative metric profiles... More appreciation for the diversity of research, and the appropriate assessment of quality thereof, within AHSS disciplines needs to be fostered within research assessment exercises... Academics and researchers in the Arts, Humanities and Social Sciences need urgently to reach agreement on what should be valued in terms of research activities, outcomes and/or impacts, and at what level (institution, department, unit, or individual). They also need to reach consensus with key policy-makers on how this work can be suitably assessed within the broader context of performance assessment in higher education. (p. 1113)

The obvious weakness of this argument, however, is the evident difficulty which Holland, Lorenzi and Hall have in specifying what might constitute a high quality research output in the arts, humanities and social sciences. Of necessity, therefore, we have to fall back on disciplinary peer review, that is, what do our colleagues and superiors think?

We may, of course—as we did in the two previous sections—again pose the question 'does it really matter'? The RAE and REF may not be wholly fair or objective exercises, and they are certainly not transparent. Unlike journal article or doctoral degree peer review, there is also no real scope for negotiation or appeal over the results. But, if we accept that available research funds should be targeted towards those who are making the best, or at least the most, use of them, is there a better system?

Conclusion

In this chapter, we have considered the use of peer review in higher education in the context of the evaluation of journal articles, doctoral degrees and institutional research performance. While the examples have been linked to my own experience and grounded primarily in the UK, their broader relevance and applicability is fairly self-evident.

The underlying question driving the discussion in the chapter has been 'is peer review fit for purpose?', and to this have been added the related questions of 'what alternatives are there?' and 'does it really matter?'

I think we have to conclude that peer review, being of long standing and fundamental to the operation of the academic enterprise, is not going away any time soon. It does have major flaws in that it is subjective and may be manipulated, but that is another way of saying that it is human. We all have preferences and biases, but—all taken together and in the long run—these should more or less even each other out.

Are there better alternatives? Well, it depends upon your perspective, and here we run straight into the qualitative/quantitative debate that has plagued social science research for decades.

In the case of the RAE/REF, it would be perfectly possible to replace peer review with bibliometric analyses—based on journal status and citation counts—which could be completed much quicker and much more cheaply. This would probably produce not too dissimilar results to peer review for the hard sciences and related fields (i.e. the disciplines that absorb the great majority of research funds). It is doubtful that this would work, or work so well, in the arts, humanities and social sciences, however, and even in the hard sciences something would be lost in terms of the appreciation of the overall field of research.

In the case of both journal article evaluation and doctoral degree assessment, a better alternative is not so clear. It would, of course, be possible to try and improve current practices: through, for example, more careful selection of article reviewers and doctoral examiners, more extensive training and the provision of more written guidance. But this would be to add significantly to the workload of those involved, who are undertaking tasks which, whether we like it or not, are really marginal to their employment, and which are either unpaid or poorly paid.

So, does it really matter? Well, obviously, yes, or I wouldn't have written this chapter. Peer review is of critical importance to the practice of higher education. It is vital that we perform it as well as we can, bearing in mind its actual and potential deficiencies. We cannot assume that all academics are already competent at it, in its various forms, but need to provide appropriate training, guidance and support. We also need to keep a watching eye on the results of peer review, allow them to be challenged and be prepared to challenge them ourselves where we believe this to be necessary.

References

Abramo, G., D'Angelo, C., & Di Costa, F. (2011). National Research Assessment Exercises: A comparison of peer review and bibliometrics rankings. *Scientometrics, 89*, 929–941.

Atjonen, P. (2018). Ethics in peer review of academic journal articles as perceived by authors in the educational sciences. *Journal of Academic Ethics, 16*(4), 359–376.

Bence, V., & Oppenheim, C. (2004). The influence of peer review on the research assessment exercise. *Journal of Information Science, 30*(4), 347–368.

Bornmann, L., Mutz, R., & Daniel, H.-D. (2010). A reliability-generalization study of journal peer reviews: A multilevel meta-analysis of inter-rater reliability and its determinants. *PLoS One, 5*(12), e14331.

Falkenberg, L., & Soranno, P. (2018). Reviewing reviews: An evaluation of peer reviews of journal article submissions. *Limnology and Oceanography Bulletin, 27*(1), 1–5.

Hewings, M. (2004). An 'Important Contribution' or 'Tiresome Reading'? A study of evaluation in peer reviews of journal article submissions. *Journal of Applied Linguistics, 1*(3), 247–274.

Holland, C., Lorenzi, F., & Hall, T. (2016). Performance anxiety in academia: Tensions within research assessment exercises in an age of austerity. *Policy Futures in Education, 14*(8), 1101–1116.

Koya, K., & Chowdury, G. (2017). Metric-based versus peer-reviewed evaluation of a research output: Lessons learnt from UK's national research assessment exercise. *PloS One, 12*(7), e0179722.

Kumar, P., Rafiq, I., & Imam, B. (2011). Negotiation on the assessment of research articles with academic reviewers: Application of peer-review approach of teaching. *Higher Education, 62*, 315–332.

Lipscombe, T. (2016). Burn this article: An inflammatory view of peer review. *Journal of Scholarly Publishing, 47*(3), 284–298.

Marques, M., Powell, J., Zapp, M., & Biesta, G. (2017). How does research evaluation impact educational research? Exploring intended and unintended consequences of research assessment in the United Kingdom, 1986–2014. *European Educational Research Journal, 16*(6), 820–842.

McKay, S. (2011). Social policy excellence: Peer review or metrics? Analysing the 2008 research assessment exercise in social work and social policy and administration. *Social Policy and Administration, 46*(5), 526–543.

Moed, H. (2008). UK Research Assessment Exercises: Informed judgments on research quality or quantity? *Scientometrics, 74*(1), 153–161.

Mryglod, A., Kenna, R., Holovatch, Y., & Berche, B. (2013). Comparison of a citation-based indicator and peer review for absolute and specific measures of Research Group Excellence. *Scientometrics, 97*(3), 767–777.

Mullins, G., & Kiley, M. (2002). 'It's a PhD, not a Nobel Prize': How experienced examiners assess research theses. *Studies in Higher Education, 27*(4), 369–386.

Peters, D., & Ceci, S. (1982). Peer-review practices of psychological journals: The fate of published articles, submitted again. *Behavioral and Brain Sciences, 5*(2), 187–195.

Pontille, D., & Torny, D. (2015). From manuscript evaluation to article valuation: The changing technologies of journal peer review. *Human Studies, 38*(1), 57–79.

Sharp, S. (2004). The research assessment exercises 1992–2001: Patterns across time and subjects. *Studies in Higher Education, 29*(2), 201–218.

Tight, M. (2003). Reviewing the reviewers. *Quality in Higher Education, 9*(3), 295–303.

Tight, M. (2019a). *Documentary research in the social sciences*. Sage.

Tight, M. (2019b). The neoliberal turn in higher education. *Higher Education Quarterly, 73*(3), 273–284.

Tinkler, P., & Jackson, C. (2000). Examining the doctorate: Institutional policy and the PhD examination process in Britain. *Studies in Higher Education, 25*(2), 166–180.

Torgerson, C. (2003). *Systematic reviews*. Continuum.

van der Heide, A., Rufas, A., & Supper, A. (2016). Doctoral dissertation defenses: Performing ambiguity between ceremony and assessment. *Science as Culture, 25*(4), 473–495.

Watts, J. (2012). Preparing doctoral candidates for the viva: Issues for students and supervisors. *Journal of Further and Higher Education, 36*(3), 371–381.

Part III

Specificities of Different Peer-Review Practices

11

Peer Review in Academic Promotion of Excellent Teachers

Eva Forsberg, Sara Levander, and Maja Elmgren

Introduction

Promotion in academia is part of the academic reward system that comprises the many ways in which institutions and scientific fields value faculty. The reward system, which includes aspects of both merit and bias, is critical in how institutions recruit, sustain, assess, and advance faculty members throughout their careers (O'Meara, 2011). Reward systems and career structures are deeply entrenched in the national traditions of higher education systems and domestic labour markets. However, processes of convergence can be observed and, in many countries, recent reforms have addressed the management of faculty careers in somewhat similar ways (Musselin, 2005). In many countries, academic careers follow a rather formalized structure with more or less clearly delineated ranks for different stages (Musselin, 2010).

E. Forsberg (✉) • S. Levander • M. Elmgren
Department of Education, Uppsala University, Uppsala, Sweden
e-mail: eva.forsberg@edu.uu.se

© The Author(s) 2022
E. Forsberg et al. (eds.), *Peer review in an Era of Evaluation*,
https://doi.org/10.1007/978-3-030-75263-7_11

Moreover, evaluations are a hallmark of scientific merit (Merton, 1968), and an evaluation machinery has now spread to almost every corner of the academic enterprise, reward systems included (Dahler-Larsen, 2015). In evaluation practices, gatekeepers maintain a powerful role in the recognition of scholars and institutions. Judgement by peers is the evaluation form *par excellence* in the academic field, although it has been challenged by managerialism (Musselin, 2013). Peer review is crucial in determining, for example, the reputation and status of scholars and the allocation of scarce resources and academic careers (Lamont, 2009).

Research merits have long been the priority in the recognition of institutions and scholars (Merton, 1968; Bourdieu, 1996). Teaching is often downplayed, appearing as a practice of less worth in academia (Van den Brink, 2010; Levander, 2017). To counteract this tendency, various systems to upgrade the value of education and promote teaching excellence have been introduced by higher education institutions on a global scale (O'Meara, 2011). Even though institutions differ in their centre of focus, most stress a multiple form of scholarship that includes the dual mission and nexus of research and teaching (Boyer, 1990; Elken & Wollscheid, 2016; Taylor, 2008; Tight, 2016). In recent decades, there has been a quest for excellence in academic scholarship, in terms of both research and teaching and public outreach. The moral qualities of academics can also be included in the evaluation of excellence (Lamont & Mallard, 2005). Nevertheless, excellence, like quality, often lacks both an external referent and internal content; it does not refer to a specific set of things or ideas (Readings, 1996). As an empty signifier (Laclau & Mouffe, 1985), excellence has gained support and general consent on the level of discourse.

Although excellence has come to serve as the meritocratic standard and currency, it is not a universally recognized, neutral, and objective gold standard. On the contrary, there is little consensus of what constitutes excellence, what it means, how it is achieved, and how it may be assessed (Lamont, 2009). Rather, it is a fuzzy socially constructed object that is contextual and relational, gaining its meaning by an array of actors in multiple practices and through various artefacts (Angermüller, 2010). In other words, the construction and conceptualization of excellence depends on where it is used, by whom, for what purposes, and in relation to what; different standards of excellence are employed in different contexts.

While there is a great deal of literature on teaching excellence in higher education—stretching from distinct conceptualizations of the phenomenon (Boshier, 2009; Boyer, 1990) to suggestions or imperatives on how to assess it in, for example, academic recruitment and promotion (Glassick et al., 1997; Paulsen, 2002; Ramsden et al., 1995)—there is less research on how it is manifested in academic promotion processes. Prior research on peer review of excellence in academia has primarily focused on research excellence in grant proposals and in manuscripts for publication in academic journals. Even though peer review has been a prominent object of study, particularly after 1990, empirical research has not addressed peer review in a comprehensive way. In particular, comparatively few studies analyse peer review in the promotion of teaching excellence and the texts that are interchanged in these processes (Sabaj Meruane et al., 2016; Batagelj et al., 2017). Thus, we know little about the manifestation of teaching excellence in peer review in distinct academic promotion systems. In this respect, this chapter provides a substantial contribution to the extant research on peer review.

In this chapter, we explore the values and beliefs that are unveiled in the promotion of academics when teaching excellence is under scrutiny. We employ empirical data collected within the research project titled Academia, Scholar Proficiency, and Career Systems—more specifically, from an inquiry into the promotion of excellent teachers to the level of 'distinguished university teacher',[1] at a broad research-intensive comprehensive university in Sweden. Due to the principle of public access to official records, the documents in promotion processes are easily accessible for research purposes. While contextual factors such as national regulations, institutions' academic profiles, and academic evaluative cultures are critical for the specific meaning reviewers ascribe to teaching excellence, the evaluation processes of academic scholarship are of interest beyond the specific context. Moreover, the tendencies of convergence of higher educational systems require an understanding of specific national circumstances (Hamann, 2019).

In many respects, peer review in the promotion of teaching excellence is similar to other evaluations of academic performances; however, there

---

[1] 'Distinguished university teacher' is the official term of this rank at the university in question.

are also significant differences. We argue that the peer review process is mainly framed by the national and institutional context, the particular career and reward system, the type of appointment (promotion), and the specific object of evaluation (teaching excellence). Moreover, the intersection between promotion, peer review, and excellent teaching affects both the peer review process and the notion of the 'distinguished university teacher'. Furthermore, the institutionalization of this promotion practice is embedded in the tension between standardization and professional judgement. Like many other evaluation practices, the promotion process is a high-stakes activity characterized by uncertainty and risk.

The chapter proceeds as follows: first, promotion to the level of 'distinguished university teacher' will be contextualized as part of the career and reward structure within the Swedish higher education sector. We then analyse the framework of regulation and division of responsibilities between the agents involved in the promotion process at the particular university under study (the 'case university'). Next, we employ guidelines, applications, and reviewers' assessments to illustrate the meaning-making of distinguished university teaching. Special attention is paid to the reviewers' judgement and to their legitimation of their judgement. In the final section, we discuss the institutionalization of the notion of an excellent teacher as manifested in the (e)valuation process of 'distinguished university teachers'.

Career and Reward Structure in Swedish Higher Education

Although most higher education institutions in Sweden are public, they vary in size, in specialization, and in the balance between the resources allocated for research and teaching. They also differ in that universities are granted general degree-awarding powers at the second and third cycle levels, while university colleges must apply for them. The same basic legislation is valid for all Swedish higher education institutions (Swedish Govt. Bill, 2009/2010, 80). Thus, there is now some diversity in the career and reward structure, although there are still major similarities regarding the most fundamental categories.

According to national statistics, senior lecturers and lecturers make up about 30 and 15 per cent, respectively, of the research and teaching staff in the higher education system. The share of academics with professorship—the highest position a teacher or researcher can achieve—amounts to roughly 20 per cent. There are also permanent positions as researchers. Within these positions, fixed-term employments are relatively common (approximately 30 per cent), including qualification positions and positions as a researcher, visiting professor, adjunct teacher, or substitute teacher. Qualification positions include associate senior lecturer, postdoc, and postdoctoral research fellow. There is an increasing trend in the number of positions requiring a PhD (UKÄ, Swedish Higher Education Authority, 2019).

The most common procedure for appointing positions at a higher education institution is that teachers are to be appointed in competition, assessed by expert reviewers who must pay the same amount of attention to the assessment of research as to the assessment of teaching expertise. Unless it is manifestly unnecessary, expert reviewers are expected to be used for the appraisal of a professor. The positions of professors and senior lecturers and the qualification position of an associate senior lecturer are regulated by the state in the Swedish Higher Education Act (SFS, 1992:1434). Beyond these, each institution can decide what teacher categories are to be employed and how their career structure and guidelines for appointment and promotion are to be designed. As elsewhere (Höhle, 2014), a shift from a chair model to a department model can be observed in Sweden. While basic criteria of eligibility for senior lecturers and professors are still established at the government level in the Swedish Higher Education Ordinance (SFS, 1993:100), decisions on more elaborated criteria, and on criteria for the employment of other types of academic positions, are made at the institutional level.

There are three basic career structures in the Swedish higher education system (as detailed in Fig. 11.1), and they employ different denominations and levels of teaching excellence. The first career structure is the traditional ranking structure, which is mainly based on research expertise. The second is directly linked to the most common teaching positions—the mandatory professor and senior lecturer—in addition to the common teaching position of the lecturer, who is not required to have a PhD. The third is an emerging career structure similar to the system of

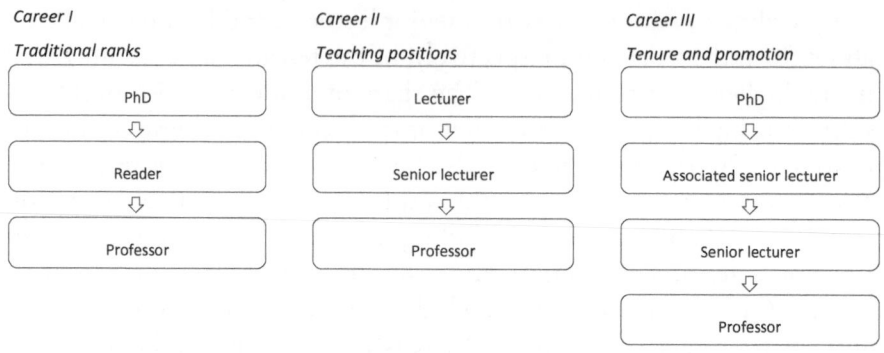

Career I

Traditional ranks

| PhD |
| ⇩ |
| Reader |
| ⇩ |
| Professor |

Career II

Teaching positions

| Lecturer |
| ⇩ |
| Senior lecturer |
| ⇩ |
| Professor |

Career III

Tenure and promotion

| PhD |
| ⇩ |
| Associated senior lecturer |
| ⇩ |
| Senior lecturer |
| ⇩ |
| Professor |

Fig. 11.1 The basic career structure in the Swedish higher education system. (Lecturer (USA: lecturer; SWE: *adjunkt*); Associate senior lecturer (USA: assistant professor; SWE: *biträdande lektor*); Senior lecturer (USA: associate professor; SWE: *lektor*); Professor (USA: full professor; SWE: *professor*); Reader (USA: associate professor; SWE: *docent*))

tenure and promotion that is employed in US universities, based on the fixed-term qualification position of associate senior lecturer.

It is possible for a candidate to be appointed to any teaching position without having held another teaching position before, although teaching expertise is required. Normally, the rank of reader is required for an appointment to professor. At many universities, a teacher may apply for a promotion from one teaching position to the next, while demonstrating the required expertise in research and teaching. In the three-track career structure presented in Fig. 11.1, there are several possible career paths. In the next section, we explore how the promotion to 'distinguished university teacher' in our case university is related to the Swedish higher education career and reward structure, and briefly comment on the promotion system and its guidelines.

Gatekeeping in the Promotion of 'Distinguished University Teachers'

In 2010, the vice-chancellor of our case university decided on a university-wide reform in which teachers could apply to become appointed as 'distinguished university teachers', at a level clearly requiring a higher level of proficiency than the level being demanded for recruitment (Guidelines for Admittance of Excellent Teachers). Thus, the admittance of 'distinguished university teachers' emerged as an additional, fourth career track (Fig. 11.2).

Criteria specifications were left to the faculty boards to decide on, in accordance with a decentralized collegial structure. Teaching qualifications should be documented and assessments should be performed by two reviewers, at least one of which must be external to the university and at least one of which must have scientific expertise in the same field as that of the candidate. The admitted teacher would receive a standardized salary increase.

Promotion processes at the university follow a formalized procedure. A requirement for admittance to 'distinguished university teacher' is permanent employment as a lecturer, senior lecturer, or professor. On the one hand, the rank of a 'distinguished university teacher' is on par with the rank of a reader. On the other hand, it is neither a part of the traditional career track nor a part of the tenure and promotion career track, both of which are based on possession of a PhD. Instead, teaching excellence is directly linked to either level of the teaching position career track (see Fig. 11.1). Each faculty board has elaborated guidelines within the institutional framework. According to these guidelines, there are no differences in the assessment of excellence based on the level of teacher position.

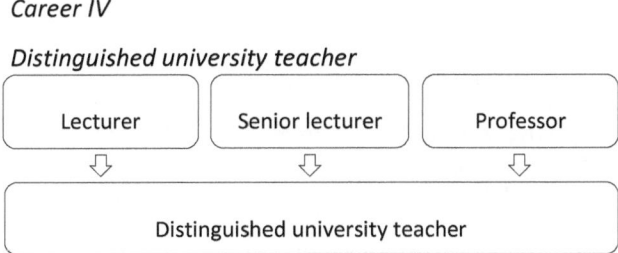

Career IV

Distinguished university teacher

Fig. 11.2 A fourth career track in the case university

The Process of Promoting Teaching Excellence

The process of promotion comprises several stages and involves different agents. Due to different structures within different scientific domains, the level of the faculty board involved may vary, and there are slight differences in the degree of delegation at different faculties. However, in general, a promotion committee prepares the recruitment and evaluation process, whereas the final decision is made by a faculty board. The promotion committees are standing committees, and the members are elected for a certain term of office. The entire process is illustrated in Fig. 11.3.

As shown in the figure, the process involves a number of gatekeepers, all of which are peers: the reviewers, the promotion committee, and the faculty board. These gatekeepers produce a number of documents: evaluation reports written by the reviewers, the proposal protocol from the promotion committee, and the final decision protocol from the faculty board/promotion committee. The faculty board plays a dual role in this process, since the board executes the local guidelines framing the whole process in addition to making the final decision, if the latter is not delegated elsewhere.

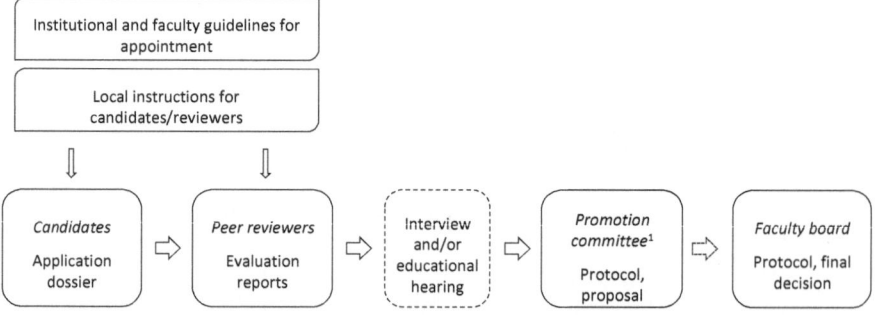

Fig. 11.3 The process of admitting excellent teachers at the university. (Lecturer (USA: lecturer; SWE: *adjunkt*); Associate senior lecturer (USA: assistant professor; SWE: *biträdande lektor*); Senior lecturer (USA: associate professor; SWE: *lektor*); Professor (USA: full professor; SWE: *professor*); Reader (USA: associate professor; SWE: *docent*)). In some cases, the faculty board delegates both the proposal and the decision to the promotion committee

The actual process starts when a candidate submits an application to the promotion committee. Then, the promotion committee selects two peer reviewers to assess the application. After the reviewers' evaluation reports are submitted to the promotion committee, the candidate may be invited to an interview and/or educational hearing. Drawing on the application, evaluation reports, interview, and educational hearing, the committee decides whether or not to nominate the candidate for admittance. Although the decision-making lies with the committee or the board, the reviewers have a crucial gatekeeping function and are key actors in this evaluation practice.

The qualification and selection of the gatekeepers are crucial to the making of the 'distinguished university teacher' in the promotion process. Faculty members on the board and committees are selected by and among scientifically qualified colleagues. In some committees, the members are themselves appointed as 'distinguished university teachers' or are considered to be especially proficient in pedagogical issues. To a varying extent, peer reviewers are chosen for their disciplinary expertise, pedagogical knowledge, pedagogical content knowledge, or expertise in the evaluation of teaching proficiency. *Pari passu* with the national emergence of the possibility of rewarding excellent teaching is the establishment of a national course programme aiming to educate reviewers in the evaluation of teaching excellence. More and more academics are attending this course, and many of the reviewers involved in these assessments at the university have taken the course.

The Mandatory Content of the Application Dossier

As shown in Fig. 11.3, reviewers must base their assessments on the information compiled in the application dossier. The dossier normally includes a cover letter, a *curriculum vitae* (CV), and a teaching portfolio (Fig. 11.4).

In the portfolio, the candidates are expected to describe and reflect upon prior experiences in areas such as scope of teaching, management and development of teaching, the teaching-research nexus, and scholarly interaction. The candidates' teaching philosophy is also relevant—that is, their reasoning about their educational aims, views on teaching, learning

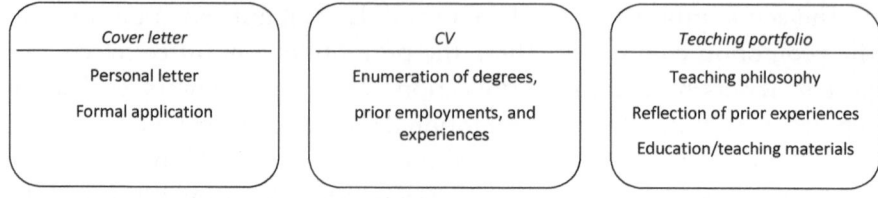

Fig. 11.4   Requested information to be included in the application dossier

theories, and so forth. It is important for candidates to provide concrete examples from practice that support their teaching philosophy and account of experiences. The portfolio should also contain various educational and teaching materials, such as a syllabus, assignments, lecture notes, and books, in order to strengthen the description and arguments made by the candidate. Moreover, a variety of testimonials, such as certificates, diplomas, student course evaluations, and attestations or affirmations from employers and colleagues, should be included in order to substantiate the excellence of the candidate. To support and guide the applicants and the reviewers' assessments, the faculty boards have developed local guidelines.

Faculty Guidelines and Candidate Applications

In this chapter, we draw on application cases from a teaching excellence reward system at a broad research-intensive comprehensive university in Sweden during 2013–2014. The data includes policy guidelines, full application dossiers, and the reviewers' evaluation reports. All three scientific domains are represented, and both admitted and rejected applications are included. The case university is divided into three scientific domains: *Humanities and Social Sciences* (HS), *Medicine and Pharmacy* (MP), and *Science and Technology* (ST). Three sets of guidelines are represented in our material, one for each domain. All guidelines include criteria, although these are elaborated in different ways regarding the aspects and examples of signs of fulfilment—that is, what kinds of indicators the applicants can point towards as evidence of their excellence. All sets of guidelines have in common an appreciation of extensive disciplinary knowledge, broad experience from teaching at various levels and in different courses, a reflective practice in which the candidate analyses her or his own teaching and its outcomes, cooperation and discussions with colleagues, and educational administration. Collaboration and academic leadership are also emphasized, but not by every faculty. Some faculties expect all aspects to be fulfilled, while other faculties regard some aspects as added value. Moreover, the ST and MP faculties emphasize the teaching-research nexus, teaching-society nexus, and student progression, while the HS faculty stresses research activities, scientific production, research seminars, and conferences. While several types of testimonials are mentioned as evidence, it is worth noting that a standard for the level of excellence is not explicit in the guidelines. Both Boyer (1990) and O'Meara (2011) emphasize the reward system as a device for institutions, departments, and disciplines to differentiate among themselves and present their practices as unique. Within the promotion system of teaching excellence, there are no signs of such organizational judgements in the guidelines.

In the applications, candidates negotiate the guidelines when addressing reviewers as 'significant others' in the promotion process (Serrano Velarde, 2018). In alignment with the guidelines, the application dossiers in our study consist of portfolios with written reflections on the applicant's educational practice, as well as attachments with various testimonies, course evaluations, and examples, normally complemented with CVs. The candidates' reflections are mostly grounded in philosophical statements, typically based on both the educational literature and experience. Various examples of teaching practice, including supervision and examination, are reflected upon, often with some literature references and attachments. Furthermore, collegial cooperation and academic leadership are covered. In many cases, reflective practice is shown through the candidate's own teaching progress and thoughts on future development. The importance of the teaching-research nexus is often discussed, with disciplinary knowledge and research insights seen as foundations for teaching, occasionally with some mention of the consequences for the design of learning activities. Reciprocal learning and the role of students in academic discussions are sometimes mentioned; for example, one applicant wrote, 'Interaction with students at both undergraduate and postgraduate levels has often meant that my research has been challenged and critiqued in interesting and at times unexpected ways'. In this chapter however, we focus on the evaluation reports written by peers.

Judgement and Legitimation in the Making of 'Distinguished University Teachers'

Recognizing excellent teaching is expedient for research on the interaction between reviewers, the object of the review (i.e. teaching excellence), and the context of the review (i.e. promotion). In addition, since attempts to promote the value of education and teaching excellence are evolving across the globe (O'Meara, 2011), early investigations may contribute to the improvement of these practices. As an emergent practice, the admittance of 'distinguished university teachers' involves insecurities regarding the prevailing norms of assessment. We may expect the various promotion texts to be more explicit about codified norms than in situations with more institutionalized evaluation practices. Informed by the sociology of valuation and evaluation (Beljean et al., 2015; Hamann, 2019), we are concerned with how value is produced and assessed in the promotion of teaching excellence. We illustrate the making of the' distinguished university teacher' in promotion evaluations by analysing the archived records of peer review reports, while considering the guidelines and candidate applications mentioned above. We distinguish between two processes that are analytically distinct yet empirically intertwined (Hamann, 2019): *the reviewers' process of judgement*, in which value and qualities are ascribed to candidates and to excellent teaching and *the process of legitimation*, in which judgements are justified and made stable.

The (E)valuation of the 'Distinguished University Teacher': The Judgement

In the (e)valuation of the 'distinguished university teacher', the reviewers ascribe value to teaching excellence in distinct ways: by explicitly pointing to qualifications and competencies found in the application dossier (referred to herein as 'existentees'); by stressing missing aspects ('absentees'); and by arguing some merits to be extraordinary ('excellencees'). The reviewers draw not only on information in the application dossier, but also on the criteria and indicators of evidence stipulated in policy. The evaluation reports reflect the guidelines in terms of structure and criteria to a rather high degree. Thus, although the wording is mostly not the same, there is an explicit and strong intertextual relationship between the guidelines and the evaluation reports. In one case in the MP domain, the list of criteria with indicators is used as a checklist and the reviewer marks which criteria are fulfilled, or not fulfilled, without providing explicit justifications.

The Scholarly Judgement of Criteria and Content

Through the reviewers' scholarly judgements, many aspects of teaching excellence are manifested. The most dominant criteria used are teaching skills, disciplinary knowledge, the teaching-research nexus, aspects of the Scholarship of Teaching and Learning,[2] a holistic perspective, development over time, collaboration, and educational leadership. *Teaching skills* refer *inter alia* to high-quality teaching, supporting students' multifarious development, constructive alignment, teaching and examination, and consideration of students' differences and diversified experiences. *Disciplinary knowledge* and *the teaching-research nexus* are deeply intertwined. Disciplinary knowledge commonly refers to the depth and breadth of the candidate's level of content knowledge, and is explicitly manifested as part of the nexus in some reports. The nexus is manifested

---

[2] The Scholarship of Teaching and Learning movement was first initiated by Boyer's (1990) seminal work, in which he proposed four different forms of scholarship: discovery, application of knowledge, integration, and teaching. He argued for the recognition and reward of all four scholarships, *inter alia* to achieve greater alignment between academic staff rewards and institutional missions (O'Meara, 2006). Based on Boyer's work, Glassick et al. (1997) stressed the importance of further assessment in order to enhance the value of other forms of scholarship in academia. Ideas about faculty conducting research on teaching and learning, excellence in teaching, the development of practice through reflection on theory and research, and experience-based knowledge on teaching (Kreber & Cranton, 2000) were later introduced as part of the notion. Furthermore, the basis for the Scholarship of Teaching and Learning is not just content and pedagogical knowledge, but also pedagogical content knowledge, as pointed out by Shulman (1986).

in various ways: through the candidate's content knowledge, undertaking of research, use of their own or others' research production, production of teaching materials (e.g. textbooks), and pedagogical and pedagogical content knowledge.

Emphasis on the *aspects of the Scholarship of Teaching and Learning* is also characteristic of this specific evaluation practice. Insights into educational research and the dissemination of practice-oriented research are included, along with examples from successful practice and development activities. In short, a research- and problem-based approach to the candidate's own teaching practice is emphasized. A *holistic perspective* refers to the candidate's ability to maintain progression throughout the teaching process and communicate the main thread to the students. The reviewers comment on the connection between current teaching and the overall education programme, and on the relationship among specific education, students' upcoming working life, and society in general. *Collaboration* and *development over time* concern communication and cooperation with students and colleagues regarding course design and development through, for example, the use of student course evaluations and discussions among peers. *Educational leadership* comprises positions and responsibilities such as (vice) head of department, director of studies, course management, and so forth.

The Scholarly Judgement of Evidence

Lists of merits (CVs), descriptions of prior teaching experiences and responsibilities, as well as commissions of trust (e.g. serving as faculty opponent or expert reviewer, and positions of authority) are common indicators to determine the fulfilment of criteria. Other commonly used indicators are testimonials such as certificates of continuing professional development (CPD) courses, records of publications and conference papers, support letters from management and colleagues, and student ratings of instructions and awards. However, testimonials *per se* are not always enough, as they must be put into context and be elaborated upon by the candidate. Furthermore, testimonials must include specific information, because if there are 'no motivations from either students or the head in the [certificate]… there are basically only statements and no material to assess or consider'. Hence, mere affirmation is not sufficient; testimonials should preferably also clearly account 'for [the candidate's] various contributions'.

Moreover, testimonials alone are insufficient evidence of excellence. The candidate's self-reflection as a teacher and reflection on educational issues in a broader sense are indispensable for the distinguished title. Reflection on one's own practice is one of the most dominant indicators of teaching excellence, both in terms of existentees and absentees. Existentees such as '[the candidate] reflects upon the relevance of the research for his teaching and how the research findings can be of use for the students' are, for the most part, acknowledged by the peers. Absentees are equally often stressed:

> *[Teaching excellence] is in part demonstrated by some student course evalua-tions, one teaching award, and affirmations from the head of department. However, I lack clearer evidence regarding e.g. well described educational considerations, discussions about how the teaching works and first and fore-most, why and with what result.*

By request of some reviewers, reflections ought to be 'developed in support of educational literature…' and based on 'extensive, practically pedagogical examples'. In turn, candidates who 'illustrate … with con-crete examples as well as [their] own arguments, theoretically connected in a very illustrative and well-thought-out way' are praised by the review-ers. Conversely, reviewers usually do not approve of excessive citations without reflection, nor of abstract reasoning without (some) references or tangible examples aligned with the rest of the portfolio. Moreover, some reviewers argue that it is 'not enough to describe a successful achieve-ment. Much more analysis and testing are required to meet the criteria for excellent teachers'. Thus, candidates are expected to problematize fail-ures and unsuccessful attempts in an investigative approach, and discuss 'what the educational problem was, how he solved it (and why), and what the outcome was'.

The same is true regarding aspects of the Scholarship of Teaching and Learning. It is not enough to participate in conferences; a discussion regarding 'the educational content of the many conferences and symposia' in which candidates partake is expected, and how this has affected the candidate's 'own and others' educational development'. It is stressed that this is not only for the sake of the portfolio, but also for the benefit of the community, as 'it would be of great value if these were distributed to colleagues through conferences, seminars, articles'. When the teaching materials support arguments such as '[the candidate] uses his own experiences from quantitative and qualitative methodology in his teaching', this is considered to be solid evidence of teaching excellence. Then again, teaching material with (according to the reviewers) misguided links to accounts in the portfolio or left uncommented may be seen as a token of a sloppy or hasty application, instead of being evidence of teaching excellence.

Although this focus of the assessments is mainly on quality, references to the level of quality are not very clear. Nevertheless, words and phrases such as 'impressive', 'excellent', 'prestigious', 'extraordinary', and 'extremely well qualified' indicate a high level of standard (excellencees). Such words are used to describe the fulfilment of several criteria and indicators, such as disciplinary knowledge, holistic perspective and reflection, educational leadership, and student course evaluations. Quantity and scope, such as 'massive teaching experience' or 'extensive experience of educational leadership from different levels', are also referred to, albeit primarily as a minimum level of qualification for teaching excellence or as being present in too-limited amounts. These are commonly referred to in a routine fashion, and are only given explicit value when related to reflection on action, developmental work, a successful outcome, and so forth.

To sum up, testimonials, reflections, and tangible examples from pedagogical practice are the most prevalent indicators of teaching excellence in our sample. However, each separate indicator is neither sufficient, nor decisive; rather, it is the combination of indicators and reflection upon them that matters.

Justification of the (E)valuation: The Legitimation

The nomination of a 'distinguished university teacher' requires not only the judgement of significant aspects of teaching excellence, but also the legitimation of these judgements. Justifications of the (e)valuations are made in several ways: through the formation of the promotion process, through the products involved and produced in the process, and through the ways in which the gatekeepers explicitly or implicitly argue their case.

Justification In-between Standardization and Professional Judgement

The promotion process is framed by institutional and faculty regulations, and is embedded in a national career structure of academic teachers. Following a process of faculty involvement, the university board decided to enlarge the reward system, and introduced the appointment of 'distinguished university teachers'. In addition to institutional regulations, each faculty board laid down rules regarding several issues that were seen to be critical for the justification of the outcome: namely, the degree of decentralization of judgement and decision, the creation of guidelines, and the subsequent balance between standardization and professional judgement.

The introduction of the promotion system was partly motivated as a way to alter the balance between the primary values of a higher education institution: research and teaching. Two parties are directly involved in the appointment of 'distinguished university teachers': actors who distribute recognition (the faculty board) and those who receive it (the appointed teachers). However, interactional third parties are also involved in the process (Sauder, 2006).

Peers, both internal and external, are involved in the formation of the promotion process and its outcome. Third-party factors that legitimate the judgements made include the promotion committees, which are composed of faculty colleagues who have been especially elected to manage and decide on the promotion of teaching excellence, and the selected peer reviewers. The question of who is considered qualified to serve as a peer reviewer or as a member of the committee is critical for legitimation. Different values and weights are given to different reviewer qualifications. The legitimacy of peers is interchangeably based on disciplinary knowledge, pedagogical knowledge, pedagogical content knowledge, and expertise in the evaluation of teaching proficiency.

The construction of comparatively elaborated and explicit faculty guidelines with criteria, indicators, and even checklists can be regarded as a way to make the promotion process relatively transparent, to standardize the process of judgement, and thus to frame the interpretation, (e)valuation, and decisions made by peer reviewers and committee members. The introduction of this promotion system at the case university created an elite/non-elite level of distinction between the basic teaching competencies required in the recruitment of teachers and teaching excellence as demanded in the appointment of a 'distinguished university teacher'. Thus, as in all ranking systems, the exclusiveness of the rank is an overall legitimation of the promotion. Although criteria and indicators are stated in the guidelines, the level of the standard of excellence is less visible and is largely left to peer reviewers to establish.

Changes in status hierarchies are never easy, and are always a matter of struggle and power. Accordingly, the employment of colleagues is no guarantee of successfully redefining the status of teaching versus research. Merton (1996) sometimes used the notion of a 'compeer' to signal that academic practice is as much competitive as collaborative. In order to legitimate the academic promotion process and shift the balance in the valuation of teaching and research, the university and faculties under study chose to seek justification in-between standardization and professional judgement of peers. Moreover, the promotion in itself is legitimated through economic remuneration for the candidates and recognition for both individuals and organizations (Hamilton, 2019). Beyond the title and possible enhancement in status, it is less clear what the rank implies for the academic work of its holder. Similar to the appointment of readers, 'distinguished university teachers' are not automatically assigned new duties. However, there are ongoing discussions on this topic in the university and there have been some signs of professional implications. Furthermore, the 'distinguished university teacher' title is not connected to a specific discipline. Thus, in a sense, it is a generic title that lacks a disciplinary foundation. Moreover, it has not been made clear whether the title is valid beyond the institution and the admitting faculty.

Justification Through Mandated Intertextuality

The justifications of the (e)valuations made by peer reviewers are mostly legitimated through explicit or implicit references to what is stated as mandatory or desirable in faculty board guidelines. Thus, the legitimation device *par excellence* in the promotion of 'distinguished university teachers' is justification through mandated intertextuality (Chen & Hyon, 2005). The reviewer report is required to interact with other documents; in particular, with guidelines and the criteria and indicators within them. These include formal judgements based on specified criteria, which allow the exclusion of candidates who are not formally qualified for the nomination (Hamann, 2019). Beside the existentees of requested merits and performances, peer reviewers frequently refer to absentees—that is, what they consider to be missing. Overall, the reports legitimate and reflect key institutional processes and values within a framework that justifies itself through its aspiration for transparency and fairness, both procedural and distributional (see e.g. Mallard et al., 2009).

Generic and referential intertextuality are both present in the peer reviews (Devitt, 1991). Although the former is implicit, it can be discerned through statements responding to formal expectations as expressed in the guidelines. The latter is explicit and is present in the form of direct references to other texts. Guidelines are referred to most often, but other texts are also invoked, such as curriculum documents, teaching materials, research publications, conference presentations, and diverse forms of testimonials, indicating evidence of an applicant's merits, performances, and achievements. Yet another form of intertextuality relates to the intersection of documents in the applicants' dossier, the faculty guidelines, and the reviewers' reports. Candidate and reviewer texts are linked through the guidelines and the candidates' anticipation and knowledge of what they believe will be especially valued by the reviewer and committees.

Justification Through Scholarly Judgement

The promotion process is marked by standardization, and mandatory intertextuality is prevalent in the reviewer reports. At the same time, the peer reviewers negotiate and interpret criteria, indicators, and standards in their (e)valuation of the candidates' records. The reviewers argue their case through different types of scholarly judgement, which are omnipresent in the reports. In line with the guidelines, reviewers mainly focus on what to judge (i.e. criteria) and on evidence of required qualification and achievements (i.e. indicators); more rarely, they focus directly on the level of teaching excellence (i.e. standards) (Centra, 1993).

The level of detail varies across *criteria* in reviewer reports. In general, criteria are briefly stated; sometimes, however, the reviewers present short rationales that further specify aspects or indicators of a criterion. Often, these rationales are closely aligned with the explanations provided in the faculty guidelines. Commonly reflected themes in the reviewers' evaluations are teaching skills, disciplinary knowledge, the teaching-research nexus, aspects of the Scholarship of Teaching and Learning, a holistic perspective, collaboration, and leadership. The reviewers appear to consider continuous change and development in response to evaluation, feedback, and reflection to be fundamental. Less frequently mentioned themes include the aims and meaning of education, student diversity, out-of-campus teaching, the use of educational media, and innovative, outstanding, or original teaching. Moreover, these latter themes are conspicuous by their absence in both the guidelines and the reviewer reports.

The most popular *indicators* employed by peer reviewers in their judgement of candidate performances are various forms of testimonials. The reviewers also make use of concrete materials related to curriculum development and courses, when these are included by the applicants. Moreover, interviews and tests on teaching competence supplement the indicators applied by the reviewers to legitimate their judgements. Observation of regular teaching or student work is not part of the assessment. In addition, no metrics are used that refer to students' retention, performances, exams, employability outcome, and so forth. Metrics related to teacher products, such as education textbooks and articles, are not used either.

Explicit references to the level of *standard* (teaching excellence) are few. According to institutional guidelines, the level of teaching excellence is 'a higher level of teaching expertise /.../ clearly distinguished from the basic level' (UFV, 2010/1842, p. 1). In turn, this 'basic level' of teaching expertise is set in the requirements for the recruitment of permanent teachers—with permanent teacher being the only category eligible for admittance as 'distinguished university teacher'. Most reviewers indicate a higher level of teacher excellence through the specific value, weight, and significance they attribute to different criteria and indicators, as well as the relation within each category and between them. As an object of (e) valuation, the level of excellence emerges in the reviewer report primarily as a relational phenomenon. The level of excellence has both a qualitative and a quantitative foundation. According to the reviewers, the diversity and amount of the candidate's experiences, skills, and achievements, in addition to the evidence substantiating these, are paramount. In comparison with the level of basic teaching competencies, the reviewers expect the fulfilment of a larger number of aspects of various criteria or themes. In some faculties, all criteria and indicators must be satisfied; in others, some criteria or indicators add value but are not mandatory. Thus, reviewers collect information with an eliminatory function, which allows the reviewers to reject an applicant. The reviewers also search for positive signs of evidence, which might add together to reach the bar of excellence (Musselin, 2002).

Furthermore, the level of excellence is manifested in arguments that consist of a link between two elements, or a chain of such links. Reviewers repeatedly address how activities are executed; merely participating is not enough, regardless of how often participation occurs. Occasionally, reviewers are explicit about the quality of the performance, adding positive descriptions such as 'impressive' and 'extraordinary' to indicate the level of excellence. Experience with different kinds of teaching covering a diverse range of activities over time is also seen as fundamental. The link between teaching content and form and student achievements (i.e. subject—teaching—learning) is present, but not especially emphasized or elaborated. A more-or-less outspoken link is constituted between teaching and disciplinary knowledge (the teaching-research nexus). The teaching-outreach link is less visible. A recurrent line of argument from the reviewers is related to the candidates' writing on teaching philosophy. The chain of links is often visualized through the identification of what reviewers judge to be left out or missing or, alternatively, to be statements without evidence. Candidates are requested to explore concrete teaching examples, preferably including student performances, course evaluations, and teaching reflections based on some kind of educational literature followed by curriculum development (activity—outcome—exploration—theorizing—feedback—change). This chain of practice-based teacher reflection operates as the prime gatekeeping function for the level of teaching excellence.

The Intersection of Promotion, Peer Review, and Teaching Excellence

Peer review emerged in modern science as a device to determine scientific quality and to allocate recognition among researchers. Nowadays, the practice of peer review has migrated and is employed in a number of evaluation practices. In this chapter, we have focused on a relatively new career track—the promotion of 'distinguished university teachers'—and identified the promotion process, involved actors, and products. The gatekeeping function of this process has been analysed in terms of judgement and legitimation. Through the former, the value and worth of different content, criteria, and indicators have been explored; through the latter, different forms of justification have been identified.

We infer that teaching competencies are ascribed with a somewhat different content and value in this specific evaluation practice, than in the recruitment of academic teachers in Sweden (see e.g. Levander, 2017; Levander et al., 2019). Reviewers relate to the criteria in policy to a greater extent, and largely highlight and discuss the many different aspects of educational proficiency, including the teaching-research nexus. In that respect, our findings do not fully support the misgivings of a disruption of teaching and research in academia due to teaching reward systems (Krause, 2009). Still, although the evaluation reports are more elaborated in these assessments, they display a rather homogeneous approach to the assessment of teaching excellence. That is, the notion of teaching excellence turns out to be very similar, irrespective of discipline, the candidate's teaching position, and/or organizational belonging. Thus, the notion of teaching excellence is rather similarly and narrowly constructed across disciplines, and the approach to its assessment is relatively uniform. This, we conclude, may be explained by the institutionalization of this specific promotion process by means of the establishment of a national course for prospective reviewers and by means of the selection of reviewers.

We have demonstrated how the career and reward structure (the context of the evaluation), the promotion process (the evaluation in itself), and teaching excellence (the object of evaluation) is decisive for the peer review practice. Accordingly, we argue that the peer review practice is mainly framed by the national and institutional context, the particular career and reward system, the type of appointment, and the specific object of evaluation. The intersection of the promotion process, peer review, and teaching excellence affects the nature of guidelines, the selection of peers, the scope and specification of the outcome, and consequently the notion of the 'distinguished university teacher'. Furthermore, we argue that there is a tension between standardization and professional judgement in the institutionalization of the promotion process.

The Promotion of 'Distinguished University Teachers': The Same, But Different

Peer review in the promotion of 'distinguished university teachers' has both similarities and significant differences in comparison with other evaluation practices of academic performance. For decades, peers have been used in external and internal evaluations of teaching programmes and exams. Thus, the evaluation of teaching and teacher competencies is nothing new, nor is the use of peers in these processes.

Reviewer judgement and justification of the level of excellent teaching add an aspect to the evaluation of teaching, however. In addition, the primary focus on educational proficiency is novel, even though disciplinary knowledge and the teaching-research nexus are expected to be taken into account. In contrast, the focus has mainly been the other way around in other kinds of promotion practices, with a particular emphasis on scientific proficiency. Furthermore, peers do not have to rank the candidates in promotion processes. In comparison with the hiring of teachers, this particular form of promotion both expands and reduces the peer review practice.

It is evident that the level of excellence is especially difficult to recognize in the promotion of 'distinguished university teachers'. When left to the peer reviewers to identify, teacher excellence has some common traits that are expressed on a rather abstract and general level in the form of expected chains or links. Nevertheless, even in this context, excellence as an empty signifier seems to have some sort of external referent, while lacking internal content (see Laclau & Mouffe, 1985; Readings, 1996); that is, the notion of teaching excellence is identified mainly as a generic phenomenon. This finding is similar to prior research on the recruitment of academic teachers in Sweden (Levander et al., 2019). The very limited focus on the products of teaching is also worth noting; metrics have a less prominent standing and original or innovative teaching is of minor importance. This finding has similarities with prior research on promotion (Hyon, 2011). However, it is a divergence from some institutional evaluations in which metrics and products of teaching hold a strong, sometimes criticized, position (Canning, 2019). Similarly, products, originality, and metrics are prominent in assessments of scientific proficiency.

The conceptualization of teaching excellence, including its assessment, is also dependent on the emergent institutionalization of this specific promotion process, which sets it apart from other peer review practices to some degree. The balance between standardization and professional judgement delimits the leeway given to peers in comparison with other academic evaluation practices. Even though the decision lies with the promotion committee, the prime gatekeeper in the promotion process is, as stated above, the peer reviewer. However, the national and institutional framing of the process causes the gatekeeping function of this particular promotion process to be more explicitly split between several actors. Faculty guidelines are definitely more elaborated when it comes to criteria and indicators, albeit silent on the level of standard. The assessment of reviewers is more bounded, as criteria and indicators are more clearly specified and fixed in advance by local gatekeepers.

A significant and powerful key mechanism in the making of teaching excellence is the selection of peers—that is, who qualifies as a peer, and why. Obviously, there are great differences in peer selection depending on whether it is teaching, research, or both that are to be assessed. When it comes to the latter, it is typical for Swedish higher education institutions to select scholars who have been recognized within the international scientific community and within the relevant inter/disciplinary domain. However, when teaching excellence is the object of evaluation, the context is constructed differently perhaps because of the national and institutional framing of education and of how education in itself is understood. As shown earlier in this chapter, peers may be disciplinary experts, have special expertise in teaching, and/or be specialized in the evaluation of teaching. Sometimes these competencies coincide, but often they do not. It is reasonable to assume that the differences in competencies have a major impact on the peer review practice and on the construction of teaching excellence within different disciplinary domains. Obviously, the peer review practice in itself, as well as its context, is decisive for the outcome. In what way and to what extent this is true needs to be further explored, preferably by means of a comparative approach.

A Latecomer and an Emergent Game Changer?

The two-staged career track (from academic teacher to 'distinguished university teacher') that is now open to all permanently employed teachers is a latecomer in the career and reward structure at our case university. It adds a rank and diverges itself from the other career paths in being open to teachers without a PhD as well. The aim of this particular promotion practice is to recognize and reward teaching excellence in order to change the balance between teaching and research, raise the status of academics engaged in teaching, and enhance the reputation of excellent teachers.

Some scholars argue that the value of rewarding excellent teaching may be jeopardized if the processes for doing so are vague, and call for more and clearer criteria, along with a congruence between criteria and indicators (Chism, 2006). Others assert that endeavours to enhance the value of teaching by means of various kinds of teaching awards are based on tokenism, and tend to counteract this enterprise rather than support it. Hence, there is an impending risk that the reward system will entail a symbolic value without leading to real changes in practice (Macfarlane, 2011). Yet others argue that organizational drift that transforms academic values may occur if 'pedagogical skills' are stressed 'at the expense of subject didactics' (Kaiserfeld, 2013, p. 174). The consequences of this particular promotion practice remain to be seen, however, and are beyond the scope of the present study.

The evaluation machinery (Dahler-Larsen, Chap. 6 in this volume) and the quality movement in academia have become such a profound part of contemporary higher education institutions that they affect academics' work in all respects. The evaluation of research as well as teaching has increased in importance and scope, and academics increasingly undertake a number of evaluation tasks each year. The time consumed for research evaluation has been estimated to equal about one month's worth of work per year for a professor (Langfeldt & Kyvik, 2011). Obviously, this impacts what we, as researchers, may expect from evaluation reports in terms of both scope and content. Furthermore, as an expert evaluation, peer review is based on professional judgement and is not expected to evoke strong formalization. It is plausible that reviewers reach a conclusion about a candidate rather quickly, based on their expertise and overall assessment. To legitimate their conclusion, they subsequently look for signs that support their judgement (Musselin, 2002). Such a logic suggests that it is less reasonable to expect a full account of the rationale of the final judgement in the evaluation reports. Hence, in order to reach a deeper understanding of reviewers' reasoning, interviews would be a promising avenue for further research. All in all, this chapter illustrates how the admittance of 'distinguished university teachers' lies at the intersection of promotion, peer review, and teaching excellence.

References

Angermüller, J. (2010). Beyond excellence—An essay on the social organization of the social sciences and humanities. *Sociologica, 2010*(3), 1–16.

Batagelj, V., Ferligoj, A., & Squazzoni, F. (2017). The emergence of a field: A network analysis of research on peer review. *Scientometrics, 113*(1), 503–532. https://doi.org/10.1007/s11192-017-2522-8

Beljean, S., Chong, P., & Lamont, M. (2015). A Post-Bourdieusian sociology of valuation and evaluation for the field of cultural production. In *Routledge International Handbook of the Sociology of Arts and Culture* (pp. 38–48). Routledge.

Boshier, B. (2009). Why is the Scholarship of Teaching and Learning such a hard sell? *Higher Education Research & Development, 28*(1), 1–15.

Bourdieu, P. (1996). *Homo Academicus*. Polity.

Boyer, E. L. (1990). *Scholarship reconsidered: Priorities of the professoriate*. The Carnegie Foundation for the Advancement of Teaching.

Canning, J. (2019). The UK Teaching Excellence Framework (TEF) as an illustration of Baudrillard's hyperreality. *Discourse: Studies in the Cultural Politics of Education, 40*(3), 319–330. https://doi.org/10.1080/0159630 6.2017.1315054

Centra, J. (1993). *Reflective faculty evaluation*. Jossey-Bass.

Chen, R., & Hyon, S. (2005). Faculty evaluation as a genre system: negotiating intertextuality and interpersonality. *Journal of Applied Linguistics, 2*(2), 153–184. https://doi.org/10.1558/japl.2005.2.2.153

Chism, N. V. N. (2006). Teaching awards: What do they award? *The Journal of Higher Education, 77*(4), 589–617. https://doi.org/10.1353/jhe.2006.0031

Dahler-Larsen, P. (2015). The evaluation society: Critique, contestability and skepticism. *SpazioFilosofico, 13*, 21–36.

Devitt, A. (1991). Intertextuality in tax accounting: Generic, referential, and functional. In C. Bazerman & J. Paradis (Eds.), *Textual dynamics of the professions: Historical and contemporary studies of writing in professional communities* (pp. 336–357). The University of Wisconsin Press.

Elken, M., & Wollscheid, S. (2016). *The relationship between research and education: Typologies and indicators. A literature review*. NIFU.

Glassick, C., Huber, M., & Maeroff, G. (1997). *Scholarship assessed: Evaluation of the professoriate*. Jossey Bass.

Hamann, J. (2019). The making of professors: Assessment and recognition in academic recruitment. *Social Studies of Science*, 1–23. https://doi.org/10.1177/0306312719880017

Hamilton, J. E. (2019). Cash or kudos: Addressing the effort-reward imbalance for academic employees. *International Journal of Stress Management, 26*(2), 193–203. https://doi.org/10.1037/str0000107

Höhle, E. (2014, September). Chair and department: Adequate models for describing Academic Career Paths? An empirical analysis in eleven European countries. *In 27th CHER-Conference in Rome*.

Hyon, S. (2011). Evaluation in tenure and promotion letters: Constructing faculty as communicators, stars, and workers. *Applied Linguistics, 32*(4), 389–407. https://doi.org/10.1093/applin/amr003

Kaiserfeld, T. (2013). Why new hybrid organizations are formed: Historical perspectives on epistemic and academic drift. *Minerva, 51*, 171–194. https://doi.org/10.1007/s11024-013-9226-x

Krause, K. (2009). Interpreting changing academic roles and identities in higher education. In M. Tight (Ed.), *The Routledge international handbook of higher education*. Routledge.

Kreber, C., & Cranton, P. A. (2000). Exploring the scholarship of teaching. *The Journal of Higher Education, 71*(4), 476–495. https://doi.org/10.1080/00221546.2000.11778846

Laclau, E., & Mouffe, C. (1985). *Hegemony & socialist strategy: Towards a radical democratic politics*. Verso.

Lamont, M. (2009). *How professors think. Inside the curious world of academic judgment*. Harvard University Press.

Lamont, M., & Mallard, G. (2005). *Peer evaluation in the social sciences and the humanities compared: The United States, The United Kingdom, and France*. Report for the Social Sciences and Humanities Research Council of Canada.

Langfeldt, L., & Kyvik, S. (2011). Researchers as evaluators: Tasks, tensions and politics. *Higher Education, 62*(2), 199–212.

Levander, S. (2017). *Den pedagogiska skickligheten och akademins väktare: Kollegial bedömning vid rekrytering av universitetslärare*. [The educational proficiency and the gatekeepers of Academia]. PhD Diss. Uppsala: Uppsala University.

Levander, S., Forsberg, E., & Elmgren, M. (2019). The meaning-making of educational proficiency in academic hiring: A blind spot in the black box. *Teaching in Higher Education*. https://doi.org/10.1080/13562517.2019.1576605

Macfarlane, B. (2011). Prizes, pedagogic research and teaching professors: Lowering the status of teaching and learning through bifurcation. *Teaching in Higher Education, 16*(1), 127–130. https://doi.org/10.1080/13562517.2011.530756

Mallard, G., Lamont, M., & Guetzkow, J. (2009). Fairness as appropriateness. Negotiating epistemological differences in peer review. *Science Technology Human Values*. https://doi.org/10.1177/0162243908329381

Merton, R. K. (1968). *Social theory and social structure*. The Free Press.

Merton, R. K. (1996). *On social structure and science*. University of Chicago Press.

Musselin, C. (2002). Diversity around the profile of the 'good' candidate within French and German universities. *Tertiary Education and Management, 8*(3), 243–258. https://doi.org/10.1080/13583883.2002.9967082

Musselin, C. (2005). European academic labor markets in transition. *Higher Education, 49*, 135–154.

Musselin, C. (2010). *The market for academics*. Routledge.

Musselin, C. (2013). How peer review empowers the academic profession and university managers: Changes in relationships between the state, universities and the professoriate. *Research Policy, 42*, 1165–1173.

O'Meara, K. (2011). Inside the panopticon: Studying academic reward systems. In J. C. Smart & M. B. Paulsen (Eds.), *Higher education: Handbook of theory and research* (Vol. 26, pp. 161–220). Springer.

O'Meara, K. A. (2006). Encouraging multiple forms of scholarship in faculty reward systems: Have academic cultures really changed? *New Directions for Institutional Research, 2006*(129), 77–95. https://doi.org/10.1002/ir.173

Paulsen, M. B. (2002). Evaluating teaching performance. *New Directions for Institutional Research, 2002*(114), 5–18.

Ramsden, P., Margetson, D., Martin, E., & Clarke, S. (1995). *Recognising and rewarding good teaching in Australian Universities*. Australian Government Publishing Service.

Readings, B. (1996). *The University in Ruins*. Harvard University Press.

Sabaj Meruane, O., González Vergara, C., & Pina-Stranger, Á. (2016). What we still don't know about peer review. *Journal of Scholarly Publishing, 47*(2), 180–212. https://doi.org/10.3138/jsp.47.2.180

Sauder, M. (2006). Third parties and status position: How the characteristics of status systems matter. *Theory and Society, 35*(3), 299–321.

Serrano Velarde, K. (2018). The way we ask for money… the emergence and institutionalization of grant writing practices in academia. *Minerva, 56*(1), 85–107. https://doi.org/10.1007/s11024-018-9346-4

SFS. (1992:1434). *The Swedish Higher Education Act Including the Act on Amendment of the Higher Education Act (2019:505)*.

SFS. (1993:100). *The Swedish Higher Education Ordinance Including the Act on Amendment of the Higher Education Ordinance (2019:276)*.

Shulman, L. S. (1986). Those who understand: Knowledge growth in teaching. *Educational Researcher, 15*(2), 4–14.

Swedish Govt. Bill. (2009/2010:80). *En reformerad konstitution* [A Reformed Constitution]. Stockholm: Ministry of Education.

Taylor, J. (2008). The teaching-research nexus and the importance of context: A comparative study of England and Sweden. *Compare: A Journal of Comparative and International Education, 38*(1), 53–69. https://doi.org/10.1080/03057920701467792

Tight, M. (2016). Examining the research/teaching nexus. *European Journal of Higher Education, 6*(4), 293–311. https://doi.org/10.1080/2156823 5.2016.1224674

UFV. (2010/1842). *Guidelines for admittance of excellent teachers*. Adopted December 6, 2011. Revised May 15, 2012. Uppsala University.

UKÄ, Swedish Higher Education Authority. (2019). *Higher Education Institutions in Sweden. 2019 Status Report*. UKÄ.

van den Brink, M. (2010). *Behind the scenes of science: Gender practices in the recruitment and selection of professors in the Netherlands*. Pallas Publications.

12

Peers in Systematic Review: Gate Keeping Understandings of Research in the Field

Tine S. Prøitz

Introduction

The introductory chapter of this book illuminates the far-reaching centrality of scholarly peers and their importance in assessing the quality of scientific work. The role of scholarly peers in a range of review processes has become institutionalised and integrated into most of the activities in academia (Musselin, 2013; Forsberg et al., 2021 in this book). An interesting aspect of this development is how the initial idea of peer evaluation and assessment of the quality of scientific work has migrated into a range of other academic contexts. In several ways, this migration can be seen to extend the role of peer review beyond the traditional turf of scientific reporting and publishing, potentially changing the premises and conduct as well as our understanding of what a peer evaluation entail.

Nowadays, we can observe peer evaluation and peer assessment as a mandatory and integrated element of scientific research such as in

T. S. Prøitz (✉)
University of South-Eastern Norway, Campus Vestfold, Norway
e-mail: Tine.Proitz@usn.no

© The Author(s) 2022
E. Forsberg et al. (eds.), *Peer review in an Era of Evaluation*,
https://doi.org/10.1007/978-3-030-75263-7_12

meta-analysis and systematic review. Peer evaluation is described as an essential element of quality assurance of the strictly defined methods and procedures of systematic review (see e.g. Slavin, 1986; Petticrew & Roberts, 2006; Gough et al., 2012; Torgerson 2003).

In this chapter, it is argued that the involvement of scholarly peers deeply embedded in the central stages of the systematic review processes has similarities with traditional peer review processes in academic publishing that aim to ensure the quality of academic work. However, the review process of a systematic review can also be distinguished from a peer review in academic publishing in how the review of manuscripts traditionally aims to formatively contribute to and ensure the quality of future publications. The systematic review process entails a peer evaluation after publishing and for purposes other than publishing. Further, peers are involved in evaluation of studies that are to be included or excluded according to the predefined criteria of the systematic review study. As such, peers in a systematic review can be regarded to make re-judgements of the quality and the relevance of already published work in accordance with the specific scope and predefined criteria for review studies.

The systematic review process places the scholarly peer within the research process where he/she becomes a central part of the scientific method, often contradicting the temporality of conventional *ex-ante* positioning of peer reviewers in academic publishing. The peer evaluator of systematic review can be considered to be in an *x-nunc* position (from now on and as long as the process goes, Jibi, 2020) limited to the frame of the systematic review in question. This positioning of the peer in systematic review represents something new and rather different from the intuitively perceived peer reviewer role. This positioning also often represents a breach of the academic principles of anonymity and distance between the researcher and the peer in situations of academic judgement.

The methods and procedures of a systematic review are examples of the migrating functionality of the peer evaluation process for assessing the quality of academic work. With its well-defined methods and procedures, a systematic review frames the involved scholarly peers in a highly specialised way. This framing partly builds on and it has characteristics similar to a peer review and partly differentiates a systematic review from conventional peer review processes, challenging our understanding of the roles of scholarly peers in assessing the quality of academic work. Thus, peers in systematic review processes are an interesting example of how a peer review framed in another context can contribute to a change in the premises and conduct of the peer role. In this chapter, these issues are illuminated by analysing the functions of peers in systematic review and discussing the roles of scholarly peers framed by other academic contexts—such as the systematic review.

The functionality and roles of scholarly peers in a systematic review must be seen in light of the rise of a general societal and policy-driven evidence movement within most fields (Hansen & Rieper, 2009). Within education, this development is reflected in the policy expectations for practitioners and professionals to use evidence when making decisions about teaching, learning and school development (Hansen, 2014; Levinsson, 2013; Sundberg, 2009; Gough et al., 2012; Levinsson & Prøitz, 2017; Prøitz, 2018). The systematic review phenomenon is grounded in ideas about methodological approaches that aim for highly detailed, universal and standardised stages of conduct (Davies & Nutley, 2000). However, with the growing knowledge base on research synthesis, variations in approaches have been acknowledged and problematised (Gough et al., 2012; Levinsson & Prøitz, 2017). Nevertheless, the involvement of peers to ensure the relevance and scholarly quality of primary research included in the systematic review studies is a stable feature across varying approaches (Prøitz 2018). In spite of the extensive and growing body of literature on various approaches to systematic reviews and ongoing debates on methodological and procedural issues, studies on the roles of the peers involved in research synthesis seem to be scarce, warranting closer analysis and discussion of the function of peers in systematic review processes.

Thus, this chapter presents an analysis and discussion of the function of peers (also called field experts, experts or peer reviewers) in scientific quality work of systematic reviews. The analysis draws on literature on traditional peer review in academic publishing and systematic reviews and a document analysis of systematic review technical reports within the field of education. The study is guided by the following questions: What are the functions of scholarly peers in a systematic review? What are their main tasks? What consequences does the analysis have for our understanding of scholarly peers in various types and processes of scientific work?

This chapter is divided into five sections. The study's thematic and research questions are presented in the first, introductory, section. The characteristics of the systematic review process are described in the second section. The analytical framework is presented in section three, followed by the method and document material described in the fourth section. In the fifth and final section, the results of the analysis are discussed and some concluding remarks are provided.

Peers in Academic Publishing and Systematic Review

For the sake of this study, it is necessary to describe the background and context of peer evaluation and assessment in both academic publishing and systematic review. Throughout the chapter, a choice has been made to use the terms *peer* and *peer work* to capture the varied actions of the peers involved in systematic reviews. Here, peer work is considered to cover all activities that scholars perform to ensure academic quality when being involved in systematic review processes as well as those performed by peers in other academic situations.

Peer review in academic publishing has been defined as 'the process by which research output is subjected to scrutiny and critical assessment by individuals who are experts in those areas' (Hames, 2012, p. 16). Simply put, the traditional peer review process requires researchers to prepare a manuscript that reports their research and submit this manuscript to a journal for publication consideration in which the peer review process is a central part of the decision.

Based on this definition, the traditional peer review occurs before publishing. The peer review processes can be traced back 300 years to the regulated consultations of publications by experts among the members of the Royal Societies of Edinburgh and London (Hames, 2012; Spier, 2002). However, peer review first became widespread in the twentieth century; today, it has grown into a massive activity in the form of 25,000 peer reviewed journals (Hames, 2012; Ware & Mabe, 2009). Editors and researchers have appreciated how the peer review process has helped strengthen scientific communication through its regulatory characteristics of control and trust in research quality (Ware & Monkman, 2008). Peer review is also, to an increasing degree, criticised for issues related to quality and fairness and abuse and bias, for being expensive, slow and conservative as well as for lacking consistency.

According to Hames (2007, 2012, p.22) a peer reviewer in academic publishing is expected to prevent the publication of bad work, check that the reported research has been carried out well and without flaws in design or method, ensure correct reporting and interpretation of results, ensure results are not too preliminary or speculative, provide editors with evidence to judge the relevance of an article for a journal, provide authors with quality and feedback, improve the quality and readability of articles and maintain the integrity of scholarly record. The expectations neither define how to recognise 'bad work' nor exemplify what is meant by 'research carried out well' nor stipulate what is meant by correct interpretations, preliminary or too speculative. To a large extent, the essential judgement of quality is left to the scholarly peer based on the individual academic understanding of quality and merits of the qualified peer.

In contrast, systematic review is a rather new invention. The development of systematic reviews can be traced back to the meta-analysis by Glass and Smith in the 1970s, regarded as a cornerstone in the rise of evidence-based medicine (Gough, 2004; Bohlin, 2011). Inspired by Anglo-American success stories of clearinghouses, centres for 'what works' and 'best evidence' programmes, European governments, researchers and private entrepreneurs have embraced the idea of evidence based practice in various fields (Hansen, 2014; Levinsson, 2013; Sundberg, 2009). This has led to an evidence-based movement calling for systematic reviews in most fields. Systematic review is grounded in ideas about methodological approaches that aim for highly detailed, universal and standardised stages of conduct (Davies & Nutley, 2000). In general, the systematic review process is defined by certain successive steps of scientific conduct. Quality assessment is a central element in most of these steps; although, the use of scholarly peers is a stable characteristic of the method, there are variations in how and when peers are involved in the review process. Mostly the systematic review process contains the following steps: formulate a research question and develop a protocol, define the studies to be included (inclusion criteria), search for studies, screen studies, describe studies (the systematic review mapping can stop at this step or continue towards obtaining the full map and research synthesis using the following steps), appraise the study's quality and relevance, synthesise the findings (answer the research question) and communicate and engage (Gough, 2007).

Based on a study of approaches presented by agencies developing systematic reviews and their review reports in the field of education, agency-specific variations in procedures of the review process were observed as a general characteristic, as was the employment of peers (Prøitz, 2015). Looking at different examples of the procedural steps used for a systematic review in education, peers can participate in the overall review teams/review groups or serve as field experts, employed by the review team of the review study to support the relevance and quality assessment phases of the review process. Often, the review protocol defining the scope of the review process is established before field experts are involved in the process, but experts can participate in approving the protocol as well as in reviewing the quality of the review steps, the review process and/or the final review report (Prøitz, 2015)

Peers participate in reviewing the protocols and methods. They offer suggestions for revisions and re-submission of protocols and methods. The same peers are also involved in reviewing drafts of the final review reports and providing feedback and suggesting revisions before approval by the review team. Peers can also be a part of advisory groups to the review team of a systematic review and participate in the evaluation of the defined quality and relevance procedure by classifying primary research in accordance with quality standards. Furthermore, peers can also contribute by assessing the evaluation process and suggesting adjustments before participating in the evaluation of the quality of the primary studies procedure. In sum, peers in a systematic review can participate in a variety of procedural steps, they can also serve as members of review teams overseeing the whole process, play an active part in the procedure and be external reviewers of the final review report.

Analytical Framework

In this section, the analytical framework of the study is presented. The peer review role is analysed and discussed by focusing on the status function of peers in the context of the systematic reviews using speech act theory motivated by the work of Searle (1995, 2005). This approach provides an analytical tool to identify what counts as peer review in systematic review processes (Searle, 1995, 2005).

In Searle's (1995) project, there is a defence of the idea of reality as independent of us as opposed to the idea that all reality is human creation. According to Searle (1995), there are objective facts in the world that are only facts because we believe them to exist. Searle (1995) calls some of these facts 'institutional facts' (e.g. money, marriage) as opposed to non-institutional facts or 'brute facts' (e.g. mountains, trees) (p. 2). Searle makes a call for the analysis of the role of language in the constitution of institutions as he considers that researchers in social science have taken language for granted and overlooked the building blocks of social reality (Searle, 2005). The creation of institutional facts is enabled by collectively accepted systems of rules (procedures, practices), by which members of a collective impose a specific status function on a phenomenon as an institutional fact, which also gives the phenomenon a specific function through agreement and acceptance. The collective assignment of status and function also involve recognition of something or someone having power by virtue of its institutional status. The creation of an institutional fact requires a collective recognition and acceptance of so-called deontic powers, e.g. rights, duties, responsibilities and obligations. A relevant example for this study is how peer evaluation in varied academic situations is based on collectively agreed upon and recognised powers, which assign the right and duty to express evaluative comments, suggest improvements and make judgement of another researcher's work with authority for quality assurance required by the peer review function.

According to Searle (1995), a collective's agreement on giving a specific phenomenon (e.g. peers in systematic review processes) a particular status function can be expressed with the logic of 'X counts as Y in context C'. Searle described the rules in these systems as having *the form of X counts as Y in C, where an object, person or state of affairs X is assigned a special status, the Y status, such that the new status enables a person or object to perform functions that it could not perform solely in virtue of its physical structure, but requires as a necessary condition, the assignment of the status.* (1995, p. 22)

Inspired by Searle, this study investigates the defined status of peers/field experts (X) and what their function 'counts as' (Y) in systematic review processes (C). Thus, Searle's (1995) logic provides a tool to analyse the collectively assigned status function of field experts in systematic review processes as described in systematic review technical reports.

Method

This study draws on a content and document analysis (Bowen, 2009; Cohen et al., 2011) of data extracted from technical reports that describe the method and procedures for determining the relevance and quality of a review involving external peers applied in systematic review processes. The technical reports provide thorough descriptions and rich information about the scope of the review in question, methods, procedural requirements and quality assurance processes where peers are potentially involved.

The review reports that were studied were selected from three different research agencies in Nordic countries (the Swedish Institute for Education Research, the Norwegian Knowledge Centre for Education and the Danish Clearinghouse). All three agencies were established during the last decade, and they form part of government-induced initiatives inspired by the ideas of evidence movement. They all have a certain focus on education and have developed their own agency-specific procedural way of conducting reviews, although they rely on well-known systematic review methods.

The documents were studied based on a three-phase process. In the first phase, 15 selected published review reports of the three Nordic knowledge agencies were skimmed (5 reports for each of the agencies, see "Selected Systematic Review Technical Reports Analysed in the Study" for a complete list of the selected reports). The aim of this phase was to identify the structural patterns of reporting because systematic review reports usually have a rather defined structure, although with an agency-specific variation (Prøitz, 2015). In this phase, mention of peers (e.g. as peer reviewers, external experts, field experts and researchers) in the different chapters, sections or paragraphs of the reports was identified in relation to the different steps of the review process to identify the functions of the peer work. In the next phase, an in-depth reading of the selected reports from each of the agencies was conducted. In this reading, a special focus on the identified passages where peers occurred was employed. In the third phase of the reading, the findings from the second phase were validated across the selected reports of each agency to see if the identified patterns of peer involvement could be characterised as a more general way of conducting the review process. This approach confirmed the existence of overall agency-specific structural patterns of reporting and the involvement of peers in the review processes. The examples presented can be considered as representations of the typical involvement of peers, as described in the 15 technical systematic review reports produced by the three Nordic agencies for reviews within the field of education.

For the analysis, characteristics of the peer status function, as described in the systematic review technical reports, are identified in the studied materials. The observed involvement of peers in the review processes is further discussed in relation to Searle's (1995) work on institutional facts and, in particular, the assignment of the status function of peers in review processes.

Examples of Peers in three Review Processes

In the following section, the document material is described in detail. The presentation of the material focuses on the use of peers in the review processes as reported in the technical systematic review reports of the three selected agencies.

Swedish Institute for Educational Research

Two external researchers were invited to participate in the review project (Skolforskningsinstitutet, 2019) in a so-called project group consisting of the two researchers and the research agency's internal staff after conducting a needs assessment (behovsinventering) among stakeholder groups and researchers in the field. A pilot study confirmed the review theme to be relevant and validated the existence of available primary studies. The role of the external experts was to 'contribute with their understanding of research within the field based on their expert knowledge' (Skolforskningsinstitutet, 2019, p. 60).

The two experts were further involved in Step 2, which was the *relevance review* of the systematic review process. In Step 1, the Skolfi staff had completed the first screening of the 9662 articles that were identified in the searches. Based on information in the titles and abstracts, 8646 studies that were considered 'clearly not relevant' were excluded. In Step 2, the external researchers individually reviewed the 1016 studies by reading the titles and abstracts. If it was unclear if a study was relevant, it was included in the next phase. Thus, the researchers excluded 815 studies in this phase. In Step 3, the two researchers individually reviewed the 201 studies left after a full reading of the text. In cases where the researchers disagreed, the studies were discussed and disagreement was resolved by consensus. In Step 3, 151 studies were excluded. In Step 4, the researchers conducted a collective relevance and quality review of 50 studies. The quality review was conducted with the support of the Skolfi guidelines for quality review (which was missing in the document, but which can be found on the Skolfi web page[1]). In this process, another 35 studies were excluded; thus, 15 articles were included in the study as being relevant and of 'good quality' (Skolforskningsinstitutet, 2019, p. 63).

The external researchers were also involved in the data and result extraction process in which they described the purpose of the studies and their results in writing on an A4 page. These writings were later used in

---

[1] Bilaga 2 Underlag för bedömning av studiernas kvalitet (Retrieved 14.12.19). https://www.skolfi. se/wp-content/uploads/2019/12/Bilaga-2-Underlag-f%C3%B6r-bed%C3%B6mning-av-studiernas-kvalitet_lek.pdf

the process as guidelines for the internal project group's understanding of the results of the different studies. The external researchers were also involved in writing the report. When the report was finalised, it was first read by internal staff and then by two different external researchers within the field that were hired only for this purpose. The first two external field experts are presented in the report with their full name, title and an extensive biography; the other two experts reading the report were un-named. The Swedish Institute for Educational Research has the most standardised description of the procedures regarding the involvement of external researchers/peers described in every report under a separate chapter with the heading: Method and Conduct.

The Norwegian Knowledge Centre for Education

The review produced by the Norwegian Knowledge Centre for Education (Kunnskapssenter for Utdanning [KSU]) is a so-called rapid review characterised as 'a format developed to do reviews quickly while at the same time ensuring the same quality criteria as for systematic review and has the same requirements for systematization and transparency' (Lillejord et al., 2018, p. 10). The rapid review was based on guidelines and tools developed by the Evidence for Policy and Practice Information and Co-ordinating Centre (EPPI) at the University College London. Search resulted in 2542 hits. All references were imported into the EPPI-Reviewer 4 programme, which was developed to handle large amounts of data. The process of screening the articles consisted of three steps. In Step 1, two un-named researchers performed the first screening by reviewing titles and abstracts according to three, predefined thematic inclusion and exclusion criteria. In that step, 1685 studies were excluded leaving 53 articles with potential relevance for the review study. In Step 2, two researchers independently reviewed the quality and relevance of the studies. The applied quality criteria are partly described by the concepts of validity, reliability and generalizability and partly by three questions to be answered: Is the research question clearly defined? Are the research method and the research design specified? Is there coherence between the research question and the results? Each point and question are to be reviewed as high (explicit and detailed description of the method, data collection, analysis and result; the results have clear support in the findings), medium (satisfying description of the method, data collection, analysis and result; the results are partly supported in the findings) or low (weak description of the method, data collection, analysis and result; results have weak support in the findings). When in doubt, the articles were presented to the project group to make a final decision. After Step 2, 33 studies were included in the research mapping for analysis. In Step 3, the articles were prepared for synthesis and the content of the studies was mapped by methods for data collection, data analysis and country. After mapping, a number of un-named researchers read the full text of the articles and every study was described in a short summary that helped clarify how it could contribute to the research question of the systematic review.

The report says little about the researchers involved in Step 2 and Step 3 of the review process and about their involvement and how they worked. Thus, some differences were found between the selected reports in terms of whether the involved peers/researchers/research group are named or whether their involvement is specified. For example, it seems that this varies with the magnitude and ambitions of the review; larger studies provide more details on the involvement particularly of research groups, but mostly in general terms. This is in contrast to the example of the Swedish report, where the full name and the individual title, experiences and qualifications of the two external researchers involved in Step 2 were presented, while the peers used in the review of the final report were not named.

Danish Clearinghouse for Educational Research (DCU)

The process of developing the research mapping is described as an example of the DCU standard procedure, which consists of using the software tool EPPI-Reviewer version 4.7.0.0 developed by the EPPI-Centre. The mapping was conducted as a collaboration between staff at the DCU and the members of the review group. The review process, including all communication between the DCU staff and the review group, was documented (Bondebjerg et al., 2016). The review process consisted of several phases; members of the review group were involved in the last phase of the process which entailed coding and review. The report describes a collaborative effort where DCU staff filled in the forms of the EPPI-Reviewer and sent them to a member of the review group. The review group member then filled in his or her review of the study, including potential corrections or additions to the DCU staff coding. In the end, the two reviewers agreed on how much evidence weight each individual study should have in the study.

The reported study identified 2409 references in the search process; 197 references were excluded due to duplication, 2122 were screened by title, abstract and full text, 2062 studies were excluded in the first screening, 150 references were coded and reviewed in the second screening. Finally, 144 studies were included. The screening and exclusions that occurred before the second screening were solely done on the basis of relevance by DCU staff. Assessing the quality of the studies was not part of the first screening process. The 144 included studies were closely read, coded and reviewed in the EPPI-Reviewer. Based on the codes, every study was given an evidence weight (high, medium or low) characterising the degree to which the individual study fulfilled general scientific standards for empirical research; thus, 63 studies were reviewed to be of high or medium evidence weight. The quality review considers if the study actually investigates the issue it is meant to evaluate, if there is coherence between the premises of the study, the data and the conclusion and if the study achieves its aim. It also includes ethical criteria in data collection and selection, and the way the relationship between the empirical data and the conclusion is described. The quality review also considers the generalisability study. The report refers to the DCU research quality guidelines, which are written in a separate document.

The three example reports illustrate varying degrees of thoroughness and transparency in the reporting of who the peers/external experts were as well as their roles and involvement in the review process. In the Swedish example, the peer/experts involved in the process were presented in a separate section with their full name and biography, while the identities of the peers/external experts employed to read the final report remained anonymous as in conventional blind peer review. In the Norwegian case, the names of the peers vary, and little information is given about the 'researchers' that were involved and the extent of their involvement, including whether the researchers were external or internal. However, in larger studies, the review group is usually presented in full with names and affiliations on the first page of the Norwegian reports. In the Danish report, the external researchers are named as part of the review group without being emphasised particularly. In all three cases, the peers/experts were involved in the later stages of the review process after the initial screening processes, which were often performed by agency staff. In the Swedish and Norwegian cases, the peers/experts were involved in reviewing titles and abstracts, as well as in the phase of reading the full text of the articles considering both issues of relevance and quality. The Danish peers/experts mainly reviewed the coding made by the DCU staff in the EPPI-Reviewer system, and they added their review of the quality of the articles.

Peers in the Systematic Review Processes

Overall, there are several features that can be regarded as characteristic of the work of peers in the systematic review processes studied. First, general principles of scientific quality such as transparency, validity and reliability for quality assurance seem to be underlying elements throughout the work. Second, the peer evaluations are framed by the scope of the systematic review, the method, the procedures, the review protocol and its defined inclusion and exclusion criteria. As such, the peer work of ensuring scientific quality is not only framed by general principles of scientific quality but also, and more strongly, defined by specific and strict principles of method and procedure. The principles can be interpreted as devices for quality assurance that both secure transparency and delimit the space available for professional judgement from going outside the scope of the systematic review. The peers in a systematic review are employed to maintain the quality of the systematic review related to the scope and methods of the review in question.

This issue can be further interpreted by the work of Searle (1995) and his ideas about how larger groups of people assign status functions, such as in this study the scholarly community more or less worldwide, have agreed on the idea of peer review as a sound way to ensure scientific quality in academic reporting and publishing in general and how this idea serve a somewhat similar function in a systematic review process. However, the logic of X counts as Y in a context C also helps identify the status function of peers in systematic review as something different from the conventional peer work when seen in relation to the frames of reference and context.

Following this line of thinking, the peer review process in, for example, academic publishing is assigned its status function through the expectations of being an anonymous/blind guarantor of academic quality based on individual scholarly merit providing academic judgement, critique and formative advice. In a systematic review, the peer can be considered to have been assigned status function as a known guarantor of academic quality and relevance based on individual scholarly merit as defined by the scope, purpose and procedures of the systematic review in question. (see Table 12.1)

The significance of the differences can be further illustrated by the difficulties that would occur if peers in a systematic review were to take on the role/status function of the peer in publishing. Their function as producers and guarantors would fall outside the focus of the review for the publishing frame, and it would most likely lose status function within the openness of framing the peer review for academic publishing. In consequence, to a large extent, peers in a systematic review are peers primarily seen in relation to the method, scope, purpose and procedure of the review. In contrast, for peers in academic publishing, the scholarly status

Table 12.1 The status function of peers in academic publishing and in systematic review

X	Count as Y (status function)	In context C
Peer evaluation	Anonymous/blind guarantor of academic quality based on individual scholarly merit and ex-ante judgement, critique and formative advise	Academic publishing
Peer evaluation	Known guarantor of academic quality and relevance based on individual scholarly merit and ex-nunc judgement defined by the method, scope, purpose and procedures	Systematic review

function can be more broadly defined for a larger field of scholarly expertise, where the peers have a certain responsibility to uphold the 'record' of the field and provide help, support and advise to authors, mainly through processes characterised by anonymity and distance between the researcher and the peer reviewer. Similarly, this status function of the peer in the publishing context would not be functional within the framing of a systematic review.

Another characteristic of the systematic review process, displayed by the presented material, is the set of procedures that are applied when the peer evaluators disagree. Consensus on the quality and relevance of reviewed work is important for the overall quality of the systematic review, and specified practices of conduct are defined in the method to reach a common agreement among peer evaluators. In the material, approaches for reaching agreement among the involved reviewers were described in the Swedish and the Danish examples. This is in contrast to the ideal of a blind review in a conventional peer review for academic publishing where disagreement between reviewers is partly left to be resolved by the author in the manuscript revision and partly to be resolved by the editorial decision and advice given to the author and sometimes by involving yet another reviewer to obtain a third opinion.

The material also displays how the peer/expert in a systematic review is mostly involved after the scope and purpose of the study has been set by the review team and also often after the groundwork of searching and screening of primary studies has been completed by the agency staff. This issue of temporality illustrates yet another and central feature of the peer work of systematic review and it is probably where peer work in systematic review distinguishes itself mostly from the traditional peer evaluation. The formative aspect of ex-ante peer evaluation is a central academic principle that underscores the ambition and importance of collegial sharing, critique, correction and revision for improvement of research. The peer evaluation in the systematic review processes does not aim for such formative purposes but it can be considered as an integrated ex-nunc judgement of another kind and for other purposes, where published studies are measured up against specified criteria of relevance and selected on the basis of being the best fit with predefined criteria, including aspects of quality. The aspect of temporality highlights issues regarding the professional judgement of the involved peers in systematic review. In Chap. 3, Vanderstraeten interestingly documents how the reviewer/author and editor roles have changed from being blurred and intermingled to becoming more specialised, standardised and pronounced as a result of historical contingencies that defines the grounds for scientific research. Looking at the peer reviewers in this chapter illuminates how experts in systematic review today also seem to work under resembling blurred lines, for example through their roles as experts evaluating published work and contributing to the development of new systematic review publications at the same time. This issue raises questions relevant for all peers regarding what the peer work is about, and with reference to the thematic of this book, what peers in varied academic contexts including systematic review are 'gatekeepers' of? In the context of systematic review, peers are not only making re-judgements of already reviewed and published research they also function as gatekeepers of the given standards, guidelines and procedures of the review method.

Concluding Remarks

In this study, we have seen how the involvement of peers in systematic review processes makes use of peers resembling those in a traditional academic peer review process to ensure academic quality. We have also seen that there are differences between the roles and the status function of peers related to the framing and purpose of a peer's 'guarantor role'. The study highlights the issues of the role and status of peers in varied academic contexts. It also highlights how the peer review role change with changing temporalities and different devices that provides different spaces for professional judgement.

As such, the analysis lays the groundwork for a debate on peers in different contexts framed by different processes with different purposes, and it questions whether a peer review is the same when the premise of the scholarly activity changes. This study also highlights the difficult question of the function of the peer reviewer in-between being the anonymous and distant person ensuring and guaranteeing scientific quality and being the one to openly and actively participate in the formation of a scholarly product while also playing the peer role. This question is highly relevant considering the more recent developments and debates on the need for more open peer review processes in academic publishing, changing the premises of the conventional activity.

With this backdrop, this study calls for a stronger framework or, potentially, a typology distinguishing between varied forms of peer work to clarify the differences between the roles of peer reviewers in different academic activities, considering the migration of the use of peers in a range of academic work.

References

Bohlin, I. (2011). Evidensbaserat beslutsfattande i ett vetenskapsbaserat samhälle. Om evidensrörelsens ursprung, utbredning och gränser. In I. Bohlin & M. Sager (Eds.), *Evidensens många ansikten* (pp. 31–68). Arkiv Förlag.

Bondebjerg, A., Jessen, A., Jusufbegovic, L., & Vestergaard, S. (2016). *Forskningskortlægning og—vurdering af skandinavisk dagtilbudsforskning for 0–6 årige i året 2016. [A research mapping of Scandinavian preschools for children aged 0–6 in 2016] København: Dansk Clearinghouse for Utddannelsesforskning.* Aarhus Universitet.

Bowen, G. A. (2009). Document analysis as a qualitative research method. *Qualitative Research Journal, 9*(2), 27–40.

Cohen, L., Manion, L., & Morrison, K. (2011). *Research methods in education* (7th ed.). Routledge.

Davies, H. T., & Nutley, S. M. (Eds.). (2000). *What works? Evidence-based policy and practice in public services.* Policy Press.

Forsberg, E., Levander, S., Geschwind, L., & Wermke, W. (2021). Peer Review in Academia. In E. Forsberg, S. Levander, L. Geschwind, & W. Wermke, (Eds.), *Peer Review in an Era of Evaluation. Understanding the Practice of Gatekeeping in Academia.* Palgrave (in this book).

Gough, D. (2004). Systematic research synthesis. In G. Thomas & R. Pring (Eds.), *Evidence-based practice in education* (pp. 44–62). Open University Press.

Gough, D. (2007). Weight of evidence: A framework for the appraisal of the quality and relevance of evidence. *Research Papers in Education, 22*(2), 213–228.

Gough, D., Oliver, S., & Thomas, J. (Eds.). (2012). *An introduction to systematic reviews.* Sage.

Hames, I. (2007). *Peer review and manuscript management in scientific journals: Guidelines for good practice.* Blackwell Publishing and ALPSP.

Hames, I. (2012). Peer review in a rapidly evolving publishing landscape. In E. Pentz, I. Borthwick, & R. J. Campbell (Eds.), *Academic and professional publishing* (pp. 15–52). Elsevier Science and Technology. http://ebookcentral.proquest.com

Hansen, H. F. (2014). Organisation of evidence-based knowledge production. Evidence hierarchies and evidence typologies. *Scandinavian Journal of Public Health, 42*(13 suppl), 11–17.

Hansen, H. F., & Rieper, O. (2009). The evidence movement. The development and consequences of methodologies in review practices. *Evaluation, 15*(2), 141–163.

Jibi, M. (2020, March). Philosophical Implications of Ex-Tunc and Ex-Nunc Testing in State Administration Disputes. In *Proceedings of the International Conference on Law Reform (INCLAR 2019)* (pp. 170–172). Atlantis Press.

Levinsson, M. (2013). Evidens och existens. Evidensbaserad undervisning i ljuset av lärares erfarenheter. Doktorsavhandling. Institutionen för didaktik och pedagogisk profession, Gøteborgs Universitet.[PhD thesis, University of Gothenburg].

Levinsson, M., & Prøitz, T. S. (2017). The (non-) use of configurative reviews in education. *Education Inquiry, 8*(3), 209–231.

Lillejord, S. Børte, K. Ruud, E., & Morgan K. (2018). *Stress i skolen—en systematisk kunnskapsoversikt* [School stress—A systematic review] Oslo: Kunnskapsoversikt for Utdanning.

Musselin, C. (2013). How peer review empowers the academic profession and university managers: Changes in relationships between the state, universities and the professoriate. *Research Policy, 42*(5), 1165–1173.

Petticrew, M., & Roberts, H. (2006). *Systematic reviews in the social sciences.* Blackwell Publishers.

Prøitz, T. S. (2015). Metoder for systematiske kunnskapsoversikter. Vetenskapsrådets Rapporter 2015.

Prøitz, T. S. (2018). Utbildningspolitikens förståelse av evidens, i N. Wahlström & D. Alvunger (Eds.), Den evidensbaserade skolan.—svensk skola i skärningspunkten mellan forskning och praktik. Natur och

Searle, J. (1995). *The construction of social reality.* Penguin.

Searle, J. R. (2005). What is an institution? *Journal of Institutional Economics, 1*(1), 1–22.

Skolforskningsinstitutet. (2019). *Att genom lek stödja och stimulera barns sociala förmågor—undervising i förskolan. Systematisk översikt 2019:01.* Skolforskningsinstitutet.

Slavin, R. E. (1986). Best-evidence synthesis: An alternative to meta-analytic and traditional reviews. *Educational Researcher, 15*(9), 5–11.

Spier, R. (2002). The history of the peer-review process. *TRENDS in Biotechnology, 20*(8), 357–358.

Sundberg, D. (2009). *Evidens i utbildningspolitiken: en kartläggning och analys av policyrelevanta kunskapsöversikter (Utbildningsdepartementets skriftserie, rapport nr 10.).* Utbildningsdepartementet.

Torgerson, C. (2003). *Systematic reviews.* Continuum International.

Vanderstraeten, R. (2021). Peer review in scientific journals: 'Disciplining' educational research in the twentieth century. In E. Forsberg, S. Levander, L. Geschwind, & W. Wermke, (Eds.), *Peer Review in an Era of Evaluation. Understanding the Practice of Gatekeeping in Academia.* Palgrave (in this book).

Ware, M., & Mabe, M. (2009). *The STM Report. An overview of scientific and scholarly journal publishing.* International Association of Scientific, Technical and Medical Publishers.

Ware, M., & Monkman, M. (2008). *Peer review in scholarly journals: Perspective of the scholarly community—An international study.* Publishing Research Consortium (PRC) Research Report. http://www.publishingresearch.net/documents/PeerReviewFullPRCReport-final.pdf

Selected Systematic Review Technical Reports Analysed in the Study

The Example Reports Used for the In-depth Reading Are Underlined

Bondebjerg, A. Jessen, A. Jusufbegovic, L., & Vestergaard, S. (2016). *Forskningskortlægning og—vurdering af skandinavisk dagtilbudsforskning for 0–6 årige i året 2016* [A research mapping of Scandinavian preschools for children aged 0–6 in 2016] København: Dansk Clearinghouse for Utddannelsesforskning, DPU, Aarhus Universitet.

Dyssegaard, C. B., Larsen, M. S., & Tiftikci, N. (2013). *Effekt og pædagogisk indsats ved inklusion af børn med særlige behov i grundskolen. Systematisk review.* København: IUP, Aarhus Universitet.

Dyssegaard, C. B., & Egelund, N. (2016). *Systematisk kortlægning om forældreinvolvering og forældresamarbejde, der kan fremme læring hos socialt udsatte børn og unge i dagtilbud og skole.* København: Dansk Clearinghouse for Uddannelsesforskning, DPU, Aarhus Universitet.

Dyssegaard, C. B., Egeberg, J. d. H., Sommersel, H. B., Steenberg, K., & Vestergaard, S. (2015). *A systematic review of the impact of multiple language teaching, prior language experience and acquisition order on student's language proficiency in primary and secondary school.* Danish Clearinghouse for Educational Research, Department of Education, Aarhus University.

Dyssegaard, C. B., de Hemmer Egeberg, J., Steenberg, K., Tiftikci, N., & citeres som Vestergaard, S. (2014). *Forskningskortlægning af håndterbare forhold til gavn for fastholdelse, øget optag og forbedrede resultater i de gymnasiale uddannelser.* København: Institut for Uddannelse og Pædagogik (DPU), Aarhus Universitet.

Lillejord, S., Børte, K., Ruud, E., & Morgan, K. (2018). *Stress i skolen—en systematisk kunnskapsoversikt* [School stress—A systematic review] Oslo: Kunnskapsoversikt for Utdanning.

Lillejord, S., & Børte, K. (2018). *Mellomledere i skolen: Arbeidsoppgaver og opplæringsbehov—en systematisk kunnskapsoversikt.* Kunnskapssenter for utdanning. www.kunnskapssenter.no

Lillejord, S., Børte, K., Ruud, E., Hauge, T. E., Hopfenbeck, T. N., Tolo, A., Fischer-Griffiths, P., & Smeby, J.-C. (2014). *Former for lærervurdering som kan ha positiv innvirkning på skolens kvalitet: En systematisk kunnskapsoversikt.* Kunnskapssenter for utdanning. www.kunnskapssenter.no

Lillejord, S., Halvorsrud, K., Ruud, E., Morgan, K., Freyr, T., Fischer-Griffiths, P., Eikeland, O. J., Hauge, T. E., Homme, A. D., & Manger, T. (2015). *Frafall i videregående opplæring: En systematisk kunnskapsoversikt.* Kunnskapssenter for utdanning. www.kunnskapssenter.no

Lillejord, S., & Børte, K. (2014). *Partnerskap i lærerutdanningen—en forskningskartlegging—KSU 3/2014.* : Kunnskapssenter for utdanning. www.kunnskapssenter.no

Skolforskningsinstitutet. (2019). *Att genom lek stödja och stimulera barns sociala förmågor—undervising i förskolan. Systematisk översikt 2019:01.* Skolforskningsinstitutet.

Skolforskningsinstitutet. (2019). *Läsförståelse och undervisning om lässtrategier. Systematisk översikt 2019:02.* Skolforskningsinstitutet.

Skolforskningsinstitutet. (2019). *Att genom lek stödja och stimulera barns sociala förmågor—undervising i förskolan. Systematisk översikt 2019:01.* Skolforskningsinstitutet.

Skolforskningsinstitutet. (2017). *Digitala lärresurser i matematikundervisningen. Delrapport skola. Systematisk översikt 2017:02 (1/2).* Skolforskningsinstitutet.

Skolforskningsinstitutet. (2017). *Klassrumsdialog i matematikundervisningen—matematiska samtal i helklass i grundskolan. Systematisk översikt 2017:01.* Skolforskningsinstitutet.

13

The Decision-Making Constraints and Processes of Grant Peer Review, and Their Effects on the Review Outcome

Liv Langfeldt

Introduction

Peer review is the foundation of quality assurance in scholarly research. Both for the social study of science and for research policy, it is fundamental to understand the processes determining the outcome of peer review – peer review sets the standard for good research and decides who gets tenure, and what kind of research is funded and published. The present study deals with the processes determining the outcome of peer review of grant proposals. The focus is on how such seemingly irrelevant

L. Langfeldt (✉)
Nordic Institute for Studies in Innovation, Research and Education,
Oslo, Norway
e-mail: liv.langfeldt@nifu.no

© The Author(s) 2022
E. Forsberg et al. (eds.), *Peer review in an Era of Evaluation*,
https://doi.org/10.1007/978-3-030-75263-7_13

or 'innocent' factors as rating scales and peer panels' ranking methods and voting systems affect the assessments of proposals, and thus what kind of projects are funded. Do such factors influence what counts as good research? Do they influence *de facto* research 'policy'?

Numerous studies of peer review focus on the reliability of, and the possible biases in, peer review, and find low degrees of agreement between referees and various kinds of bias: academic and institutional status, nationality, gender and research field of the applicant, as well as different kinds of cognitive bias, are all found to affect the outcome.[1] Other studies focus on criteria and find a common 'language' in evaluation of research quality, certain criteria that researchers pay attention to.[2] The combination of findings of low degrees of agreement between referees and of a common set of criteria for assessments of research quality indicates that while there is a certain set of criteria that reviewers pay attention to – more or less explicitly – these criteria are interpreted or operationalized differently by various reviewers.

There are no clear norms for assessments, and there may be a large variation in what criteria reviewers emphasize – and *how* they are emphasized. The determinants of peer review may in this way be accidental: for example, who reviews what research and how reviews are organized may determine outcomes, and this process may be open to various kinds of bias. As criteria have no standard operationalization or interpretation, there are ample possibilities, for instance, to choose interpretations that promote the personal favourites of the reviewers.

It should be noted that the *concept* of bias is seldom discussed, but interpreted in various ways. Some studies that find disagreements among reviewers interpret these as some sort of 'cognitive particularism' (Travis & H.M. Collins 1991), or 'confirmatory bias',[3] while others interpret disagreements as 'real and legitimate differences of opinion among experts

---

[1] Reviews of such studies are found in Campanario (1998a); Campanario (1998b); Chubin & Hackett 1990); Cicchetti (1991); Cicchetti (1991); Daniel (1993); Wood (1997). The literature is far from conclusive on the various kinds of bias.

[2] See for example Chase 1970; Dirk 1999; Gulbrandsen & Langfeldt 1997; Hemlin 1993.

[3] 'Confirmatory bias' means that reviewers are biased against research contrary to their theoretical perspective. The concept 'cognitive particularism' is used by Travis & Collins (1991) about the same kind of phenomenon: decisions based upon membership in a scientific school of thought.

about what good science is or should be' (Cole et al. 1981, p. 885).[4] Such divergent interpretations reveal a lack of common understanding not only of the notion of bias, but also about what are legitimate considerations when assessing research. One view may be that 'grant applications should be judged on universalistic criteria, such as the scientific merit of the proposal' (Travis & Collins 1991, p. 325), and that 'school of thought' is a particularistic criterion. This implies that peer review should use uncontroversial criteria, and not take a stand in ongoing debates. When scholars disagree, such 'scholarly-neutral' assessments may not be feasible.

Another point of view is to see the reviewers as central actors in the definition and redefinition of 'good research'. In this view, low inter-reviewer agreement on a peer panel is no indication of low validity or low legitimacy of the assessments. In fact, it may indicate that the panel is highly competent because it represents a wide sample of the various views on what is good and valuable research (see Harnad 1985). Broad representation of divergent judgements and open debate about criteria and assessments are then desirable, and focussing on how various models manage these concerns is consequently important.

The problems of handling disagreements between reviewers, and the need to understand the effects of various peer review models, are increased by today's high refusal rates of grant proposals. When a very small proportion of projects are funded, the effects of grant review as censorship against certain kinds of research (cognitive bias) or researchers (for example, institutional bias or gender bias) may be high. Organizational factors may increase or reduce the problems of cronyism and conservative assessments, as well as increasing or reducing arbitrary outcomes.

The main characteristic of peer review – that quality criteria have no standard operationalization, and that judgements depend on the 'intimate craft knowledge' (Ravetz 1971, p. 274) of the reviewers – is the main problem for students of peer review; biases are hard to prove for outsiders.[5] The aim of the present study is to understand and explain the

---

[4] The studies I have mentioned (Travis & Collins, 1991; Mahoney, 1977; Cole et al., 1981) all, more or less explicitly, explain the outcome of review by the reviewers' scholarly points of view. Travis and Collins also found that the research community is concerned about the effects of the reviewers' scholarly points of view, and that review panels tried to take action to modify such effects when reviewing applications based on statements from mail referees.

[5] The approach of Wennerås & Wold (1997) should be noted. By comparing bibliometrics with grant review marks for scientific competence, they found that female applicants had to have pub-

decision-making processes of grant review and their policy effects. There are no attempts to identify 'biases' on the basis of quantitative correlation between organizational factors and outcome. The review processes of different review units are contrasted to gain insight into the factors decisive for the overall direction of review outcome in terms of policy effects, not in terms of measuring specific biases. Consequently, the study does not aim at conclusions about whether peer review is 'reliable' or 'biased', nor at defining such terms in relation to peer review. The aim is a more general understanding of the decision-making processes of grant review, and how organizational constraints influence review outcome. The constraints in focus are: review guidelines; rating scales; the review panels' ranking methods; disciplinary *versus* multi-disciplinary panels; mail reviews *versus* panel reviews; and budgets. The kind of 'bias' in focus is the effect on the overall direction of the outcome, in terms of the weight put on various research policy objectives: scholarly pluralism; innovative research; the strengthening of weak research fields; geographical distribution of funds; and priority to female applicants. These effects are analysed through the weight put on such concerns in review documents, panel discussions and panels' ranking lists.[6]

Data Sources and Methods

This study includes grant reviews for The Research Council of Norway (RCN) for 1997/98 in 10 different fields: economics; history; social anthropology; philology; interdisciplinary social science and humanities; clinical medicine; pre-clinical medicine; biology; environment and development research; and mathematics. Fields were selected to include all the different review models of RCN, and a broad variety of research fields.

Data sources are (619) applications and review documents, direct observation of the panel meetings, and interviews with (25) panel members. Fieldnotes were taken at the panel meetings, and interviews were

---

lished and been cited much more than male applicants in order to get the same rating on grant review. See also Abrams (1991) and Campanario (1993). One should also be aware of the shortcomings of citation analysis when studying biases of peer review, see Nissani (1995) and Luukkonen (1991).

[6] A more detailed account is available in Norwegian (Langfeldt 1998).

taped, transcribed and analysed with the help of the NUD*IST software for qualitative data analysis.[7] The interviews with panel members were semi-structured and dealt with the objects of distributing research grants, review criteria, the different points of view on the panel, the decision processes, the effects of panel discussions on the panel members' assessments, views on various models for grant review, the role of the research council staff, and the relations between the panel and the research council.

One selected panel refused to be observed and was substituted by another panel (the stated reason for refusal was that they did not want their meeting disturbed by an outsider). Panel members were given general written information about the object of the project (to study the policy implications of different models of grant review).

The review documents, the fieldnotes and the interview transcripts were analysed with regard to the emphases on different criteria for assessing applications, and the arguments used for the ranking of applications (all observation, interviewing, transcriptions, categorizing and analyses were done by the author). Both for the assessment criteria and the ranking arguments, the main categories used for the analysis included (with examples of criteria/arguments in brackets):

– the applicants' prior merits (publications, citations, originality, solidity, experience in the research field, position/reputation of the applicant/group or institution);
– the project descriptions (methods, clarity, originality, up-to-dateness);
– the expected value of the projects (scholarly/scientific value, expected use/applicability for specific audiences or in general);
– distributional policy (research field, institution, region, gender);
– research policy objectives (building up research competence within specific fields/'needs' of the fields, [large] multi-disciplinary projects, international collaboration, scholarly breadth/pluralism/diversity, national importance of the field);
– other considerations (budget and budget obligations, maximizing the panel budget, applicant's prior/other grants).

---

[7] 'NUD*IST' stands for 'Non-numerical Unstructured Data, Indexing, Searching and Theorizing'.

Analysis of the Grant-Review Practices of RCN

The Various Models

The Research Council of Norway (RCN) practises several different models of grant review, and is therefore especially suited for the study of implications of different models. The present Council is a merger of the five previous Norwegian research councils (one council for basic research and four sector councils). Today's models for reviewing grant applications are partly adopted from the old councils, partly results of reforms after the merger. The processes of allocating *general grants* ('responsive mode' funding) in four divisions were studied:[8]

- *Medicine and Health Division*: There were 4 medium-sized peer panels (10 members) that reviewed applications in their respective area. There was no mail review. Each of the panel members marked all applications on a fine-graded scale (1.0–4.0) and tables of these individual marks and average marks were set up before the panel meetings (available only to the chair of the panel). Panel decisions were based on discussion, average marks, and the chairman's discretion.
- *Culture and Society Division* (the social sciences and humanities): There were 15 small discipline-based peer panels (3–5 members) that reviewed applications. A review of each application, with marks on a 4-graded scale, was written by one of the panel members before the panel meeting. Advisory mail reviewers – selected by RCN-staff and panel chair in collaboration – might be used, but seldom were. Panel decisions were based on discussion, negotiation and/or majority rule.[9]
- *Science and Technology Division*: Here the administrative staff ranked the applications based on mail review reports. There were usually two

---

[8] General grants or responsive mode funding refers to the funding of independent researcher-initiated projects (contrary to time-limited grant programmes in restricted research/problem areas). In addition to the divisions studied, RCN consists of two sector-oriented divisions: the Industry and Energy Division, and the Bioproduction and Processing Division. These divisions are not included in this study, as they have no budgets for independent researcher-initiated projects.

[9] This model has recently been changed. From 2001, there will be 3 multi-disciplinary panels and, in addition, mail reviews on all applications.

reviewers per application, selected by RCN staff from a pool of (Scandinavian) reviewers within the field. A 5-graded scale was used, and there were extensive guidelines for reviewers. There were three multidisciplinary advisory panels that commented on the staff's ranking of the applications related to their area, but these panels had no concrete influence on the outcome. When ranking the applications, staff used the average marks from the (two) mail reviews and criteria for priorities given by the Division Board. When more fine-graded ranking was needed to reach a decision, staff had discretion to interpret reviews.

• *Environment and Development Division*: There was one medium-sized peer panel (9 members) with representatives of a broad scale of disciplines, which reviewed all applications for general grants. There was one advisory mail review per application, selected by RCN staff. The criteria for the selection of the mail reviewers were not specified, and were unclear to the panel members, as they were not involved in this process. The responsibility for each application was divided between the panel members. The panel members' assessments were presented orally at the panel meeting. Panel decisions were based on discussion, negotiation and/or majority rule. The mark set by the mail reviewer was altered by the panel for 51% of the applications.

In all divisions, the formal decisions on grants were taken by the Division Board.[10] The panels' ranking of the applications, and the RCN-staff's ranking in the case of Science and Technology, were advisory to the Board, but in reality these rankings were the final outcome. The influence of the Board was more indirect, by appointing the review panels, by allocating the budgets between panels, and by setting the review criteria and guidelines. It should be added that there are substantial differences in funding rates. In the studied units of review, from 12% to 51% of the applications were funded (see Table 13.1). Each panel is given a budget for new applications before their panel meeting. This budget might be

---

[10] Each Division Board has 7 members, mostly with a majority from research institutions. There is one proposal submission deadline each year, so that all proposals in a field are reviewed simultaneously. The exception is the Science & Technology Division, which accepts proposals twice a year for some of the grant categories.

Table 13.1 Overview of Studied Proposals and Grants (RCN 1997/98)

Subject/panel	Proposals (number)	Applied sums in relation to RCN budget (percent possible to cover)	Percent of proposals judged as 'clearly fundable'	Percent of proposals funded	Percent of female applicants funded
Economics	32	14.0	62.5	34.4	50.0
Anthropology	25	12.2	52.0	28.0	6.3
Interdisciplinary social sciences & humanities	23	9.3	39.1	17.4	0.0
Philology	50	9.3	54.0	14.0	6.5
History	52	6.4	48.1	11.5	10.0
Pre-clinical medicine	101	8.7	24.8	30.7	34.6
Clinical medicine	86	16.9	3.5	26.7	24.0
Environment and development	122	18.3	46.7	34.5	33.3
Biology	91	25.5	67.0	37.4	33.3
Mathematics	37	39.8	86.5	51.4	33.3
Average	62	15.4	43.9	29.7	21.8

Note: The table gives an overview of studied cases along some quantifiable dimensions. The data on which conclusions are drawn, are illustrated in the text, not in this table.

Projects including both sexes are not included in the calculations

adjusted by the Division Board as it judges the lists from the various panels, or it might be reduced or increased for *all* panels as, at the time of the panel meetings, the budget proposal has not yet been accepted by the Parliament.

The main criteria for selecting the members of the review panels were research competence and coverage of the panel's field, and a fair representation of regions and gender. Rules for handling conflicts of interest were common for all divisions: in the case of any affiliation to an application (for example, an applicant from the department of one of the panel members), the involved panel member leaves the meeting during the discussion and the ranking of the application. The formal written criteria for review varied between the divisions. Translated quotes from the guidelines are given in the Appendix (below: pp. 839–41), to illustrate the differences. In addition to these review criteria, the guidelines for review included policy directions (see the section on 'Considerations Other than Research Quality in the Assessment', below: pp. 828–31).

Overview of Proposals and Grants

General grants (responsive mode funding) at RCN are organized into 30 review units/fields. This study encompasses 10 of these units of review. Table 13.1 shows the amount of proposals, budget restrictions, grading, proportion of successful proposals, and proportion of successful female applicants for each of the studied units.

As illustrated in the first column of Table 13.1, there was a large variation in *number of proposals* between the review units. The panel for interdisciplinary social science and humanities reviewed 23 proposals for the studied year, whereas the panel for environment and development research reviewed 122 applications. Comparing the columns '*applied sums in relation to RCN budget*' and '*% of proposals funded*', we see that the latter is substantially higher than the former. This is because project budgets are cut before funding, and give room for funding a larger number of proposals.

The column '*% of proposals judged as "clearly fundable"*' tells us more about the different rating scales and directions for rating, than about differences in the quality of proposals. Within the social sciences and humanities, there was a four-graded scale of which the best mark was 'clearly fundable'. Within the sciences and within environment and development research, there was a 5-graded scale of which the *two* best marks were 'clearly fundable'. Within medicine, there was a fine-graded scale (1.0–4.0) of which 1.0–1.9 was 'clearly fundable'. Medicine differed from the other divisions in that there were no restrictions against funding a proposal that was not marked 'clearly fundable' (for instance, a proposal getting 2.2 might get funds). The result of the demand for the mark 'clearly fundable' to be funded in social science and humanities was little differentiation in marking: a large proportion of the proposals get the best mark (which was needed to be part of the priority discussion of the panels). In conclusion, the percent of proposals judged as 'clearly fundable' in the different review units does not tell us much about differences in quality of the proposals. For instance, the low percentages of clearly fundable proposals in medicine reflect other directions for review, and more differentiated rating.

The column '*% of female applicants funded*' shows that proposals from female applicants had a somewhat lowwer chance of getting funded than have the proposals in general. This is most evident in anthropology, interdisciplinary social science & humanities, philology, and mathematics. It should be added that numbers of female applicants varied and were particularly small in economics (4), interdisciplinary social science/humanities (4), and mathematics (3). As the number of funded proposals also is small, statistics for several years are needed to draw any conclusions about discrimination of female applicants. Analysis of prior funding (obligations for the particular year) showed that anthropology, interdisciplinary social science/humanities and philology had a *majority* of grants to female applicants.

The Quality Criteria in the Assessments

There are notable differences between the RCN-divisions regarding the content of the studied reviews. With one important exception, the clearest differences in the weight given to the various quality criteria do *not* seem related to the differences in directions and lists of criteria given to the reviewers. At the same time there are substantial differences between the review units *within* the divisions – which operate under the same directions – further substantiating that directions to reviewers are of limited importance. Examples illustrating the assessments (of highly-ranked applications), and the differences within the divisions, are given in Table 13.2.[11]

The most striking differences (see Table 13.2) between the reviews of the different RCN-divisions regard the weight given to criteria related to

---

[11] All the quotations here are translated by the author.

Table 13.2 Review criteria

RCN division	Illustration of typical criteria (from the panel members written statements and the mail reviews)	Differences between review units within the division
The Medicine and Health Division	'The group has an impressive productivity with many articles in highly ranked journals.' 'The group and the PI are well qualified for the project and have a rich production of scientific publications and PhD's the later years.' 'The described methods seems to be the best' 'The group has developed unique methods'	More weight on applicants' prior merits in the pre-clinical panel than in the clinical panel, whereas the clinical panel was more concerned about weaknesses in methods and design.
The Culture and Society Division	'The project is scholarly interesting and gives an impression of thoroughness.' 'The application is based on a thorough discussion of scholarly perspectives and shows their relevance for the planned analysis.'	Less weight on the project description and more weight on the research milieu in the economy panel than in the other panels.
The Science and Technology Division	'The PI is an outstanding scientist' 'The applicant publishes in high ranking biological journals' 'The project is presented with a convincing quality' 'This is a central research field within mathematics'	More detailed assessments of the project descriptions within biology than within mathematics. Somewhat more weight on the scholarly importance of the field within mathematics than within biology.
The Environment and Development Division	'The project has high scientific relevance and practical-political importance.' 'The PI has unquestionable competence' 'The applicant is updated on the literature and the theoretical debates'	Only one review unit.

the project description within the Social Sciences and Humanities, the weight given to criteria related to the prior merits of the applicant (and his/her group) within the Medicine and Health Division, and the special emphasis put on the centrality and the importance of the research field in the reviews within the Science and Technology Division.[12] The reviews in the Social Sciences and Humanities were more concerned about criteria related to the project description than those in the other divisions. There is nothing in the guidelines for reviews indicating such different emphases (see the Appendix). On the contrary, if any of the divisions' guidelines may be said to put more emphases on criteria related to the project description, this would be the Medicine and Health Division, which put these criteria first and clearly set up more detailed criteria than did the Culture and Society Division. The reviews in the medical sciences, however, put *less* weight on the project description and more weight on prior research merits and reputation than did the reviews in the other divisions.

Reviews within Mathematics and Biology were more concerned about the centrality and the importance of the research field than were reviews in other divisions. This special emphasis can be related to a specification which is found only in the guidelines of the Science and Technology Division:

> *What importance does research in this area have for the development of the field (the area of research in question may no longer be of importance or, at the other end of the scale, be new and rapidly expanding)?*

The first part of this question is found both in the Science and Technology guidelines and in the Environment and Development guidelines, whereas the specification in parentheses is only found in the Science and Technology guidelines. In this case then, differences (even in brackets) in guidelines had a clear impact on the content of reviews.

As shown in Table 13.2, there are also differences within divisions. Review units that operated under the same guidelines differed clearly with regard to emphases on the various quality criteria. Within Medicine

---

[12] The weight on the various criteria is studied by an overall analysis including fieldnotes from the panel discussions and the interviews with panel members, in addition to the use – and context for use – of the criteria within the review documents.

and Health, more weight was placed on applicants' prior merits in the pre-clinical panel than in the clinical panel. Within Culture and Society, there was less weight on the project description in the economics panel than in the other panels. Within Science and Technology, there were more detailed assessments of the project descriptions within biology than within mathematics.

In conclusion, weight on quality criteria differed both between and within the RCN-divisions, regardless of guidelines.

Considerations Other than Research Quality in the Assessments

When applications were given equal rating on scholarly quality, considerations related to policy objectives and distributional policy affected the ranking. As with the quality criteria, these considerations differed both within and between the divisions.

The Divisions' Boards, to varying degrees, gave directions on distributional policy and research policy objectives to be included in the review:

- The Board of the *Medicine and Health Division* gave a limited set of policy directions. These concerned the Board's view on the priority on gender and two of the funding modes: 'applications for postdocs are to be given priority', 'the number of female postdocs should be increased', and 'senior fellowships have low priority within the Medicine and Health Division'. In addition, research needs and priorities had their separate points in the review form: 'Field with special national needs for new knowledge; Field with special national conditions for doing research; Field with special national needs for building up new competence; Field with good alternative funding sources'.[13]
- The guidelines within the *Culture and Society Division* summarized the policy priorities in this way: 'Support research recruits in fields with a

---

[13] After finishing the review of applications, the panels were asked to produce a document commenting on special needs in the fields. This document might of course be used when reviewing next year's applications, but was foremost a document giving policy input to the Board – describing the situation in the research fields, arguing for more money, and the like.

need for recruits; Support projects that lead to scholarly innovation of research fields and groups; Support projects with a potential for internationalization; Support projects that strengthen the recruitment of female researchers in fields with a low percentage of females, and projects that may lead to more females obtaining tenure'. They were also told to consider the needs for geographical distribution, for increased activity in particular fields, and for scholarly breadth *versus* depth. In sum, the guidelines allowed special arguments for most kinds of applications. It was still emphasized that most weight should be put on the need for research recruits.

- In the *Science and Technology Division,* the administrative staff ranked applications based on mail reviews and the Division Board's policy priorities. The Board's priorities included internationalization, research recruits, the national importance of the research fields, scholarly diversity, as well as distribution on gender and institutions, and the applicants' prior/other resources.[14]
- The *Environment and Development* panel was given no policy directions. However, the Division Board had more direct budget control over priorities, as it divided the budget into separate sums for fellowships for research recruits, for larger group projects, and for ordinary projects.

The considerations *taken* were only partly related to these directions. An example of implemented directions is the need for research recruits, which was given high priority within most units of review. The priority given to female applicants is an example of a priority *not related to the variation in directions.* Within the two divisions with general directions for priority to female applicants (the Culture and Society Division and the Science and Technology Division), explicit priority to female applicants was rare: the sample included priority to one female researcher within mathematics and a general priority to female applicants within economics. Except for the economics panel, the studied panels within the Culture and Society Division saw no need to be concerned about the

---

[14] These are the priorities listed by the RCN-staff in a document informing the advisory committee about the priorities to be taken.

gender of applicants (see the section on *Overview of Proposals and Grants*, above). With regard to mathematics, there were disagreements on the topic. In the staff's ranking of applications, explicit priority was given to one of the three female applicants within mathematics. The advisory panel that commented on the ranking of proposals in mathematics wanted to give priority also to another female applicant, who was number one among those not recommended for funding. This was argued against in the administration's recommendations, which was accepted by the Division Board:

> *The opinion of the administration is that an individual fellowship is a funding mode that should primarily be assessed by the quality criterion.... Based on this criterion. . . the administration maintains its initial recommendation.*

When contrasting this limited concern with the funding of female applicants resulting from general directions to give priority to female applicants with what happened within divisions that did *not* give any general directions for priority to female applicants, we find that the situation is much the same: within Medicine, there are cases of priority to females within *all* kinds of applications, not only to the post-docs specified in the directions; within Environment and Development, the panel gave explicit priority to a female applicant, although the panel was given no directions about such priorities.

Priorities related to geographical distribution also differed, regardless of directions. Such priority was given within Medicine, where there were no such directions, but not within Culture and Society, that had such directions. Priorities other than those specified in the directions also include distribution over different research fields within Environment and Development, and priority to small/weak research fields and considerations of needs/quality in terms of prior funding within Medicine. For instance, research field was a major concern when the Environment and Development panel ranked applications for fellowships, and the five candidates ranked above the cut-off line were all from different disciplines: sociology, history, biology, geography and agriculture.

We will now turn to a discussion of the *conditions* for taking considerations other than research quality into account.

The Effects of the Various Ranking Processes

The rough-rating scales and open decision-making processes within the Culture and Society Division and the Environment and Development Division gave ample room for research policy considerations, such as scholarly pluralism and support to innovative projects. In these two divisions, the clearest examples of processes in favour of support to innovative projects were observed: enthusiastic panel members managed to change the panels' views on projects that first were seen as too risky, peripheral or immature. In one case (in one of the Culture and Society panels), the chair and a new panel member had divergent opinions about an application. On the first day of the meeting, when all applications were graded, the chair said it should have the next best mark, which meant no grant. The new member wanted to give it the best mark, which meant it might get a grant (as it would be included in the ranking process the following day):

> *Chair: The applicant does not substantiate the reasons for doing what she wants to do, and the reasons are not evident.*
>
> *New member: I disagree with your reading of the project description. The research of [name of another researcher] seems to show that this is well substantiated. I see this as a springy project [potential for jumping high or long].*

The new member was supported by another panel member, and the application ended up with the best mark. In the ranking process the following day, the new member stated what should be the three top-ranked applications from the point of view of 'springiness', and managed to get the disputed application placed among the three at the top, and consequently granted (his three proposed top candidates ended up as the three top candidates).

Medicine and Health's ranking processes, on the other hand – with a larger number of panels members all giving marks to all applications individually, a fine-graded rating scale and decisions based on the average of the individual marks – promoted more thoroughness and predictability. The room for explicit considerations other than quality review was rather modest compared to the other divisions, though larger budgets per panel secured the possibility of funding a broad spectrum of research. The decision-making processes on the two observed medical panels differed: one panel put more weight on average marks, whereas the chair's discretion was more central in the other panel. In the panel depending most heavily on average marks, it was left to the individual panel member to adjust his/her marks after the panel discussion (and a new average was calculated if someone did). This process was more conservative with regard to letting the discussion influence the outcome than the processes on the other panel. This discussion of an application was typical:

> *Chair: The average mark is 2.3, the variance is generally low. [Panel member 1] gives it 2.9?*
> *Panel member 1: The applicant has little experience. . . .*
> *Chair: [Lists applicant's prior merits]. I think this is a fundable project.*
> *Panel member 2: It is somewhat incomplete.*
> *Panel member 3: I don't think so. . . .*
> *Panel member 2: The design is not stringent, it is somewhat imprecise. . . .*
> *Panel member 1: Yes, the planning is incomplete.*
> *Chair: . . . No one changes his mark, so 2.3 is upheld.*

On the other panel, the chair had the power to adjust marks after panel discussion (though in some cases the chair had to remake his decisions as panel members objected), and he also initiated proposals to move applicants up and down the 'final' list, so as more effectively to include various policy objectives. The example below illustrates this rôle of the chair (the first panel member was primarily responsible for the review):

> *Panel member 1: The applicant is well qualified. The project has some formal shortages, but it is important to establish this kind of project in Norway. I say 2.1.*
> *Panel member 2: I gave 1.9. There are some clear frustrations in the application about the conditions for doing research. They have got unique competence and from a strategic point of view this has top priority. It is a challenging project. . . The ambitions may be too high.*
> *Chair: The applicant is perfectionist and might manage. Do we say 2.1? The average was 2.2.*
> *Panel member 3: At the minimum.*
> *Chair: Then we say 2.0.*

The processes in this panel gave considerably more leeway for policy and distributional concerns than those of the former panel, which relied more on the average marks. In the former panel, the initial average mark was changed for 34% of the applications, whereas in the panel where the chair used his discretion to adjust marks, the initial average mark was changed for 48% of the applications.

In the Science and Technology Division there were only two reviewers per proposal, and no panel involved in the comparisons and ranking of the proposals.[15] In general, the RCN-staff that ranked the proposals seemed more concerned about getting the ranking 'right' with regard to the scientific merit of the application, than the panels were (see, for example, the panel's concern with gender priorities in mathematics, discussed above). When more applications had the same average mark, the staff looked to the content of the reviews and the graduation marks of research recruits, whereas review panels were more concerned about distributional policy.

Only 30% of the studied applications within Science and Technology got the same mark (on the scale from 1 to 5) from the two mail reviewers. This means that outcomes may depend heavily on which reviewers are picked for the particular proposal (randomness). The individual reviewer had a significant role in determining the outcome for the single applications. In this way, the composition of the pool of reviewers may be important for the scholarly pluralism in the overall outcome of review. The large number of reviewers in the pool, and the disagreements between the reviewers, indicate some scholarly pluralism in the pool, and should also give leeway for scholarly pluralism in the final outcome. On the other hand, few reviewers per application and no scholars to compare and rank the whole portfolio of applications give ample room for randomness – which may or may not be moderated by the competence, discretion and

---

[15] At the time of the study, there were three advisory committees that commented on the ranking without seeing applications or reviews, and they had no concrete influence on the ranking. These committees were later abolished, and one advisory panel for all applications for responsive mode funding within the Division was set up. For this panel, applications and reviews are available during the panel meeting. The new panel's influence on the outcome has not been studied. It is still the administration that ranks applications and makes the recommendations for funding that are finally approved by the Division Board. There are now 3 mail reviewers for some categories of applications.

guidelines of the RCN staff who rank the proposals. Among the interviewed members of the advising panels within the division, there were some doubts about the system. The most critical expressed it like this:

> *Now it is one staff-member that does it all. I am convinced that we [the advising panel] are more able to do it.... We cannot have a system that relies on the [one] staff-member being good.*

In conclusion, the models studied support different outcome profiles: leeway for scholarly pluralism and innovative/risky projects within the Culture and Society Division and the Environment and Development Division; thoroughness and predictability – and consequently more conservative assessments – within the Medicine and Health Division; and randomness and possibly scholarly pluralism within the Science and Technology Division. The effects of some of the organizational factors are further explained in the next section.

Important Organizational Factors Affecting Ranking

Distributional policy and research policy objectives may or may not be decisive for the outcome of grant review. A central finding is that the budget available and the rating scale both affect the degree to which such considerations are taken. Ample budgets and a rough-rating scale give much more room for policy priorities than do tight budgets and fine-rating scales. When there are funds only for a few highly-selected projects, the wide range of policy objectives that are given in (some of the RCN divisions') guidelines are impossible to fulfil. The panels do not want policy priorities to overrule research quality assessments, and instructions to include policy concerns have little to say in such a context. There is simply 'no room' for distributional policy. With ample budgets, on the other hand, there is room for funding more than a small number of 'obviously best' applications and, with a rough-rating scale for research quality, the panel ends up with several applications with identical marks. In such a situation, the panels seemed glad to have a set of policy priorities to help reaching decisions: criteria to rank the group of identically-rated possibly-funded applications. The existence of a set of identically-rated possibly-funded proposals seemed a central condition for giving priority to research with special needs (strengthen weak fields), and for taking the distribution on various subfields into consideration (pluralism). Also, original/innovative projects that do not easily compete with projects formulated within established traditions and methods, have better chances for funding when there is room for more than the proposals getting top rating. Tight budgets and fine-rating scales, on the other hand, tend to strengthen established research and give less pluralism in funded research. The panel discussion of an unorthodox application (that did not receive any grant) within a field with high rejection rates illustrates that budgets may also be a direct argument against risk projects:

> Panel member 1: I doubt this project, I don't think it will succeed.
> Panel member 2: It has got charm, it tries to do away with the force of gravity!
> Chair: With a better budget we could have taken the chance on a wild card a year.

In addition to budgets and rating scales, *the decision-making process itself* is found to be important for the scholarly pluralism in funded research. The ranking of applications depends heavily on the method applied. Methods implying that all panel members get their favourite candidate funded secure pluralism far better than methods eliminating proposals to which a majority of panel members do not give priority (given some scholarly pluralism on the panel). On the other hand, the first set of methods let single panel members decide outcomes, and are thus open to accidental circumstances.

Table 13.3 illustrates a situation where three different methods of ranking give very different outcomes. With method 1, each panel member has as many votes as there are applications to be funded. With method 2, each panel member has one decisive vote. With method 3, applications not to be funded are excluded by majority votes. The table illustrates the logic of methods that were observed at the panel meetings when ranking applications with identical marks. In the meetings, the decision-making processes were far from as simple, explicit and structured as in Table 13.3.

Table 13.3 The results of different methods for the ranking of applications a-f, of which there is room for three within the budget

| Ranking | Panel members' (X, Y and Z) rankings | | | The result of | | |
	X	Y	Z	Method 1	Method 2	Method 3
1	a	b	f	b	abf	bde
2	b	d	b	e		
3	c	e	e	a		
4	d	a	d	fdc	cde	a
5	e	f	c			c
6	f	c	a			f

Table 13.4 Combination of elements of methods 2 and 3, and other methods, in one of the Culture and Society Panels

Eight applications for fellowships had received the best mark. Three of these were to be granted. One of the four panel members proposed they should first eliminate the weakest applications. Three candidates for elimination were proposed, discussed and eliminated seemingly by consensus. Five applications of which three would be granted remained. The chair asked for the 'first choice' of the panel members. Because of conflicts of interest, one of the five applications was excluded from the first phase of the following discussions (and ranked as number 5 when the involved panel member later on left the room). The following discussion included the pros and cons of the various candidates, the needs of different fields and institutions, solidity versus originality, and gender. Not all panel members stated any explicit opinion about what was their first choice. The two candidates that had been explicitly stated as someone's first choice were given the two first places. The third place was decided by a vote (of 3 against 1) between the two remaining candidates.

An example from the meetings is given in Table 13.4 – an example in which elements from various methods were combined.

The situation in Table 13.3 is simplified to include only three panel members and six applications, of which only three may be funded. Only application 'b' is funded with all three methods of ranking:

- *Method 1*: All members propose 3 candidates for funding; that is, they have 3 votes each. Application b gets 3 votes and first rank; e gets 2 votes and second rank; a, c, d and f get 1 vote each. Third rank is left to member X's favourite a, because this member has only got in one of his 'votes' for rank 1 and 2, while the two other members have got two of their 'votes' among the first two on the funding list.

- *Method 2*: All members propose 1 candidate for funding; that is, they have 1 vote each. The three applications that receive one vote are funded. These are a, b and f.
- *Method 3*: Elimination. Candidates to be eliminated from funding are voted on in the order they are proposed. The outcome depends on the chosen voting order. Here the supposed voting order is f, c, a. These three are all eliminated with votes two against one, and the three applications remaining for funding are b, d and e.

A likely reaction when seeing the methods spelled out and analysed as in Table 13.3 is that serious funding agencies should and would never allow such arbitrary outcomes, and that the RCN findings must be exaggerated or exceptional. It should be noted that the table explains the underlying logic of methods that were adopted *ad hoc* by the observed panels. The methods had no formal status, and no stated rationale. The panels had to make a decision in one way or another, and found a way to do it. There are also more general arguments for the extent of the problem. Social choice theory has shown that there are fundamental problems in the aggregation of preferences. Voting methods *do* affect outcomes, and the choice between methods is problematic as there are no simple clues as to which method is the better (from the point of view of a set of fundamental requirements).[16]

The implications of the method chosen may be substantial. Given that the panel members' different rankings (at least partly) represent conflicting scholarly norms and interests, method 2 gives funding to projects scoring very differently in this regard. The favourite candidates of member X and Z are funded, although these applications are at the bottom of other panel members' lists. If agreement on high ranking indicates that the projects are uncontroversial regarding research questions, scientific methods, and so on, while disagreement indicates controversial research and risk-projects, method 2 may fund controversial research and

---

[16] See, for example, McLean (1987), which discusses various voting systems, and also presents a proof of Arrow's Theorem, the classic mathematical presentation of this dilemma.

risk-projects, while method 3 tends to fund uncontroversial and safe projects. In this way, the panels' choice of ranking methods has far-reaching implications on the chances for various kinds of research to be funded.

Conclusions

The most unambiguous conclusions to be drawn from this study regard the effects of guidelines, rating scales, budgets and ranking methods.

The guidelines given to the panels had little effect on the criteria they emphasized,[17] whereas mail reviewers were more consciously attempting to write reviews in accordance with the guidelines. Put more clearly, it seems that panels do as they like, whereas mail reviewers do as they are told – or, at least, mail reviewers phrase their reviews more in accordance with the guidelines, to make sure they have influence on the ranking of proposals. The criteria emphasized in the review documents, in the panel discussions and by the interviewed panel members were studied. The panel reviews within *medical science* were those most focused on applicants' prior merits, whereas the panels within the *social sciences* and the *humanities* reviews were more focused on criteria related to the project description. The focuses of these panels ought to be the *opposite* of these, according to their guidelines. The mail reviewers within the *sciences*, on the other hand, wrote reviews more in accordance with their guidelines – with specific focus on the centrality and importance of the research field (for example, stating that this is outdated research, or that this research will have central future importance), which was a specific concern of the guidelines within this RCN division.

The differences in reviews found *within* the divisions (that is, differences within units given the same guidelines) underline the limited effects of guidelines, both for panel reviews and for mail reviews. Furthermore, geographical priority, priority to female applicants and several other policy concerns were taken by panels contrary to their guidelines. The guidelines of some divisions told panels to include such concerns, but the

---

[17] But, as explained, guidelines seem to have had substantial effect on how grading scales were used (see the explanations of Table 13.1, above).

panels did not include them; whereas other divisions did not give any instructions on such concerns, but the panels did include them.

Whereas the guidelines, which are supposed to influence reviews, did so to a limited extent, factors that are *not* supposed to influence outcomes were found to be much more important. The size of the budgets and the kind of rating scale applied were found to affect such policy concerns as the funding of fields with special needs, the distribution on the various sub-fields, priority to female applicants, and geographical priority. Also, original and controversial projects seemed to have better chances with ample budgets and rough-rating scales. With a *rough-rating scale* for research quality, the panel ends up with several applications with identical marks, and with *a good budget*, the panel may fund more than a small number of 'obviously best' applications. Such a situation, with identically-rated projects that *may* get funds, was a central condition for peer panels to rank applications on the basis of policy objectives. With the opposite kind of situation, with funds only for a few highly-selected projects that were ranked on the basis of a fine-rating scale, there was no room for supplementary policy priorities. The members of the review panels did not want policy priorities to overrule the research quality assessments. Tight budgets and fine-rating scales may therefore easily strengthen established research fields, and give less pluralism in funded research.

The ranking method applied by the panel may be decisive for the outcome of review. Some methods imply that all panel members get their favourite candidate funded. If the panel members' different ratings represent conflicting scholarly norms and interests, and there is some scholarly pluralism on the panel, such methods ensure some scholarly pluralism. This kind of method also involves more randomness as, in reality, *single* panel members decide outcomes. On the other hand, there are methods that *eliminate* proposals to which a majority of the panel members do not give priority. It is argued that these methods tend to support uncontroversial and safe projects, as agreement on high ranking of a proposal indicates that the projects are uncontroversial regarding research questions, scientific methods, and the like, while disagreement indicates controversial research and risk-projects. In this way, the panel's choice of ranking method may have far-reaching implications on the chances for various kinds of research to be funded.

In sum, the organization of grant review is found to influence what counts as a good and relevant grant application: (1) guidelines seem to have limited effects on review panels; but (2) the review outcome is found to be highly dependent on (a) rating methods, (b) rating scales and (c) budgets. Each of these factors may have far-reaching implications on the funding chances of applications for different kinds of research. Various organizational factors also interact and reinforce each other. As candidates for further study on how the organization of grant review affects outcomes, two hypotheses – including a broad set of organizational factors – should, on the basis of the present study, be emphasized:

- Ample budgets, rough-rating scales, heterogeneous panels and open decision-making processes give leeway for scholarly pluralism and innovative/risky projects.
- Tight budgets, fine-rating scales, average marks and majority decisions tend to strengthen established research and give less pluralism.

Another finding is that administrative ranking gave less room for policy concerns than panel ranking, as administrative staff were more concerned about ranking on the basis of scientific merit. There is no obvious reason for this effect of administrative ranking, and further study may show that the effect of administrative ranking depends on the kind of administrative staff involved, and what instructions they are given (see Rip 1994, pp. 16-17).

It should be noted that there is an inherent tension between the different aims of research councils: good and reliable peer review on the one hand, and various policy aims on the other. Those review models that score highly on thoroughness and reliability do not score highly with regard to encouraging controversial projects or securing greater scholarly pluralism, and vice versa – leaving those trying to improve grant-review processes in a constant dilemma. Consequently, unambiguous recommendations for the design of grant review cannot be made, except that conflicting concerns should be balanced consciously.

At the beginning of this paper, two different views on peer review bias were presented, views on whether 'school of thought' is (or is not) a legitimate consideration when assessing research. Regardless of one's standing on this question, one might embrace the intuitive supposition that processes properly aggregating marks on the scientific merit of the applications yield less bias than a process of discussions and negotiations.[18] The findings of this study might be used to challenge that view. Let us say we accept that the possibilities of ranking applications uniquely on the basis of neutral universalistic criteria of scientific merit are limited (which must be said to be one central implication of decades of studies of peer review), adopting the view that 'school of thought' is an inherent and legitimate basis of peer review, and that reviewers are central actors in the definition and redefinition of 'good research'. This position opens up the view that peer-panel discussions and negotiations involving policy objectives may be a better way to avoid biases, than processes simply aggregating marks on scientific merit. Review panels that are concerned about the distribution of funds for research directions/'schools of thought', gender, institutions, position, and so on, may reduce such biases. The present study shows that peer panels are willing to take up such concerns, and in some cases do so more than do their administrative staff. However, this willingness is restricted by budgets and indirectly by rating scales. Such seemingly irrelevant factors affect the review outcome, as they decide whether panels only deal with 'scientific merit', or also include discussions and negotiations on distributional policy and other research policy concerns.

Appendix: The Review Criteria in the Different RCN Divisions

The RCN Divisions' guidelines listed the following review criteria:

Medicine and Health Division:

- The project: Updated within the field, nationally and internationally; Whether the project will produce important new knowledge; Whether the research questions are clearly presented; Whether the methods are suited; Whether the objectives are clearly presented; Plans for dissemi-

---

[18] For an example of this approach, see NIH 1996.

nation and contact with users; Whether the project is part of national or international collaboration.

- The principal investigator or person applying for fellowship: Education; scientific qualifications.
- The research *milieu*: Whether the *milieu* takes part in national/international collaboration and/or multidisciplinary/ crossdisciplinary collaboration; Student flow.

Culture and Society Division:

- The principal investigator or person applying for fellowship: Education; scientific qualifications.
- The research *milieu*: The reputation of the research *milieu*; The *milieu's* circle of scholarly acquaintances; Whether the *milieu* has sufficient resources to accomplish the project; Whether the project fits the scholarly activity of the *milieu*; Student flow; Publication rates of the *milieu*.
- The project: Updated within the field, nationally and internationally; The research question and its delimitations; Theory and methods of the project; Whether necessary ethical concerns are taken care of; Plans for dissemination and contact with users; Budget and progress plans; The objectives of the research.

Science and Technology Division:

- The applicant (individual applicant or person applying for a group): What is the applicant's contribution to the field in recent years? Does the applicant have sufficient knowledge, experience or potential to contribute significantly to this field? Has the group where the applicant works the necessary resources to carry out the project? The ability and resources of the applicant may be compared to other recognized research groups in the field.
- The scientific quality and importance of the project: How does the application score in relation to well-defined goals, the methods and theory, scientific originality? Can the project be done within the stated time-span? Is it probable that the project will lead to new knowledge or significant progress in the field? What importance does research in

this area have for the development of the field (the area of research in question may no longer be of importance or, at the other end of the scale, be new and rapidly expanding)?

- Usefulness and contribution to the field and the research community: Will the project recruit researchers, in terms of PhDs, postdoctoral fellows, or in other ways strengthen the national research infrastructure? May the project contribute to the internationalization of Norwegian research in the field, and if so, in what way? To what degree may it be expected that this research will have bearing outside its field, for example on the development of new technology or knowledge that may contribute to the solution of important societal problems?

Environment and Development Division:

- The applicant/research *milieu*: What is the applicant's contribution to the field in recent years? Does the applicant have sufficient knowledge, experience or potential to contribute significantly to this research (education, scientific qualifications)? Has the group where the applicant works the necessary strength (e.g. student flow, publication rate) and resources and/or infrastructure to carry out the project?
- The scientific quality and importance of the project: How does the application score in relation to well-defined goals, the methods and theory, scientific originality? Can the project be done within the stated time-span and budget? Is it probable that the project will lead to new knowledge or significant progress in the field? What importance does research in this area have for the development of the field? Are necessary ethical concerns taken care of? Plans for dissemination and contact with users? Also see the demands of the project description that are expressed in the application form.
- Usefulness and contribution to the field and the research community: Will the project recruit researchers, in terms of PhDs, postdoctoral fellows, or in other ways strengthen the national research infrastructure? May the project contribute to the internationalization of Norwegian research in the field, and if so, in what way? To what degree may it be expected that this research will have bearing outside its field, for example on the development of new technology or knowledge that may contribute to the solution of important societal problems?

References

Abrams, Peter A. (1991). The Predictive Ability of Peer Review of Grant Proposals: The Case of Ecology and the US National Science Foundation. *Social Studies of Science,* Vol. 21(1), 111–32.

Campanario, Juan Miguel (1998a). Peer Review for Journals as It Stands Today, Part 1, *Science Communication,* Vol. 19, 181–211.

Campanario, Juan Miguel (1998b). 'Peer Review for Journals as It Stands Today' Part 2, *Science Communication,* Vol. 19, 277–306.

Chubin, Daryl E. & Hackett, Edward J. (1990). *Peerless Science.* New York: State University of New York Press.

Chase, Janet M. (1970). Normative Criteria for Scientific Publication, *The American Sociologist,* Vol. 5, 262–65.

Cicchetti, Domenic V. (1991). The Reliability of Peer Review for Manuscript and Grant Submissions: A Cross-Disciplinary Investigation, *The Behavioral and Brain Sciences,* Vol. 14, 119–86.

Cole, Stephen; Cole, Jonathan R. & Simon, Gary A. (1981). 'Chance and Consensus in Peer Review', *Science,* Vol. 214(20 November), 881–86.

Campanario, Juan Miguel (1993). Consolation for the Scientist: Sometimes it is Hard to Publish Papers that are Later Highly-Cited. *Social Studies of Science,* Vol. 23(2), 342–62.

Daniel, Hans-Dieter (1993). Guardians of Science: Fairness and Reliability of Peer Review. Weinheim, Germany: VCH Verlagsgesellschaft.

Dirk, Lynn (1999). A Measure of Originality: The Elements of Science, *Social Studies of Science,* Vol. 29(5), 765–76.

Gulbrandsen, Magnus & Langfeldt, Liv (1997). *Hva er forskningskvalitet? En intervjustudie blant norske forskere.* Oslo: Norwegian Institute for Studies in Research and Higher Education, Report No. 9/97.

Harnad, Steven (1985). Rational Disagreement in Peer Review, *Science, Technology, & Human Values,* Vol. 10(3), 55–62.

Hemlin, Sven (1993). Scientific Quality in the Eyes of the Scientist: A Questionnaire Study, *Scientometrics,* Vol. 27, 3–18.

Langfeldt, Liv (1998). *Fagfellevurdering som forskningspolitisk virkemiddel. En studie av fordelingen av frie midler i Norges Forskningsråd.* Oslo: Norwegian Institute for Studies in Research and Higher Education, Report No. 12/98.

Luukkonen, Terttu (1991). Citation Indicators and Peer Review: Their Time-Scales, Criteria, and Biases. *Research Evaluation,* Vol. 1, 21–31.

Mahoney, Michael J. (1977). Publication Prejudices: An Experimental Study of Confirmatory Bias in the Peer Review System, *Cognitive Therapy and Research*, Vol. 1, 161–75.

McLean, Iain (1987). *Public Choice*. Oxford: Basil Blackwell.

NIH (1996). *Report of the Committee on Rating Grant Applications*. Washington, DC: National Institutes of Health, Office of Extramural Research.

Nissani, Moti (1995). The Plight of the Obscure Innovator in Science: A Few Reflections on Campanario's Note. *Social Studies of Science*, Vol. 25(1), 165–83.

Ravetz, Jerome R. (1971). *Scientific Knowledge and its Social Problems*. Oxford: Clarendon Press.

Rip, Arie (1994). The Republic of Science in the 1990s, *Higher Education*, Vol. 28, 3–23.

Travis, G.D.L. and Collins, H.M. (1991). New Light on Old Boys: Cognitive and Institutional Particularism in the Peer Review System, *Science, Technology, & Human Values*, Vol. 16(3), 322–41.

Wennerås, Christine & Wold, Agnes (1997). Nepotism and Sexism in Peer-Review. *Nature*, Vol. 387(22 May), 341–43.

Wood, Fiona Quality (1997). *The Peer Review Process*. Canberra: Australian Research Council, National Board of Employment, Education and Training, Commissioned Report No. 54.

14

Typecasting in the Recruitment of Full Professors

Sara Levander, Eva Forsberg, Sverker Lindblad, and Gustaf J. Bjurhammer

Introduction

The recruitment of full professors is critical for the formation of academia. Usually, it is closely connected to the fulfillments of the universities' core missions, and thus, has long-term effects on the future academic profile of departments and higher education institutions (HEI). The latter holds true both in terms of the profile of the professorship per se, and insofar as professors commonly are involved in the strategic recruitment work and the recruitment processes of other academic staff members at the department (Klawitter, 2015). Thus, the recruitment of professors impacts not only the HEI and department, but also individual academic's careers. Hence, it is a political and highly symbolic act (Musselin, 2010).

S. Levander (✉) • E. Forsberg • G. J. Bjurhammer
Department of Education, Uppsala University, Uppsala, Sweden
e-mail: sara.levander@edu.uu.se

S. Lindblad
University of Gothenburg, Gothenburg, Sweden

© The Author(s) 2022
E. Forsberg et al. (eds.), *Peer review in an Era of Evaluation*,
https://doi.org/10.1007/978-3-030-75263-7_14

327

However, the professorship is paramount not only for the prosperity of the HEIs, but especially so for the establishment, development and communication of the discipline. In a Humboldtian sense the professorship unites the two functions research and education, and is therefore a fundamental part of the disciplinary building (Hofstetter & Schneuwly, 2002). In fact, as part of a discipline's self-perception the professorship may be considered an integrated part of the discipline itself (Lindberg, 2006/2007). Although disciplinary teaching and research traditionally have been the core activities incumbent on the professoriate, the academic profession has undergone profound changes during the last decades, and academic work is no longer solely about teaching and research (Fumasoli et al., 2015). Consequently, expectations put on the professoriate have shifted over the years. This is also reflected in job advertisements for vacant professorships: in addition to teaching and research, candidates are progressively expected to meet additional criteria to be appointed as professors (e.g. the ability to attract external funding, to cooperate within and outside academia, to conduct international work, and to occupy a specialized academic profile) (Klawitter, 2015).

Since the professorship is a particularly interesting index for the emergence and re/formation of a disciplinary field (see e.g. Hofstetter & Schneuwly, 2002), the gatekeeping practices involved in the appointment of the professorship are of utmost importance to study. This is the task that this chapter takes on. We are especially interested in the conditions that frame these gatekeeping practices. Here we focus on one of the initial and formal aspects of the typecasting process (Hamann & Beljean, 2019), that is, the gatekeeping practice were candidates are valued and evaluated based on how well they fit a certain type, or academic profile, as outlined in job advertisements. Job advertisements are the result of "a two-part labor supply construction process" (Musselin, 2010, p. 56). During this phase in the process, the breadth of the field and how specialized or vague the job advertisement should be is determined. This step helps to organize and guides the subsequent matchmaking process, in particular the legitimacy of assessments and decisions made by peer reviewers and committees (Hamann & Beljean, 2019; Musselin, 2010). We use a few cases to illustrate how the intellectual and social organization of the field of education science(s) [pedagogiska kunskapsområdet] is manifested in the recruitment of full professors. By means of the advertisements, categories indicating what the candidates are expected to do, within which area of expertise, and so on are outlined. Thus, we may expect that boundaries for the disciplines of the education science(s) and their disciples are drawn up, negotiated, and established. The field of education is especially interesting for the study of the intellectual and social organization of a field since it is—in comparison with some other disciplinary fields within the social sciences—a highly heterogeneous field (Forsberg & Sundberg, 2018) with diversified job tasks.

The Field of Educational Science(s)

In the recruitment of professors in education, the ways in which the educational science(s) are organized and carried out are important aspects and a complex matter in several respects. A way to capture this complexity is presented by Hofstetter and Schneuwly (2002) who situate education as a field of study with a threefold basis with different interests—in relation to the academy and scientific demands, in relation to the teaching profession, and in relation to society and expectations on educational systems in terms of economy, culture, and so on—often transmitted by policy-making. These different interests often produce dilemmas for the field of educational sciences and by that also varying expectations on professors in this field.

Educational ideas and reasoning have a long history in academia—mostly in Philosophy. But as a distinct part in the system of higher education and research, educational science(s) has a shorter life, often related to the emergence of the social sciences and welfare state organization. Internationally the study of education can be characterized by three clusters of knowledge traditions; *academic knowledge traditions*, *practical knowledge traditions* and *integrated knowledge traditions* (Whitty & Furlong, 2017).

Academic knowledge traditions are depicted in terms of 'singulars' and 'regions'. Singulars have their own intellectual field of texts, practices, and rules of entry. They are protected by strong boundaries and hierarchies. In some singulars, for example mathematics, the knowledge structure is relatively unified and hierarchical, and they have well-agreed procedures for testing new knowledge. Other singulars, for example sociology, have a more eclectic knowledge structure in which different subgroups adopt different methodological and epistemological assumptions. Regions on the other hand, are made up of a number of singulars that are re-contextualized into larger units, and they operate both in the intellectual field of disciplines and in a field of practice. Despite the fact that they consist of different sub-disciplines they function to some degree as a single discipline, because they are held together by their engagement with a specific field of practice.

Practical knowledge traditions are closely linked to the world of practice. They share a common interest for professional knowledge and enhancing the educational practice, rather than developing discipline-based knowledge. Some of them have a long history, such as *The normal college tradition* that dates back to the end of the 17th century, while others, such as *The networked professional knowledge tradition*, stress the importance of treating practitioners as the main source of the creation of professional knowledge. Other examples of practical knowledge traditions are *Education as a Generic*, that is, performance-oriented movements to enhance generic competences and standards of practice, *Personal liberal education* and *Craft knowledge*.

Integrated knowledge traditions are neither primarily academic nor primarily practical in their genesis, and they do not consider the links between theory and practice as something to be achieved. Rather, these links are considered as central to the process of knowledge production itself. These traditions are limited in number, and their status in academia is debatable. Examples of integrated traditions in the study of Education are *Pedagoġija (Latvia), practitioner enquiry/action research, clinical practice*, and *learning sciences*. The necessity of searching for the development of integrated traditions is pointed out in the literature; to create powerful professional knowledge it is needed to bring together disciplinary knowledge and other external knowledge with professionals' reflective practice and practical theorizing. (Whitty & Furlong, 2017)

Although there are commonalities independently of national context, the emphasis and impact of these traditions differ somewhat across nations depending on local demands and prerequisites. The institutional organization also differs across nations (Whitty & Furlong, 2017). Organizationally, educational science(s) may be at home in the universities, but it may also proceed 'in between' universities and educational systems (Keiner, 2019).

While some countries have a unified institutional structure, that is, the university, others make formal divisions between different types of universities. Yet others have a highly fragmented institutional structure including distinct specialist institutions based on different knowledge traditions.

In Sweden, educational research has to a very large extent been institutionally organized in the discipline of Pedagogik (Rosengren & Öhngren, 1997). Today the organization of the field looks different. We find a shifting basis for educational research in terms of relevance: first in relation to science, then to policy-making, and then over to professional practice and efficiency. A point of departure for describing the development of educational science(s) in Sweden is the making of professorships in education as a science (Pedagogik) and by that Education as a scientific discipline. After a long debate, Pedagogik was introduced in 1908 at Uppsala University and a few years later at the University of Lund. An important reason for this introduction was to create a scientific basis in the training of teachers based on, for example philosophy (why education?) and psychology (how to make students learn?). In terms of relevance, the important point was that studies in education had a scientific ground. Since then, demands of relevance have been of vital concern in changes of education as a discipline in Sweden. Particularly scientific relevance has been a dominating demand (Foss Lindblad & Lindblad, 2016). This was later followed by education science as a basis for extensive reforms of primary and secondary education after World War II, and thus a matter of policy relevance (see the Research Council evaluation of educational sciences in 1995). However, this was combined with increasing demands for professional relevance. First in terms of application of scientific findings as in educational technology and in the making of a sectoral organization of educational research in the 1970s, and later by a practical turn in the 1990s with expansion of didactics and in the making of a clinical educational science, such as pedagogical work. To this fairly straightforward history of the educational science(s) in Sweden, education has increasingly been subject to other disciplines and to cross-disciplinary ambitions, as manifested in the turning of the Swedish Research Council resources for educational research (Benner, 2009). Several other disciplines, mostly within the social sciences and the humanities now receive around half of the resources for educational research from the research council (Foss Lindblad & Lindblad, 2016). This is accompanied by a growing number of academics working in the system of higher education and research having a PhD based on educational research. Thus, the competition for research funds is even fiercer.

Recruitment Procedures—A National Context

Each national higher education system has its own specific procedures for recruiting professors, and the position entails different duties and rights, expectations and accountabilities depending on the particular system. Although there national distinctions and varying details of the pathway to the professorship across nations and HEIs (Angermuller, 2017), the use of peer review is customary and a common denominator is the earning of the PhD as an initial step (Hamann, 2019). As illustrated in Chap. 13 in this volume, this is true also in Sweden.

Until 1993, the majority of professors in Sweden were appointed as chairs or with authorization [fullmaktsprofessorer]. These types of professorships meant a fairly high level of autonomy, and a very high level of employment security. In contrast to today when professors are appointed by the university, they were appointed by the government. They were also leaders of the department and responsible for the discipline (e.g. Pedagogik). However, in 1977, when people other than the professors could be elected to head of department, the professors' hold over the department was diminished. In part, this was related to the general perception that the professoriate was an overly privileged group of academics. During the 1990s and onwards, the system has successively been transformed into one where department heads primarily serve as administrators and decision-makers. This transformation was part of what is often regarded neoliberal reforms of higher education, with changes in governance and reduced collegial decision-making, which led to competitions between universities and departments over financial resources and students. In turn, this has had important implications for positions and tasks for professors. In addition to changed power relations, the professors' duties nowadays primarily encompass research, teaching and, increasingly so, public outreach. Due to altered funding regimes, a good deal of the professors' time is currently dedicated to attract research funding.

The recruitment process is regulated by national law and ordinance, and the use of peer review in academic recruitment has been an institutionally established practice since 1876. While the government stipulates the criteria of eligibility and assessment, most higher education institutions have, as in many other countries, developed supplementary local guidelines and regulations to rule these processes. The eligibility for appointment to full professor in Sweden is the demonstration of both research and teaching expertise, and as much attention shall be given to the assessment of teaching expertise as to research expertise (SFS, 1993:100/2010:1064). In the discipline of education (pedagogik), the double competence was often stressed as a criterion of eligibility in the recruitment of professors during the twentieth century (Lindberg, 2006/2007). This meant candidates had to have qualifications equivalent to a double docent competence—double in the sense of being competent in at least two distinctly different scientific areas—to be considered eligible. Education was regarded such a wide and comprehensive discipline that it required research experience from at least two distinct disciplinary sub-areas to take on the responsibility of representing such a discipline.

The entire recruitment process involves several steps. First (1), there is the initial decision to recruit, usually made by the faculty board. This decision is context dependent, for example, the academic profile of the HEI, and involves negotiations at the faculty and departmental level. Subsequently (2), the vacant position is publicly advertised. The advertisement is essential for the entire process since it conveys desirable competences stipulated by the HEI and thus guides the subsequent steps. Third (3), applicants apply for the position, which is subsequently followed by (4) the selection of external peers by the recruitment board. Customarily two reviewers, external to the HEI in question, are appointed. The reviewers are assigned the task to assess the candidates' qualifications in terms of teaching expertise, research expertise, and expertise in academic leadership. Traditionally, the external reviewers are selected based on their expert knowledge in the area or subject of the current position. However, due to the expansion of positions and review tasks during the last decades, today, reviewers are increasingly faced with the task to assess candidates who are not in the same research field as themselves. Rather, it is becoming more and more common that reviewers are expected to assess candidates in fields that are "overlapping", "related" or "adjacent" to the field of the reviewer (Kaltenbrunner & de Rijcke, 2019, p. 873). In a heterogeneous field such as the educational sciences, this is even more so. The reviewers' evaluation reports (5) form the basis of the recruitment board's nomination of whom to hire (6). The final decision (7), for professorships, lies with the Vice-Chancellor. Figure 14.1 illustrates this process, and highlights the step under scrutiny here.

The recruitment process lies in the intersection between institutional desires and academic values, and each step in the process involves negotiations and decisions on managerial and/or collegial levels. These decisions are made by distinct actors, who all serve as gatekeepers of different kinds and at different stages in the process. For instance, before the advertisement is published heads of departments, deans, members of the recruitment boards, and staff from human resources departments have been involved in decisions on recruitment, profile, desired qualifications etc.

Typecasting in the Recruitment of Full Professors

The analysis is framed by Hamann and Beljean's (2019) concept of typecasting in recruitment of academics. The concept is most commonly associated with typecasting processes in the art, theatre, and film

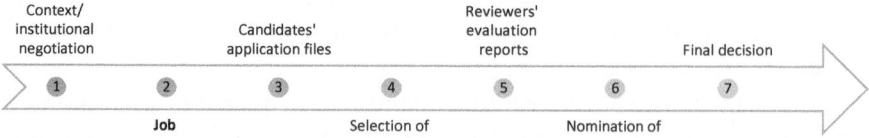

Fig. 14.1 The process of recruiting full professors in Sweden

industry, where it generally refers to the process when an actor becomes strongly identified with a specific character, or characters having similar traits. Among actors, typecasting often has negative connotations as it indicates an actor's limitation or lack of talent (Wojcik, 2003). In this chapter, the concept is based on research on typecasting in labor markets (Zuckerman, 2005), and refers, as stated, to the gatekeeping practice were candidates are valued and evaluated based on how well they fit a certain type outlined in the advertisement, that is, the preferred academic profile. It is part of controlling access to desirable positions in a specific scientific field. In this process of sorting out the wheat from the chaff, candidates are "screened according to and matched with specific categories that are considered relevant for the job at hand" (Hamann & Beljean, 2019, p. 15). The job advertisements mobilize these categories and serve as a backdrop for the subsequent match-making process. To the extent that reviewers explicitly refer to documents other than the application files, they typically do so in relation to the advertisement. In the match-making process, the reviewers are highly dependent on whether the advertisements are 'open' and vaguely written or if they are specified and precise (Musselin, 2010). The former strategy opens up the field of candidates to elicit and choose the best one, while the latter targets the population of candidates.

The advertisements used here are derived from recruitment processes of full professors in the field of educational sciences at three higher education institutions (HEIs) in Sweden: two universities (University I and II) and one university college (University College). All positions issued during the period 2011–2018 are included, which in total gives seven distinct positions—one at the university college, two and four at the two universities respectively—in Pedagogical work, Child and youth studies, Education, Sociology of education, Education focusing on special needs education, and Curriculum studies focusing on the Swedish language. While the entire corpus includes job advertisements, application files, external reviewers' evaluation reports, decision protocols and policy documents guiding the process, the advertisements constitute our prime source of data. Although there are minor divergences in terms of structure and headings, scope and level of detail, the advertisements follow a similar structure and cover a similar content. In contrast to the advertisements studied by Hamann and Beljean (2019), the advertisements by far exceed more than a few sentences; they range from two to up to four pages.

The Intellectual and Social Organization of Educational Science(s)

When we now turn to the typecasting process in the recruitment of professors, we could expect that the field no longer has a basis in one single discipline, and that there are different ways to state what is expected from a professor. The professorships in our sample encompass both general (Education, Pedagogical work, Child and youth studies, Sociology of education) and specialized professorships (Curriculum studies focusing on the Swedish language, Education focusing on special needs education). Some of these are differentiations from Education, some also have roots in other subjects or disciplines, and yet another is a specialization within Education.

The Emergence of a Heterogeneous Field

Our cases show that the field is characterized by some common features as well as by variation. We can see several of the knowledge traditions pointed out by Whitty and Furlong (2017) reflected in the professorships, and their interweaving with the social world and the professional field of reference (Hofstetter & Schneuwly, 2002) differ. In essence, there is a dominant interest in the teaching profession and teacher education. In a majority of advertisements this is manifested in terms of "particularly there will be job tasks that focuses teachers' professional knowledge", "the research shall meet the challenges in education of special teachers in preschool and school" or "…practical experience from preschool" being considered an additional qualification for the position. Focus is on the formal school system, particularly preschool and school, and again on teacher education. In this sense, we see similarities to the practical knowledge traditions identified in prior research (Whitty & Furlong, 2017). In spite of this common trait, and despite our small sample, our cases also reflect the heterogeneity pointed out in the literature (Hofstetter & Schneuwly, 2002; Forsberg & Sundberg, 2018). Although there is an emphasis in interest towards the formal school system, taken together, the advertisements demonstrate an interest in the entire school system (preschool to higher education) as well as other professional practices and everyday life. In part, this is of course related to the fact that the sample comprises positions in disciplines with a long academic tradition, as well as more recently established subjects with a special focus. Hence, the objects of knowledge vary from children's and young peoples' literacy or assessment of teaching content, to the philosophy and politics of education, socialization, and *Bildung*. While the focus on certain substantive educational topics such as early years and assessment in some of the professorships would indicate a position as a 'region', statements like "[the discipline] is based on more general traditions in sociology and historical science" could be an indicator of a 'singular' tradition (Whitty & Furlong, 2017). The absence of such specifications (as in e.g. Education) may also imply a 'singular' position.

The Selection of Peers

The heterogeneity of the field is reflected also in the selection of peer reviewers. Around half of them are professors in Education, the remaining in Pedagogical work, Subject didactics, Sociology, History of economics, Special needs education, Science of literature with focus didactics and Bilingualism with focus on Swedish as second language. This would, on the one hand, suggest the field is a multidisciplinary based field, or a *region* (see Whitty & Furlong, 2017). On the other hand, the fact that professors in Education (all of which holds a PhD in Education) are selected to review candidates for positions other than in Education would rather indicate that Education as a discipline is understood to encompass the other (sub)disciplines, and thereby being a *singular* (see Whitty & Furlong, 2017). Indeed, it might simply imply that there is a lack of reviewers in the specific subject or discipline at hand, and that reviewers in disciplines adjacent or overlapping to the field of the reviewer are engaged (see Kaltenbrunner & de Rijcke, 2019).

The Mobilization of Job Categories

The mobilization of job categories and stipulation of desired qualifications is a prerequisite for the reviewers to be able to make the right fit. In one sense, the job categories are similar across cases: teaching, research, and collaboration. These categories encompass different and more or less specific duties and responsibilities. Similar to previous findings (Klawitter, 2015) it is clear that the contemporary professor not only is expected to teach and conduct research, but also to perform a range of distinct job tasks. These involve various administrative responsibilities, academic leadership including conflict managing, relational work such as internal collaboration, public outreach, service, as well as attracting research funding.

The emphasis of job categories differs, however, depending on type of professorship and discipline. Teaching and research are dominant in all advertisements, even if the establishment and development of these activities are in focus for the more specialized professorships or professorships tied to the establishment of new subjects on third cycle. A noticeable difference is that public outreach or collaboration within academia is hardly mentioned in the professorships from University II, while this is a highly stressed category in the other advertisements, particularly so for the professorships in Pedagogical work. This could be because Pedagogical work is a relatively new, explicitly practice oriented and interdisciplinary subject that is dependent on collaboration across disciplinary boundaries inside and outside academia to be able to develop. One of the professorships in Pedagogical work was, as mentioned, tied to the establishment of Pedagogical work as a third cycle subject which might have accentuated this need for cooperation in the building-up phase even more. Such an inference is contradicted however, by the fact that the professorship in Education focusing on special needs education at the University II also was tied to the establishment of a new subject on third cycle, which rather would suggest these differences vary by type of HEI. The fact that collaboration is not explicitly accentuated as a job task in the advertisements at University II might seem odd at first, but one interpretation could be that this is regarded such an obvious part of the professor's duties that it does not need explicit articulation. The teaching-research nexus is treated in a similar way; it is an explicitly and highly stressed job task at the

University College and University I, but not at all mentioned as a duty for the professor at University II, albeit stipulated as a criterion of assessment.

Moreover, the job descriptions show divergences in level of detail. To varying extent, we find reflections of the two distinct strategies by which advertisements can be constructed (see Musselin, 2010). Most detailed descriptions are found at the University College, and least at University II. That is, most detailed descriptions are found at the youngest HEI, where the professorship is tied to the establishment of a new subject on third cycle level. Least detailed are the two general professorships from the oldest university (University II). And again, a more general difference between the advertisements from University II and the others is that they contain rather short standardized formulations in respect of job tasks. This may reinforce the assumption that the process of recruiting full professors is an established practice at this university in which there is consensus on what to expect from a holder of this position. It also implies that more newly established disciplines want to guide both candidates and peer reviewers more in what they desire.

Taken together, these findings suggest that type of HEI is highly influential on the design and content of the advertisements.

The Intersection Between Institutional Demands and Scientific Values

Although all advertisements contain articulations like "contribute to develop the subject child and youth studies" focus lie primarily on departmental job tasks and duties such as "have a key role in the further development of cooperation between various specializations at the department and thereby regenerate and reinforce scientific profiling" or "[c]ontribute to the establishment of a premeditated graduate school". At University II, the departmental expectations are more pronounced insofar as they expect the future candidate to undertake administrative and managerial duties such as "head of department or director of studies".

The departmental expectations may also be perceived in terms of whether the professor is expected to primarily build or develop an activity or practice. In this regard, there is a slight distinction between the professorships that are tied to the establishment of a new subject on third cycle and the other professorships. Unsurprisingly, for the former there is an emphasis on building, even if development also is stated. While it is most emphasized and elaborated for the professorship tied to the establishment of the subject Pedagogical work, it is briefly stated for the professorship in Education focusing on special needs in that the professor "will play an important part in the establishment and development of the subject area".

Eliminatory Criteria, Signs and Retrospective Judgments

Prior research has shown that reviewers look for eliminatory criteria (criteria of eligibility) and positive signs (criteria of assessment) when making their assessments (see Musselin, 2002). For this match-making process, the advertisements serve as a backdrop and are of great significance for the reviewers and their work in finding the right candidate. A common eliminatory criterion stipulated in the advertisements is "demonstrated" or "documented" scientific and teaching expertise, in accordance with the higher education ordinance. Here too, there are differences in scope and level of detail, depending on subject/discipline and/or HEI. Most advertisements elaborate what they mean by scientific and teaching expertise (and this vary somewhat), but one only refers to "the higher education ordinance" and "[the University's own appointment regulations]". Interestingly enough, while "independent research activities within [det barnpedagogiska forskningsfältet]" and "extensive and documented empirical research on children and preschool" are stipulated as eliminatory criteria for the professorship in Child and youth studies, none of the other advertisements stipulate knowledge in a specific field of research as an eliminatory criterion, and none stipulates a specific subject or discipline for the PhD. Another interesting divergence is University II's specification of "personal qualities required to do the job satisfactorily" as an eliminatory criterion.

While subject or discipline is sparsely articulated in relation to eliminatory criteria, it is more emphasized as a positive sign. Experience from research in the same field as that of the position is to a varying degree stressed for all professorships. Other positive signs are the ability to attract research funding, to establish and develop a research environment, and to collaborate across disciplinary or institutional boundaries, as well as experience from successful doctoral supervision, experience from national and international research networks and international publication in high quality journals. As pointed out earlier, University II has quite short and "open" advertisements in terms of future job tasks. In terms of assessment criteria, the situation is reversed. They are relatively elaborated, albeit standardized across cases, and they include criteria for the assessment of administrative and management expertise. The latter includes for example "the capacity to manage operations and personnel, to make decisions, to take responsibility, and to motivate others and provide them with the prerequisites to efficiently achieve common goals". It also refers to "[t]he ability to coordinate the group and to help create commitment, participation, and job satisfaction, as well as the ability to manage conflicts". Common for all professorships is that both eliminatory criteria and signs are expected to be based on retrospective judgments. That is, the reviewer shall assess the candidates' prior achievements, rather than ability and potential to perform future activities.

Conclusive Notes

Typecasting as a Device of Peer Review in the Formation of Education Science(s)

In this chapter, we have studied one of the initial stages in the typecasting process in the recruitment of full professors in the field of education science(s). We have argued that job advertisements serve as a backdrop for the typecasting process in finding the right fit. Advertisements are the results of numerous collegial and managerial deliberations and negotiations at institutional and departmental levels. We have demonstrated the possibility to discern a re-/formation of a scientific field and a professorship by studying job advertisements. From the viewpoint of the knowledge traditions pointed out by Whitty and Furlong (2017), our findings suggest that the professorship of education science(s) in late modernity is tied to both practical knowledge traditions and academic knowledge traditions ('singulars' and 'regions'). However, even if we were able to discern indications of these traditions, and a varying intersection of them in some of the professorships, the indicators are divergent to some extent and manifest articulations in this respect are few. Documentation from the preparatory work, and the reviewers' evaluation reports might contain more elaborated articulations in this regard.

Peer review in academic recruitment and promotion is characterized by a high level of intertextuality (see e.g. Chen & Hyon, 2005). Reviewers' evaluation reports are no exception; reviewers explicitly refer to prior documents in the process, such as policy documents, job advertisements, and candidates' application files. The reviewers in our sample do set out from the advertisements when doing their assessments. We have pointed to several distinctions in the advertisements, and these distinctions create different conditions for the reviewers and the subsequent match-making process. The shorter and the more opaque advertisements are more is the space, responsibility, and power given to the reviewers. This highlights the question of what qualifies a peer, also in view of that prior research has indicated that reviewers' qualifications and own academic profile influence their evaluations (Levander, 2020).

Traditionally, peers have been selected on the basis of their expert knowledge in the subject area or discipline. Based on the selection of reviewers in our sample, there is reason to further examine the boundaries of the subjects or disciplines in the field of education science(s) and what consequences those boundaries might have for the (continuing) formation of the disciplines in the field and the field as whole. The qualification of those who serve as gatekeepers in a discipline is one possible way to undertake such studies (see Sugimoto et al., 2011).

The Professor of Education Science(s) in Late Modernity

Although there are many different manifestations of disciplinarity (Sugimoto & Weingart, 2015), in this chapter, we have focused on one particularly interesting index for the emergence and re/formation of a disciplinary field—the professorship (Hofstetter & Schneuwly, 2002). We have demonstrated that the formation of the professorship lies in the intersection of academic values and institutional norms, although an emphasis is put on the latter. We argue that in specific cases presented here the formation of the professorship and education science(s) in late modernity is both contingent and context dependent. Type and traditions of the HEI as well as the academic profile of the department and the maturity of the subject influence what is expected of the holder of the position. In fact, these aspects seem to have a greater influence on the advertisements and the formation of the professorship than the discipline per se. There are indicators that the educational science(s) in the typecasting of professorships in Sweden may be regarded a hybrid structure (see Schriewer & Keiner, 1992). On the one hand, educational science(s) emerges as a mode of reflection *within* the education system, on the other hand as a mode of reflection *on* the education system (see Keiner, 2019).

Irrespective of discipline and HEI, we can see a narrowing of authority for the professor and at the same time an increase of responsibilities. The professors of today need to have a broad competence in a variety of job categories (see Hamann & Beljean, 2019). It is simply not enough to be an outstanding researcher and excellent teacher, the professor also needs to be a successful fundraiser and appreciated collaborator.

To conclude, in this chapter we have presented findings from a rather small-scale study of professor recruitment in Sweden. The phenomena studied here need further academic attention, preferably by means of full cases in various national contexts and a comparative approach. The framing of this study and the findings pointed out may well serve as an entrance for such future studies.

References

Angermuller, J. (2017). Academic careers and the valuation of academics: A discursive perspective on status categories and academic salaries in france as compared to the U.S., germany and great britain. *Higher Education, 73*(6), 963–980. https://doi.org.10.1007/s10734-017-0117-1

Benner, M. (2009). *Kunskapsnation i kris?: politik, pengar och makt i svensk forskning*. SISTER.

Chen, R., & Hyon, S. (2005). Faculty evaluation as a genre system: Negotiating intertextuality and interpersonality. *Journal of Applied Linguistics, 2*(2), 153–184. https://doi.org/10.1558/japl.2005.2.2.153

Forsberg, E., & Sundberg, D. (2018). Formeringen av det pedagogiska kunskapsområdet—mot ett forskningsprogram. *Pedagogisk forskning i Sverige, 23*(5), 5–20.

Foss Lindblad, R., & Lindblad, S. (2016). Higher education and research in a steady state–on changing premises and practices for educational research in Sweden. *Nordic Journal of Studies in Educational Policy, 2016*(1), 32371.

Fumasoli, T., Goastellec, G., & Kehm, B. M. (2015) (red.). *Academic work and careers in Europe: Trends, challenges, perspectives* (Vol. 12). Springer.

Hamann, J. (2019). The making of professors: Assessment and recognition in academic recruitment. *Social Studies of Science*, 1–23. https://doi.org/10.1177/0306312719880017

Hamann, J., & Beljean, S. (2019). Career gatekeeping in cultural fields. *American Journal of Cultural Sociology*. https://doi.org/10.1057/s41290-019-00078-7

Hofstetter, R., & Schneuwly, B. (2002). Institutionalisation of educational sciences and the dynamics of their development. *European Educational Research Journal, 1*(1), 3–26.

Kaltenbrunner, W., & de Rijcke, S. (2019). Filling in the gaps: The interpretation of curricula vitae in peer review. *Social Studies of Science, 49*(6), 863–883. https://doi.org/10.1177/0306312719864164

Keiner, E. (2019). 'Rigour','discipline' and the 'systematic': The cultural construction of educational research identities? *European Educational Research Journal, 18*(5), 527–545.

Klawitter, M. (2015). Effects of institutional changes on requirements for vacant professorships in Germany. *Working Papers in Higher Education Studies, 1*(2), 1–19.

Levander, S. (2020). Construction of educational proficiency in academia: Peer review of educational merits in academic recruitment in sweden. *Education Inquiry*, ahead-of-print(ahead-of-print), 1–18. https://doi.org/10.1080/20004508.2020.1843234

Lindberg, L. (2006/2007). Disciplinen och den dubbla kompetensen. Noteringar i anslutning till ett kvalifikationskrav. Studies in Educational Policy and Educational Philosophy. E-tidskrift 2006:2/2007:1.

Musselin, C. (2002). Diversity around the profile of the 'good' candidate within French and German universities. *Tertiary Education and Management, 8*(3), 243–258. https://doi.org/10.1080/13583883.2002.9967082

Musselin, C. (2010). *The market for academics.* Routledge.

Rosengren, K., & Öhngren, B. (Eds.). (1997). *An evaluation of Swedish research in education.* HSFR.

Schriewer, J., & Keiner, E. (1992). Communication patterns and intellectual traditions in educational sciences: France and Germany. *Comparative Education Review, 36*(1), 25–51.

SFS. (1993:100/2010:1064). Högskoleförordningen. [Higher education ordinance]. Ministry of Education and Research.

Sugimoto, C. R., Ni, C., Russell, T. G., & Bychowski, B. (2011). Academic genealogy as an indicator of interdisciplinarity: An examination of dissertation networks in library and information science. *Journal of the American Society for Information Science and Technology, 62*(9), 1808–1828. https://doi.org/10.1002/asi.21568

Sugimoto, C. R., & Weingart, S. (2015). The kaleidoscope of disciplinarity. *Journal of Documentation, 71*(4), 775–794. https://doi.org/10.1108/jd-06-2014-008

Whitty, G., & Furlong, J. (2017). *Knowledge and the study of education: An international exploration.* Symposium Books.

Wojcik, P. R. (2003). Typecasting. *Criticism, 45*(2), 223–249. https://doi.org/10.1353/crt.2004.0005

Zuckerman, E. (2005). Typecasting and generalism in firm and market: Genre-based career concentration in the feature film industry, 1933–1995. In Jones, C. & Thornton, P. (Ed.), *Transformation in Cultural Industries* (Research in the Sociology of Organizations, Vol. 23, pp. 171–214). Emerald Group Publishing Limited.

15

Assessing Academic Careers: The Peer Review of Professorial Candidates

Björn Hammarfelt

Introduction[1]

Academic careers have two characteristic features, which have to be considered when studying their evaluation: the content of work (e.g. research done) plays an important role, and the research community has a great deal of influence when evaluating academic careers. The dependence on colleagues means that the reputation of an academic is dependent on their recognition among a wider community of peers. The primary means for gaining a reputation among colleagues is through publications, and the recognition of a researcher is largely dependent on their writings. In fact, reputation and recognition gained through publications has been a

---

[1] This chapter is a revised, updated and shortened version of an article published in *ASLIB Journal of Information Managment* under the title: "Recognition and reward in the academy: Valuing publication oeuvres in biomedicine, economics and history" (2017). https://doi.org/10.1108/AJIM-01-2017-0006

B. Hammarfelt (✉)
University of Borås, Borås, Sweden
e-mail: bjorn.hammarfelt@hb.se

E. Forsberg et al. (eds.), *Peer review in an Era of Evaluation*,
https://doi.org/10.1007/978-3-030-75263-7_15

crucial merit for career advancement in academia since the birth of the research university in the late eighteenth century (Josephson, 2014). Generally, it is assumed that the competition for positions in academia has increased over the last decades, and while idioms like 'publish or perish' are usually reiterated rather carelessly, there appears to be some substance to the claim about the increasing pressure to publish (Van Dalen & Henkens, 2012).

Academic researchers are continuously evaluated on the basis of their publication record, either as part of informal assessments or in the form of more regular systems of evaluation. A formal evaluation, which may have significant consequences for the individual career, takes place when applicants for an academic position are evaluated on the basis of their research merits, teaching and administrative skills. This chapter looks at discipline specific evaluation practices in three fields; biomedicine, economics and history. The material consists of reports (sakkunnigutlåtanden) commissioned by Swedish universities when hiring new professors. Independent referees hired to evaluate and compare candidates author these texts. The approach here is not so much to study what constitutes 'value' in these evaluations, rather the focus is on how 'value' is enacted with special attention to the kind of tools—judgements, indicators and metrics—that are used. A selection of 45 assessment reports from four major universities in Sweden are used to study how publications are valued in this context. Commonly, the number and quality of publications are two main criteria through which research quality is evaluated. However, more exact studies of how research quality is defined in the context of evaluating candidates for academic positions are quite rare (Hemlin & Montgomery, 1993; Nilsson, 2009; Hammarfelt & Rushforth, 2017), and research on conceptions of research quality has foremost been focused on the peer review process of grants (see e.g. Langfeldt, 2001; Lamont, 2009; Van Arensbergen et al., 2014) rather than on academic careers. Moreover, the literature on academic careers tends to focus on structural aspects such as differences between national career systems (Musselin, 2009) or systematic discrimination based on gender (Steinpreis et al., 1999), while actual evaluation procedures have attracted less attention.

In focusing on how contextual information, such as information on the status of the publication channel, or externalities (e.g. bibliometric measures), are brought in to evaluate candidates this study engages in the current debate on peer review and indicator use in research assessment (Wouters et al., 2015). Externalities are defined as features such as publication channel, age of the texts, reviews, bibliometric indicators and prizes, which can be assessed without evaluating the epistemological claims made in the actual text. Recent research has shown how indicators are employed as 'judgement devices' (Karpik, 2010; Hammarfelt & Rushforth, 2017) when evaluating research. The journal impact factor has been identified as one frequently used such device which is integrated in the field of biomedicine where it also affects epistemological considerations (Rushforth & Rijcke, 2015). The present study broadens the perspective introduced in these studies by engaging with contextual information about publications that might be used in similar ways, but which must not directly involve the use of bibliometric indicators. Thus, the purpose of this study is to provide a more detailed understanding of how 'research quality' is defined and constructed in the context of evaluating the publication oeuvres of candidates for academic positions.

Three fields of research—biomedicine, economics and history—were deliberately selected to highlight distinctive disciplinary valuation practices, although similarities in-between fields will also be emphasised. These fields were chosen on the basis of their being large high status fields both within and outside academia.

The chapter starts with a short outline of research on peer review and perceptions of scientific quality. The subsequent section introduces the theory of judgement devices suggested by Karpik (2010), and the analytical frame developed by Whitley (2000). Material and methods are thereafter presented and the recruitment system in Swedish academia is briefly explained. The findings are structured on five main themes identified in the material: authorship, publication prestige, temporality, reputation within the field and boundary keeping. The concluding section summarises and discusses the implications of this study.

Picking the Best: Peer Review in Assessment Procedures

Conceptualisation of 'scientific quality' in the context of peer review is a reoccurring topic in the literature. A noticeable strand within this area is studies that look at the work of grant panels, and how notions of quality are negotiated in this context. Seminal works, like Lamont's (2009) study of peer review, show how field specific quality criteria are negotiated in multidisciplinary panels. Following in this tradition, several studies examine how judgements are made and negotiated in panels evaluating research grant applications (Langfeldt, 2001; Roumbanis, 2017). The present study distinguishes itself from these approaches in several ways; it concerns itself with intra-disciplinary peer review, it looks at peer review that is done remotely (not in panels) and it uses reports, not interviews or ethnographic observation, as its primary material.

Conceptualisations of research quality when evaluating and ranking candidates for academic positions has been much less studied, perhaps due to difficulties in gathering empirical material on procedures for evaluating candidates. Hemlin and Montgomery (1993) looked at assessment reports concerning candidates for 31 professorships in the humanities, the social sciences, medical sciences and natural sciences. They found considerable overlaps in how quality was judged across research fields, for example, mentions of methods, 'problems' and 'results' were frequent and 'stringency' and 'novelty' were deemed as important attributes for high quality research across all domains. Yet they also found differences which could be explained by the division between 'hard' and 'soft' sciences.

The qualitative and comparative approach developed by Nilsson (2009) is of relevance for the present study. By studying assessment reports across three disciplines, physics, political science and literature, over a time-period of 45 years Nilsson depicts how notions of quality have developed over time. However, while she chose to select a few reports for each year, the present study gathers instead a larger number of contemporary reports in order to get a deeper understanding of how conceptualisations of quality are expressed when evaluating careers. A similar approach, but with a focus on teaching merits, is Levander's (2017) study of how pedagogical merits are evaluated. A notable finding in this study is that research merits—often in the form of publications—usually have greater impact on the final ranking compared to other accomplishments.

Hammarfelt and Rushforth (2017) analysed the use of metrics in assessment reports in biomedicine and economics. Their findings indicate that both disciplines use metrics rather extensively to assess candidates, but the type of use is dependent on the organisation of the field and on specific disciplinary publication patterns. The study showed how bibliometric indicators are used as 'judgements devices' to differentiate between candidates. The focus of the present study is more expansive as it incorporates a broader set of externalities used in the evaluation of the quality of publications.

Analysing Referee Reports

The methodology adopted in the current study is best described as a qualitative content analysis where quotes, rather than statistics, are used to illustrate findings. Three fields—biomedicine, economics, and history—which, to some extent, represent three 'cultures' (social science, natural science and the humanities) have been selected for analysis. Hence, the overall design of the study and the selection of fields assume that disciplinary differences might be a fruitful approach for studying how academic worth is judged. Yet, in order to avoid a simple confirmation of rather established conceptions of differences across disciplines special attention has been paid to details, which may contradict this neat separation of fields.

Fifteen external referee reports from each discipline were randomly selected from four universities in Sweden (Lund University, Umeå University, University of Gothenburg and Uppsala University). A total of 45 reports, each comprising about 1–38 pages, was deemed large enough to provide a variety of different types of reports, while maintaining the possibility for a detailed analysis of the arguments made in each report.[2]

---

[2] The number of available reports at these universities over the period (2005–2014), 18 for history, 54 in economics and 132 in biomedicine, provided a further limitation to the number of reports that could be included in this study.

Material from a ten–year period, 2005–2014 was collected. Although these are official documents that are accessible to anyone according to 'offentlighetsprincipen' (the principle of public access to public records), it was decided to anonymise both referees and applicants. All reports were therefore coded based on year, field (biomedicine: bio, economics: eco and history: his) and university (Lund University: LU, University of Gothenburg: GU, Uppsala University: UU, Umeå University: UMU). Many of the reports, especially in economics and history, were written in Swedish or other Scandinavian languages and quotes used in the analysis were translated to English by the author.

The common routine for recruiting academic personnel in Sweden can briefly be described in six steps: (1) a decision to recruit is made by the head of the department or the dean, (2) a description of the position and the qualifications needed to acquire the position is drafted and the job opening is advertised, (3) applications from possible candidates, containing a CV, selected publications, and a description of pedagogical merits are submitted, (4) external referees, are chosen to access and sometimes even rank candidates, (5) these assessments together with interviews and trial lectures by the leading candidates are used to form a final ranking of candidates (usually by a recruitment board), (6) based on this ranking the formal recruitment decision is made by the relevant authority (e.g. department head or dean).

My focus is specifically on stage 4 when CVs and a selected number of publications (usually around ten) are sent to external referees (so-called sakkunniga) who are assigned the task of making unbiased evaluations and ranking candidates. Reviewers usually make judgements on all merits, including teaching and administration, but research merits, and specifically publications, continue to play a key role in the final ranking (Levander, 2017). The usual structure of these documents can be summarised as follows: first, a general introduction presenting the assignment, followed by detailed descriptions of each candidate and concluded with a ranking of applicants.

The methodology chosen has similarities with directed content analysis, also called deductive content analysis (Hsieh & Shannon, 2005; Mayring, 2000) in that the analysis is guided by the theoretical frame provided by Whitley's theory on the organisation of research and Karpik's concept of 'judgement devices'. Initially this theoretical viewpoint facilitated a focus on intellectual and social aspects of academic careers expressed through the evaluation of publication oeuvres using externalities. After a first reading of the documents five main themes, authorship, publication prestige, temporality, reputation within the field, and boundary keeping, were identified as the main evaluative categories. However, as will be evident in the material, these categories are in no way mutually exclusive, and neat separations are not to be expected.

Judgement Devices and Intellectual Structure

When evaluating candidates, referees face the task of assigning value to specific research accomplishments and producing a ranking of applicants. This task is difficult because each academic career is distinctive and multidimensional. Such unique and not easily compared entities are termed 'singularities' (Karpik, 2010). Examples of singularities are literary works or a medical doctor and when comparing and evaluating such 'goods' consumers often make use of so-called judgement devices. Judgement devices provide external support for making and legitimating decisions, and their use in academic recruitment was first suggested by Musselin (2009). Musselin's study pointed to a more general use of judgement devices, but for the more detailed and comparative approach taken here it is important to consider the different types of devices identified by Karpik: *appellations*, *cicerones*, *confluences*, *networks* and *rankings*. Two of these are less applicable in the context of academic valuation; *networks* describe how the personal network of a buyer (friend and family) influences their choices, and *confluences* relate to who buyers navigate in a physical space, for example in a store. Appellations and rankings have previously been identified as particularly useful for understanding evaluation procedures in referee reports (Hammarfelt & Rushforth, 2017). *Appellations* can be defined as a type of certification or brand, for example, prestigious journals or publishers that assign value to products (articles/books). *Rankings*, on the other hand, assign value by a hierarchisation of products based on specific criteria. *Rankings* can be further divided into 'expert rankings' (e.g. prizes and diplomas awarded by juries) and 'buyers rankings' (top ten products and bestseller lists) (Karpik, 2010, p. 46). A third judgement device, which is relevant for this study, is what Karpik calls *cicerones*, authorities in the form of guides or critics, which help consumers in making their choice.

The use of judgement devices can be further understood in relation to the social and intellectual structure of research fields (Hammarfelt & Rushforth, 2017). Thus, Whitley's (2000) study of the intellectual and social organisation of research fields is used in order to analyse how judgement devices are employed in specific disciplinary contexts. How the three selected fields, biomedicine, economics and history, are depicted in Whitley's framework is summarised below (Table 15.1).

Whitley introduces two main axes that can be used to describe intellectual fields: mutual dependency and task uncertainty. Mutual dependency measures the degree to which a researcher is dependent on colleagues, while the degree of task uncertainty reflects agreement on the goals of research and the methods used. Whitley then continues by separating technical and task uncertainty and functional dependency and strategic dependency thus allowing for an intricate description of fields through 16 possible characterisations. Whitley's theories are here used for the specific purpose of providing an analytical lens through which disciplinary differences in assessing careers become visible.

Table 15.1 Characterisation of research fields using Whitley's typology

Field	Characterisation according to Whitley (2000, p. 158f.)
Biomedicine	*Professional adhocracy* is characterised by combining reduced technical task uncertainty with high strategic task uncertainty. There is considerable standardisation of skills and technical procedures. No single group dominates when defining scientific criteria and various groups influence the field in terms of funding and employment.
Economics	*Partitioned bureaucracy* combines high technical task uncertainty with low strategic uncertainty, and high strategic dependency. These are rule-governed fields, and hierarchically organised fields, where theoretical elaboration and analytical abilities carry greater value than empirical investigation.
History	*Fragmented adhocracy* combines high task uncertainty with low degrees of mutual dependency. In these fields, research is personal and weakly coordinated, common-sense language is used when communicating results, and specialisation is formed around empirical objects.

Merits and Their Assessment: Five Main Themes

The findings are structured around five main themes: authorship, publication prestige, temporality, reputation within the field and boundary keeping. These themes emerged through an iterative categorisation of topics when analysing the reports. While this structure is useful for presenting the results in a systematised manner, it should be emphasised that such an arrangement is a simplification of a broader narrative. Moreover, many themes intermingle throughout the material and this is also visible in the analysis. As the current study has a focus on the evaluation of researchers as authors, it is logical to begin the analysis by scrutinising the notion of (co-)authorship across the three fields.

Authorship, or the Reading of By-Lines

It is well-established that notions and practices surrounding authorship differ considerably between research fields, which is reflected in that the average number of authors per publication varies from one or two in many humanities fields to tens- or even hundreds in the biomedical- and the natural sciences (Marušić et al., 2011). Naturally, these authorship practices have consequences for how collaboration in the form of joint publications is evaluated in the context of publication oeuvres. Moreover, research fields differ in their focus either on individual publications, or on the oeuvre as a whole. As Hemlin and Montgomery (1993) suggest, the medical and natural sciences tend to have a greater focus on the whole oeuvre, while the assessment of individual publications are the prime method through which research is assessed in the humanities.

Collaboration in the form of co-authorship is rarely touched upon in history, probably because it is quite infrequent, but there are instances when referees find it difficult to separate individual contributions and posit this as a potential problem: "it is not always easy to separate the role and responsibility of the two authors." (His UU 2013-1, p. 3). However, on other occasions co-authorship might point to distinct qualities and due to its rarity it can be seen as a merit, rather than a problem:

"[co-authorship]…shows her ability to work and think together with other researchers and authors" (His LU 2011-1, p. 8). Overall, however questions regarding co-authorship are few and co-authored pieces are uncommon.

The presence of several authors in the by-lines is more frequent in economics, and typically, two or three authors write the majority of papers, although there are instances of longer by-lines. In these instances of 'multiauthorship', the value of a publication becomes unclear, as the role of the individual is hard to distinguish:

> *This resembles laboratory sciences where all those involved in a large project are included as authors. […] The joint authorships make it a bit hard to pinpoint individual contributions, but xxx's publication list includes several articles and papers written by him or with only a few co-authors, so clearly there is a fair amount of independent work.* (Eco GU 2007-3, p. 5)

Who you publish with matters, and papers co-authored with senior colleagues are generally viewed with a bit of scepticism: "As the other top candidates, xxx has a stellar publication record. However, it is a slight disadvantage that all his best papers are joint with senior co-authors" (Eco GU 2014-3, p. 7). Similar judgements are made in biomedicine, where too many publications with your former supervisors are seen as an indication of being too dependent: "She has not yet established herself as an independent researcher which is illustrated in that her former supervisor is co-author on 15/16 publications" (Bio GU 2006-1, p. 7). The author order, which has been found to play a central role for credit assignment in medicine (Biagioli, 1998), is consistently referred to in the reports. Generally, it is first and foremost last authorships that are counted when publication oeuvres are valued, and being middle authors counts for very little: "The results have been published in 41 multi-authored original publications, but most with the applicant in somewhat anonymous positions in the author sequences of the articles" (Bio LU 2011-1, p. 14). Prestige is instead attached to the first and the last position and the author order also signals degree of independence: "He clearly demonstrates independence with several publications as last or main author" (Bio UU 2014-11, p. 1) and last authorships also signify leadership: "He is frequently the senior author on his publications in recent years, indicating that he is clearly the leader behind the research line" (Bio LU 2005-6, p. 4). Hence, the ability to interpret author by-lines, and give credit based on this reading is a key competence when evaluating bio-medicine, and the arrangement of authors as well as the reading of authorship order is highly standardised.

Hence, the reading and interpreting of author bylines is an established practice in biomedicine. The evaluation of multi-authored publications is less straightforward in economics as this quote illustrates, "It is always difficult to evaluate a candidate who publishes with many co-authors, especially when they are very senior" (Eco UU 2013-1, p. 4). In history, co-authorship is still more of a curiosity rather than a problem, and the single author is the norm. Independence from senior researchers is not an issue discussed in evaluating candidates, which is expected given that research in history, according to Whitley (2000), is personal, weakly co-ordinated and highly specialised even early in the career.

Publication Prestige and the Importance of Articles in 'Top-Journals'

The type of publication channel that is assessed, and how it is valued var-ies considerably; monographs are the most prestigious publication chan-nel in history, while journal articles are the most important merit in biomedicine and economics. Book chapters are not uncommon in eco-nomics, but in general they have less status than journal publications: "xxx has a series of articles in books about economic development but lacks scientific merits in the form of journal publications, which are needed to compete for the position" (Eco 2008-4, p. 2). Usually, evalua-tors in economics and biomedicine put considerable emphasis on publi-cation channels, and papers in highly reputable journals are much valued. Publishing in more general high status journals is considered an impor-tant achievement in both fields, particularly in economics:

Xxx has maintained high productivity since the PhD defence in 1998, and has an impressive productivity. However, publications in more general journals would have helped to spread the results to other researchers. (Eco GU 2008-4, p. 3.)

Xxx shows relatively high productivity but his research has not yet reached the best journals. (Eco GU 2008-4, p. 4)

Overall, the ability to publish in top journals emerges as the most important criteria for valuing careers in economics, and top journals, or highly ranked journals are mentioned in almost all reports. Sometimes it is a clearly distinctive factor: "I chose to rank first xxx because she is the only who has a top-5 publication" (Eco UU 2013-1, p. 1). A similar view is expressed by this reviewer:

A university that aims to compete at the first or second tiers in Europe should expect its full professors to show the ability to publish at least a few articles in the best journals in the field. Publishing a paper in a top finance journal requires a degree of effort, awareness of the latest thinking in the field, and excellence, which any number of articles in journals below second tier could not match. (Eco UU 2006-1, p. 5)

Apart from highlighting the significance of papers in top journals, as outlined above, these quotes also indicate the hierarchal structure of the field, where top institutions and top journals can easily be identified (Fourcade et al., 2015). A logical consequence, as noted in the quote above, is that top researchers should publish in the best journals, and the highest ranked universities should employ them. As argued by Hylmö (2018, p. 295) these top journals "merge with an understanding of something like a disciplinary core", and in order to be accepted researchers need to adapt to an "established disciplinary style of reasoning". While hierarchies exist across all disciplines, it is probably warranted to claim that there is greater agreement on top journals or best universities in economics compared to many other fields. The hierarchal organisation of economics, which according to Maeße (2017), is further accentuated by resources and academic capital being concentrated to a few 'top' institutions, has direct consequences for how individual researchers are evaluated.

Top journals, or high impact journals, have a distinct role in biomedicine, while other types of publications, including dissertations matter less when evaluating research. Similar to economics, reviewers of candidates for positions in biomedicine tend to discuss the status of the journal in which an article appears, and the names of prestigious journals, or in Karpik's terms, brands, to support their judgements:

For several years, he has published regularly as the corresponding author in excellent journals such as Chemistry and Biology, J. Biol. Chem, Blood, Biochemistry. He is also co-author of papers in prestigious journals such as Science and Nature. (Bio UU 2008-1, p. 2)

The 'market standing' of these 'brands' are then often confirmed by the implicit and explicit use of the Journal Impact Factor (JIF):

He has published 27 papers and most of these are in high impact journals such as EMBO journal, Science, Journal of Clinical investigation, PNAS, JBC and Journal of Physiology. (Bio LU 2005-6, p. 6)

The JIF is used here as a judgement device that informs and supports assessment. Similarly to how journal rankings are employed in economics, JIF functions as a device which provides a shortcut to evaluating research; for example, a paper published under the brand 'Nature' automatically benefits from the reputation of the journal. Relating to Whitley's characterisation of biomedicine we can also regard the use of the JIF as a form of standardisation, which supports decision making in a situation where several different groups have to reach agreement when evaluating scientific quality.

Journal articles, especially if they are peer reviewed, are a strong merit in the field of history and journals with good reputations are appreciated: "a considerable number of her publications have appeared in renowned series or journals" (His GU 2014-1, p. 9). However, the skills associated with writing and publishing monographs is still highly valued: "The research is both in-depth and original, but its merits are devalued by the fact that xxx has not published any larger monographic work since the doctoral thesis in 1991" (His UU 2013-1, p. 3). The importance of the text's length is further accentuated by the use of the number of pages as one of the few 'metrics' mentioned in evaluation reports in history:

The dissertation is long (622 pp.) [...] The study is a large (579 pages) and is detailed research... (His GU 2014-1, p. 5)

Scientifically xxx is relatively well qualified with two monographs, and one longer article of 61 pages as well as a comprehensive report of 271 pages (His GU 2013-1, p. 17).

The use of numbers for measuring the length of publications is noteworthy as referees in history otherwise tend to rely on narrative accounts, which do not make use of quantitative data or metrics. Hence, the length of the publication is clearly an important factor when evaluating publications in history. Moreover, while the dissertation plays a minor, and in biomedicine, a negligible role when evaluating candidates, the assessment of doctoral theses, almost exclusively in the form of a book, take up a considerable part of the evaluation report. In part, this relates to the temporal horizons through which research is assessed.

Temporality. The Importance of a 'Positive Trajectory'.

When reading the reports it becomes evident that the temporal foci of reviewers are quite distinctive in each discipline. As noted above, historians tend to spend considerable time describing and valuing the dissertation, which in many instances is stated as being the candidates' strongest research merit. Many descriptions start out with a lengthy description of the dissertation work of the candidate, and the importance of the doctoral thesis is underlined: "xxx greatest scientific merit is his dissertation." (His GU 2007-1, p. 16) or in the case of a professor who is an author of several monographs: "The dissertation, which is of high scientific quality, is xxx strongest scientific qualification" (His UmU 2012-2, p. 5). Still, of course, career or in this case, publication trajectories, also matter in history, as expressed by this reviewer: "His research does not show any clear progress" (His LU 2011-1, p. 13).

Dissertations are, with no exceptions, published as monographs, and many of them receive prizes, or other awards, which are important merits. Hence, for younger researchers and even for more experienced scholars the dissertation is a persistent yardstick by which they are judged, and looking at the origin of an academic career will always be relevant. Particular emphasis is put on not only methods used or findings presented in the dissertation, but also on language and presentation. Thus similar to views of Hemlin and Montgomery (1993), aspects like writing style and reasoning are highlighted. In history, first impressions last—if not forever—for a very long time.

The dissertation plays a lesser role in economics and biomedicine, and here focus often lies on recent work. The dissertation in these fields is a starting point for a career, and rarely its high point. Evaluations of candidates in economics often go one-step further and evaluate research that has not been formerly published (e.g. pre-prints). Similar practices can be found in biomedicine and history where drafts or book manuscripts under consideration are included in the evaluation. However, in economics, forthcoming work is given greater weight compared to both history and biomedicine, and this difference can partly be explained by the tradition among economists to publish pre-prints ahead of formal acceptance. Yet, there are also suggestions that economics as a field is forward looking, and interested in being not only a descriptive but also a predictive science; " [economist] 'live 'in the now', and see trajectories from the present forward', while sociologists have the reverse intellectual attitude, looking at the present as the outcome of a set of past processes" (Fourcade et al., 2015, p. 109, citing Abbott, 2005). The forward-looking focus is reflected by many reviewers not only making judgements on research done, but also predicting which researchers have positive trajectories. This can in turn influence how researchers are compared:

> As they have different expertise, it is hard to rank them. xxx and yyy have a richer publication record, but zzz is at an earlier stage in his career and on very positive trajectory. (Eco GU 2014-3, p. 1)

> Xxx has clearly improved his scientific qualifications over the last years, and there is reason to believe that he will publish well also in the future. (Eco GU 2008-4, p. 8)

In biomedicine, successful publication careers are partly defined by how fast a candidate moves from being first author to last author. Moreover, the number of publications is of great importance for evaluating the direction of the career trajectory: "His list of publications reveals a remarkable and unexplained decrease in scientific productivity during

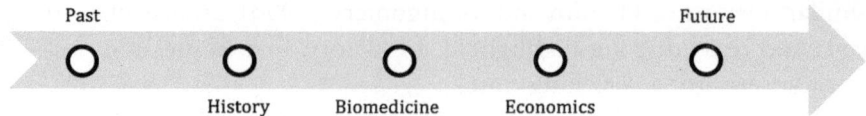

Past   Future

History   Biomedicine   Economics

Fig. 15.1   Schematic overview of temporal focus when evaluating research quality

the last six years" (Bio LU 2011-1, p. 13). Here publications are evaluated as part of an oeuvre, rather than as single works: "It is not only rarely seen, but also stimulating to evaluate such a consequential research career" (Bio LU 2011-1, p. 8). Overall, it is evident that these three disciplines employ slightly different temporal horizons when evaluating research. These can be schematically illustrated on a timeline (Fig. 15.1).

Overall, many of the evaluation reports build on an assumption of what might be defined as an 'ideal trajectory' of the academic career. Thinking in terms of trajectories is a fundamental feature of western modernity (Appadurai, 2013, p. 223f.), and this logic is apparent also when evaluating academic research (Felt, 2017; Hammarfelt et al., 2020). In this case, publications, (co-)authorship, and indicators are used to position and compare individual careers against an 'ideal trajectory'; a trajectory which is partly field specific. Yet, as shown in the next section, the type, amount and the temporal frequency of publication is not enough for evaluating a candidate, as reputation within the discipline carries great weight when evaluating publication oeuvres.

Racing for the Prize: Reputation Through Awards and Citations

The reputation that a publication of a scholar has gained within the discipline is an important criterion for assessing scientific merits. Often are external information, such as reviews, prizes, citations or similar, brought in to form and substantiate claims. As we will see different forms of 'indicators' representing the reputation of a scholar are introduced depending on the discipline. These indicators are all said to represent the recognition and impact that a particular publication or an oeuvre has gained in the research community.

Prizes, peer review assignments, membership in associations and editorships are all important signs of recognition in history, and appreciation in form of reviews is quite often mentioned in connection to monographs. The finding that reviews play an important role for assessing reputation is in line with previous research suggesting that reviews might be seen as an indicator of impact (Zuccala & van Leeuwen, 2011). Prizes, often for dissertations and books, are repeatedly used to present the reputation of a scholar. While national (Swedish) organisations are most visible, we also see that international engagements in projects, review-assignments and associations are highly valued. Candidates that exclusively publish for a Swedish audience are often criticised by reviewers, which might indicate that the criterion of 'international reach' has gained in importance in comparison with Hemlin and Montgomery's (1993) study.

Prizes and book reviews serve in many ways the same role for historians as citations do in biomedicine and economics. These are used to showcase the recognition that particular publications have gained in the community:

The dissertation was awarded with the Geijerprize and is still her strongest merit. (His GU 2013-1, p. 13)

Xxx has established herself as a leading researcher in her area. Which among other things is made visible in the reviews of her dissertation that have been published in international journals. (His GU 2007-1, p. 16)

Prizes can be seen as type of endorsement, which in Karpik's vocabulary might be defined as an *expert ranking*, while the authority of reviews builds on the embodied and softer form of expertise in the form of critics or guides, or what Karpik (2010) terms *cicerones*.

In economics citations in specific publications, or to the whole oeuvre, are often used to measure the impact, and indirectly the reputation of researchers. For example, it can be stated, "they have both made an impact on the profession, for instance both have a fair number of citations" (Eco GU 2008-5, p. 1), or similarly, it can be formulated in this way: "A search in Google scholar gives 197 hits which suggests an average/high visibility

Table 15.2 Judgement devices used to assess the recognition of publications in the discipline (*type of device according to Karpiks typology*)

	Biomedicine	Economics	History
Externalities	Reputation of journal (*appellation*)	Reputation of publisher/journal (*appellation*)	Reputation of publisher/journal (*appellation*)
	Impact factor (*appellation*)	Impact factor (*appellation*)	Prizes (*expert ranking*) Reviews (*cicerone*)
	h-index (*buyers ranking*)	Citations (*buyers rankings*)	
	Citations (*buyers ranking*)	h-index (*buyers rankings*)	
	Prizes (*expert ranking*)	Prizes (*expert ranking*)	

in the scientific community" (Eco UmU 2012-1, p. 1). Similar statements are made in biomedicine, with the difference that the amount of citations per author and paper can be considerably higher than in economics: "His main author papers include papers with notably high citation rates (up to <1000 citations), demonstrating his ability to publish visible cutting edge research" (Bio UU 2008-2, p. 2).

Overall, we find that a range of judgement devices is used across these fields, with significant overlap between them, however it is important to note that the extent of use differs considerably between fields (Table 15.2).

Prizes, for example, are rarely mentioned in biomedicine and economics (one instance each) but frequently used when evaluating careers in history. Similarly, it is evident that these fields have distinct practices when it comes to defining and defending their borders.

Boundary Keeping and the Shielding of Academic Markets

External reports serves not only the purpose of assessing the merits of candidates, but these texts also make distinction between those that can be recognised as peers, and thus eligible for a position, and those that do not belong to the community (cf. Levander, Forsberg, Lindblad & Bjurhammer, *this volume*). The disciplinary boundaries shield the market, and otherwise highly competent candidates have little chance to compete if they are deemed as 'outsiders'. Usually reviewers refrain from making an assessment of such candidates: "scientific and pedagogic merits are primarily from the field of art history and he can therefore not be included on the short-list." (His GU 2014-1, p. 11) or they make qualifications: "If his main and nearly exclusive research and publication area (…) is seen as belonging to the field of history, he would have a very strong and internationally qualified record" (His GU 2014-1, p. 15). Similar statements are also made in economics, "xxx is not an economist. All his publications are in non-economics journals" (Eco UU 2013-1. p. 5), or "The work shows good familiarity with the research area, but it is outside mainstream economics. This is shown also by the fact that xxx has no publications in general economics journals" (Eco GU 2007-3, p. 4).

Overall, it is evident that economist and historians are strict when it comes to upholding boundaries to other fields, but while publishing in key economic journals is enough for being recognised as a peer in economics, formal training as a PhD is a key qualification in history. This is probably due to relatively porous boundaries to other fields such as art history, economic history and history of ideas. The focus in biomedicine is more on specific competencies and whether the candidate will fit into a particular research profile or lab and, as suggested by Whitley (2000, p. 161), a single group does not control the labour market in biomedicine. Using Whitley's theoretical frame it can be suggested that formal institutional origin—for example, being trained as a historian—seems to play a decisive role in determining disciplinary borders in fields where agreement on research procedures or goals are less useful for defining the core of the discipline. Fields with a certain agreement on methods and procedures, might instead, as in the case of economics, define 'membership' as having the skills needed to contribute to the advancement of the field.

Discussion

The peer review of academic careers is a complex and demanding task, also for experienced reviewers. Careers, even when summarised in publication oeuvres, are multifaceted and not directly comparable. While disciplinary norms, and 'judgement devices' in the form of externalities may be of great help to reviewers, the many uncertainties and disagreements in the ranking of candidates are norm rather than the exception. Importantly, reviewers not only have to be experts on current research in their field, but they must also be well acquainted with current assessment procedures and evaluation criteria's used. As expressed by Hammarfelt and Rushforth (2017, p. 178), "it is knowing *how* and *when* to deploy indicators which should be considered the marker of expertise in such evaluative contexts."

Still, the three fields under study all emphasise similar aspects when evaluating candidates and these can be summarised in five themes: authorship, publication prestige, temporality of research, reputation within the field and boundary keeping. These aspects are evidently the structure of all the reports, and a generic narrative form can be distilled from across all disciplines, making it accessible for practitioners that are familiar with the form but not experts on the evaluation procedures of specific fields.

While the criteria through which publications oeuvres are evaluated are fairly similar, the emphasis placed on these criteria varies greatly. Questions concerning co-authorship are prominent in biomedicine but less emphasised in economics. The reputations of publication channels in the form of highly ranked journals or journals with high impact matter a great deal in economics and biomedicine, while monographs and the length of publications are important for historians. Ways of assessing the impact of these publications in a community of peers differs; citations are quite often utilised in biomedicine and economics, while prizes and book reviews are used as 'indicators' of impact in history. Borders to other neighbouring disciplines are keenly defended in history and economics. Biomedicine is more porous. Overall, these results seem to support the notion that disciplinary differences do have great influence on evaluation procedures.

The evaluative procedures identified in these documents can then be further understood through Karpik's theory of judgement devices. On an abstract level, it seems that the dominance of *appellations* in the 'standardised' field of biomedicine, *rankings* in the 'hierarchically' organised discipline of economics, and the influence of *cicerones* in the 'individualistic and weakly co-ordinated' field of history align well with the structure and organisation of these fields. However, it is worth emphasising that there also are several instances where the connection between disciplinary structure and evaluation procedures is less obvious, and judgement devices in the form of *appellations* and *cicerones* are found across all fields.

One feature, which is not easily incorporated in this arrangement, is how temporal aspects come to influence evaluation. It might in fact be argued that temporal dimensions cut across all other dimensions, and that 'trajectorial thinking' is an integral feature when evaluating research. Indeed, the findings of this study indicate that research fields use distinct temporal horizons when evaluating research, which partly relates to epistemological factors. The ambition of economics to be a forward-looking field which tries to predict the future influences how research is evaluated, and the same applies to the field of history where past achievements, and especially the origins of academic careers are emphasised. Overall time-perspectives seem to have a significant influence over how research is valued (Hammarfelt et al., 2020), yet temporality has so far been little discussed in the literature on research evaluation.

A common fallacy in recent debates on how to evaluate research is the assumption that agreement on the criteria for evaluating research means that there is a general consensus on how these criteria should be applied. However, as this study has shown, the repertoire of indicators and externalities, that are brought in to make and substantiate claims about the quality of research is distinctive for each field. The valuation of co-authorship or publication channels is field specific, as is the time-horizon from which research is evaluated. Overarching systems for evaluating research employed by nations or institutions, are by their very nature limited to using a very broad and crude set of indicators, and the measures used rarely reflect how scientific quality is defined within specific fields. The objective of this study is not to overcome this inherent tension between field specific evaluation repertoires and more generic peer review procedures. Rather, it illustrates that while a somewhat general agreement might exist on what constitutes research quality across fields, the actual tools and devices used to make these criteria tangible and comparable are distinct and not easily generalised.

The evaluation of applicants for academic positions based on their publication record is nothing new, and similarly we should not assume that different 'short-cuts', or judgement devices used for evaluating publication oeuvres is a late-modern innovation. As far back as the late eighteenth century concerns were expressed regarding the practice of over emphasising opinions expressed in well-respected journals when evaluating candidates for academic positions (Josephson, 2014, p. 36). Similarly, it should be emphasised that the practice of reading texts and assigning scientific value to content, structure, style, findings and relevance of research is still an important, and in many cases the dominant form, of evaluation across all three fields. This kind of 'classic', or perhaps idealised, peer review is, despite the availability of a range of indicators and metrics, still the primary practice used for evaluating candidates. So, in the context of evaluating candidates for academic positions it might be misleading to emphasise tensions between the use of indicators or other externalities, and 'pure peer review'. Rather, the use of judgement devices, for example bibliometric indicators, should be seen as integrated within a larger set of evaluative practices.

References

Appadurai, A. (2013). *The future as a cultural fact: Essays on the global condition*. Verso.

Abbott, A. (2005). The idea of outcome in US sociology. In G. Steinmetz (Ed.), *The politics of methods in the human sciences: Positivism and its epistemological others* (pp. 393–426). Durham, NC: Duke University Press.

Biagioli, M. (1998). The instability of authorship: Credit and responsibility in contemporary biome dicine. *The FASEB Journal, 12*(1), 3–16.

Felt, U. (2017). Under the shadow of time: Where indicators and academic values meet. *Engaging Science, Technology, and Society, 3*, 53–63.

Fourcade, M., Ollion, E., & Algan, Y. (2015). The superiority of economics. *The Journal of Economic Perspectives, 29*(1), 89–113.

Hammarfelt, B., & Rushforth, A. D. (2017). Indicators as judgment devices: An empirical study of citizen bibliometrics in research evaluation. *Research Evaluation, 26*(3), 169–180.

Hammarfelt, B., Rushforth, A. D., & de Rijcke, S. (2020). Temporality in academic evaluation: 'Trajectoral Thinking' in the assessment of biomedical researchers. *Valuation Studies, 7*(1), 33-63.

Hemlin, S., & Montgomery, H. (1993). Peer judgements of scientific quality: A cross-disciplinary document analysis of professorship candidates. *Science & Technology Studies, 28*(1), 19–27.

Hsieh, H. F., & Shannon, S. E. (2005). Three approaches to qualitative content analysis. *Qualitative Health Research, 15*(9), 1277–1288.

Hylmö, A. (2018). *Disciplined reasoning: Styles of reasoning and the mainstream-heterodoxy divide in Swedish economics.* Diss. Department of Sociology, Lund University.

Josephson, P. (2014). The publication mill: The beginnings of publication history as an academic merit in German universities, 1750–1810. In P. Josephson, T. Karlsohn, & J. Östling (Eds.), *The Humboldtian tradition: Origins and legacies* (pp. 23–43). Brill Academic Publishers.

Karpik, L. (2010). *Valuing the unique: The economics of singularities.* Princeton University Press.

Lamont, M. (2009). *How professors think: Inside the curious world of academic judgment.* Harvard University Press.

Langfeldt, L. (2001). The decision-making constraints and processes of grant peer review, and their effects on the review outcome. *Social Studies of Science, 31*(6), 820–841.

Levander, S. (2017). Den pedagogiska skickligheten och akademins väktare: Kollegial bedömning vid rekrytering av universitetslärare. Diss. Acta Universitatis Upsaliensis, Uppsala.

Maeße, J. (2017). The elitism dispositif: hierarchization, discourses of excellence and organizational change in European economics. *Higher Education, 6*, 909–927.

Marušić, A., Bošnjak, L., & Jerončić, A. (2011). A systematic review of research on the meaning, ethics and practices of authorship across scholarly disciplines. *Plos One, 6*(9), 1–17.

Mayring, P. (2000). Qualitative content analysis. *Forum: Qualitative Social Research, 1*(2), Art. 20.

Musselin, C. (2009). *The market for academics.* Routledge.

Nilsson, R. (2009). God vetenskap. Hur forskares vetenskapsuppfattningar uttryckta i sakkunnigutlåtanden förändras i tre skilda discipliner. Diss. Acta Universitatis Gothoburgensis, Gothenburg.

Roumbanis, L. (2017). Academic judgments under uncertainty: A study of collective anchoring effects in Swedish Research Council panel groups. *Social Studies of Science, 47*(1), 95–116.

Rushforth, A., & Rijcke, S. (2015). Accounting for impact? The journal impact factor and the making of biomedical research in the Netherlands. *Minerva, 53*(2), 117–139.

Steinpreis, R. E., Anders, K. A., & Ritzke, D. (1999). The impact of gender on the review of the curricula vitae of job applicants and tenure candidates: A national empirical study. *Sex Roles, 41*(7–8), 509–528.

Van Arensbergen, P., van der Weijden, I., & van den Besselaar, P. (2014). The selection of talent as a group process. A literature review on the social dynamics of decision making in grant panels. *Research Evaluation, 23*(4), 298–311.

Van Dalen, H. P., & Henkens, K. (2012). Intended and unintended consequences of a publish-or-perish culture: A worldwide survey. *Journal of the American Society for Information Science and Technology, 63*(7), 1282–1293.

Whitley, R. (2000). *The intellectual and social organization of the sciences.* Oxford University Press.

Wouters, P., Thelwall, M., Kousha, K., Waltman, L., de Ricjke, S., Rushforth, A., & Franssen, T. (2015). *The metric tide. Literature review. Supplementary report I to the independent review of the role of metrics in research assessment and management.* HEFCE.

Zuccala, A., & van Leeuwen, T. (2011). Book reviews in humanities research evaluations. *Journal of the Association for Information Science and Technology, 62*(10), 1979–1991.

16

Bureaucratic, Professional and Managerial Power in University Tenure Track Recruitment

Tea Vellamo, Jonna Kosonen, Taru Siekkinen, and Elias Pekkola

Introduction

Finnish higher education policy has followed the global trend of providing more autonomy to universities, including independent personnel policies. In the 2010 higher education reform, the status of Finnish universities was changed from public bureaus to foundations and corporations under public law. Meanwhile, the status of university personnel was changed from civil servants to employees. Universities were granted independent status as employers, and they were empowered with their own

T. Vellamo (✉) • E. Pekkola
Tampere University, Tampere, Finland
e-mail: tea.vellamo@tuni.fi

J. Kosonen
University of Eastern Finland, Kuopio, Finland

T. Siekkinen
University of Jyväskylä, Jyväskylä, Finland

© The Author(s) 2022   **371**
E. Forsberg et al. (eds.), *Peer review in an Era of Evaluation*,
https://doi.org/10.1007/978-3-030-75263-7_16

human resource management (HRM) practices independent from state human resource policies (Kivistö et al., 2019; Siekkinen et al., 2016).

During the 1990s and 2000s, in most Finnish universities, personnel and financial decision-making were gradually transferred from collegial (tripartite) multimember bodies to rectors and deans. This resulted in the managerialisation, centralisation and professionalisation of decision-making in personnel affairs. Simultaneously, tenure track systems were introduced. Consequently, the role of internal academic bodies was weakened while that of institutional managers and external scientific evaluation was strengthened (e.g. Pekkola, 2014; Siekkinen, 2019). At the same time, with their new organisational form, universities were becoming more goal-oriented with unified strategies, more managerial central coordination and control, and building organisational identities related to these practices (Pietilä, 2015).

In this chapter, we are interested in tenure recruitment at a technical university. We analyse how tenure track recruitment in similar fields reveal differences between a technical foundation university and a multidisciplinary corporate university. For this, we compare the tenure track recruitment process of Tampere University of Technology (TUT) and the University of Tampere (UTA). TUT and UTA are an interesting pair to compare as they represent very different kinds of universities organisationally and discipline-wise. The two universities merged in 2019 and their different recruitment processes are now being formed anew. Since 2010 most recruitments have still been open vacancies; less than 10% of the recruitments at TUT were international tenure track recruitments, whereas at UTA the ratio was less than 5% (statistic from 2010–2014, Välimaa et al., 2016). In UTA, tenure track recruitments were limited in number, but they were open in varied fields from education and medicine to game culture.

We specifically focus on the recruitment process and criteria, the different powers within the process and the way they are related to the organisational identity of the technical foundation university as compared to the multidisciplinary university. The analysis was conducted on the practices of the two universities prior to their merger, examining documents from 2011 to 2017. Concerning the recruitment process and criteria, we ask the following question:

- How are bureaucratic, managerial and professional powers manifested in the tenure track recruitment processes?

It is interesting to analyse the tenure track recruitment process in similar fields, and therefore we chose our cases from Computer Science (TUT) and Information Technology (UTA). As the fields are similar, the way these different powers are present in the recruitment process may be related to the organisational form and organisational identity of the two universities. This hypothesis is supported by (currently scarce) empirical evidence on the differences between Finnish comprehensive universities and technical universities, which suggests that the management culture in technical universities is more managerial (Pekkola, 2011, 2014) and the identity of its staff members is more entrepreneurial (Vellamo et al., 2019). We would anticipate that these aspects are visible in the recruitment process as well. Our analysis may contribute to increasing transparency in recruitment processes by disclosing information on the agents, their power balance, their criteria and their evaluation and decision-making processes.

Selection and Recruitment of Candidates: A Regulative Perspective

All Finnish universities are regulated by the same legislation concerning their personnel, primarily defined by labour law, which, however, does not regulate the selection of employees or the evaluation of competency. All universities are regulated by the Universities Act (558/2009), according to which universities define the qualification requirements of staff and the procedures for recruitment in the university rules (Section 31). Professors can be recruited either through an open vacancy or by invitation. Invitation without public notice of vacancy is an exception and can be utilised only if the invited professor is an academically distinguished person who indisputably fulfils the qualification criteria or if the position is non-permanent (cf. Pietilä, 2017; Universities Act 33 §). In addition to traditional vacancy-based recruitment (open vacancy), universities may decide on tenure track procedures. However, external evaluation of the candidates' qualifications is required by legislation in both cases, although the university has freedom in choosing the evaluators (Universities Act 33 §). Generally, decisions regarding the competency and selection procedures of employees fall within the scope of university autonomy, and the law grants the same autonomy in the recruitment processes for both universities under public law and foundation universities. However, the foundation universities were the first to adopt the tenure track process, introduced later to some of the universities under public law.

Labour law does not obligate universities to make administrative decisions on personnel selection, and thus they are not required to justify such decisions. However, a university must be prepared to demonstrate why the appointed person was regarded as the most qualified for the task. Universities are also obliged to demonstrate the non-discriminatory nature of recruitment under the Non-discrimination Act (1325/2014) and the Act on Equality between Women and Men (609/1986). An employer may not discriminate against applicants based on age, gender, or other similar personal characteristics. For example, the Finnish Ombudsman for Equality reminded UTA in 2011 that the university must always carry out a comparison of merits when there are both male and female applicants (13.6.2011 TAS432/2010, dnro 473/2009).

Whether selection for academic tenure is seen as falling under labour law or being an administrative decision, the university must factually compare the applicants' merits. The job description, confirmed in advance by the employer, plays a central role in this comparison. The comparison should be based on objectively demonstrable merits, and the merits under comparison should be apparent in the application documents (HE 19/2014 vp, p. 73–76). Typically, the following merits are compared in universities' selection procedures: research, teaching and societal services (Clark, 1987). According to Levander et al. (2019), administrative proficiencies are sometimes included in tenure track evaluation. Different merits can be given different weights, especially concerning the job description.

It should be noted that the purpose of regulations targeting universities is not to restrict the employer's right to choose the most suitable and best person for the position; rather, to ensure decisions are made on a non-discriminatory basis and on comparison of merits (Bruun & von Koskull, 2012). The subjects of discrimination and merit evaluation in recruitment are controversial; there is a certain level of resistance to addressing gender and equality issues outright in the recruitment process as recruitment has always been considered solely based on merits (van den Brink et al., 2010). In this view, merit is not problematised as an objective criterion. There are studies on the paradox of meritocracy and the difficulties of defining and quantifying merit as well as on discrimination in recruitment processes and merit evaluation (Castilla & Benard, 2010; Nielsen, 2016; van den Brink et al., 2010). Despite claims that "in true meritocratic systems everyone has an equal chance to advance and obtain rewards based on their individual merits and efforts, regardless of their gender, race, class, or other non-merit factors," meritocratic organisational values have been shown to favour males over equally qualified females and other under-represented minorities (Castilla & Benard, 2010, p. 543). Recently there has been a tendency to measure, rationalise and access academic activities, despite the unquantifiable character of academic results and work such as publications or teaching (Musselin, 2007, p. 11).

Recruitment and Tenure Track Systems in Finland

The tenure track system originates from, and has been mostly adopted by, universities and colleges in the United States, where tenure was initially intended to promote academic freedom from external accountability in exchange for serving the greater public good (Finkin, 1996; Kezar & Sam, 2011; Rhoades, 2010), but the tradition of academic freedom associated with tenure has cracked lately as, especially public universities, have decreased the number of tenure positions and increased other, more flexible contracts (Ehrenberg, 2012; Siekkinen, 2019). In Finland, however, the still recent tenure track is viewed as a privilege granted to those who undergo a peer review process to prove themselves as scholars, even though criticism has been leveraged by the Finnish Union of University Professors (Pietilä, 2015).

In many European countries, tenure track implementation is related to internationalisation, competition, profiling and the evaluation of (research) performance. Two recent trends are particularly important in this context. First is the increasing competition among European universities to become top-level academic institutions. Second, because of this competition, universities are now trying to recruit the best scholars internationally, which has increasingly globalised the academic labour market (Mohrman et al., 2008; Pietilä, 2015; Regets, 2007; Välimaa et al., 2016), a trend that has also led to universities attempting to become more comparable to other academic institutions. When attracting international scholars, for instance, having a career progression model that is familiar across national borders is particularly important (Arnhold et al., 2018).

In Finland, the introduction of the tenure track system has been justified by competition by and comparability with other higher education systems (Kivistö et al., 2019). It has been implemented in Finnish universities in an attempt to increase attractiveness among international applicants. The system was introduced, coincidentally, alongside the new university law in 2010 and was first adopted by Aalto University, followed soon by others. However, so far, only Aalto uses tenure track as its dominant recruitment model (Kivistö et al., 2019), while other Finnish universities fill most positions through other recruitment methods. There are organisation-specific differences in tenure track recruitment, whose model has been developed over the last decade (Kivistö et al., 2019; Pietilä, 2017; Välimaa et al., 2016), the main difference being related to the entry and exit phase, as well as promotion (Arnhold et al., 2018). In the tenure track, the possibility of progressing to full professorship is defined with set targets and a schedule; in the open vacancy model, the only possibility to advance is to apply for another position in an open call. Positions in teaching and research outside the tenure track can be varied, from fixed-term researcher positions to teaching specific lecturers or full permanent professorships. This has led to a situation where many academics are stuck in a position without advancement opportunities. Accumulating merits might also be difficult in other positions, whereas in tenure track positions it is part of the job description. Many entry-level open vacancy positions are fixed term, and the evaluation of the candidates is based on requirements of the task and not potential, as in the tenure track. The more senior fixed-term vacancies are considered as positions from which the employee either retires or resigns. The major difference can be summarised in that tenure track positions include the aim of becoming more merited and advanced within the position, whereas in other tasks there is no such definition inherent to the position. Tenure recruitment should thus focus on the potential of the candidate, whereas open vacancy focus on the merits and the requirements defined in the call for the particular position. From a legal perspective, the differences are related to the question of how potential can be evaluated. The evaluation of set requirements can be more easily questioned or justified based on set criteria (See Kivistö et al., 2019; Siekkinen et al., 2016).

Both recruitment processes usually involve external evaluation, although there may be some exceptions in filling an open vacancy. The evaluation typically considers teaching, research and societal service which are differently emphasised and valued across universities and by different people within the university. Macfarlane (2005) argues that some academic responsibilities are emphasised, while others are neglected in the hiring and promotion criteria. Although external evaluators do not make the decisions, their evaluations may have a significant impact (Pekkola, 2014). Those making the recruitment decisions for academic positions in the university serve as institutional gatekeepers, and their role in emphasising the different evaluation criteria is crucial in the process (Levander et al., 2019; Merton, 1973). Different actors, such as academics, administrators and high-level management, exert different kinds of power in the process. Our interest is not in evaluating the actual qualifications of the applicants, rather in examining how they have been evaluated in the recruitment process to theoretically define what kind of power these instances represent.

Collection and Analysis of Data

To examine the processes, we obtained data on the tenure track recruitment 2011–2016 in specified fields. Due to the limited scope and focus on a particular field, we acquired only five case studies as examples of the recruitment process in the two universities at different times during the period.

We collected the following documents for each case analysed:

- Setting the open tenure (establishing the need)
- The recruitment call
- Evaluations of qualifications (including a trial lecture)
- External expert evaluations
- Memos of the working group evaluating teaching skills, research merits, and candidate suitability
- Summary of the applicant's evaluation
- The dean's proposals for the board, with justifications
- The board decision

We excluded the application documents and their annexes from our examination, as we did not aim to evaluate the candidates as such. Other proposals, materials, and decisions related to the process and the selection decision were also used to support the analysis. The material available for each tenure recruitment case was similar, but there were differences between the universities and in some cases; for example, applicant summaries were more detailed than in others.

The analysis was conducted by examining the documents using the chosen theoretical approach of the three different powers—bureaucratic, professional and managerial—and defining the instances of use of the different types of power.

As the data set consists of few cases, no broad conclusions can be drawn, but we believe the cases are representative of their organisations at the given time. We looked at three cases from TUT, 2011–2016; these represent a change in the TUT tenure process (as they are from different times for the same department). The development of the tenure process is also manifested in the institutional-level documents and instructions we used as background material. We examined two tenure track recruitment cases from UTA 2013–2015. These five cases represented all the tenure track recruitments in the specified fields since tenure track had been introduced in these universities.

Influence and Actors in Recruitment: Bureaucratic, Managerial and Professional Powers

Following Pietilä (2017) and van den Brink et al. (2010), we approach tenure track recruitment as a "site of political struggle." We define the different powers that compete in the recruitment process as *bureaucratic*, *managerial* and *professional*. Their goals, and the people who exert them, differ.

Bureaucratic power (Weber, 1978) may be defined as the legal power related to rules and regulations. Administration follows the administrative tradition, primarily adhering to the process and definitions of the university and unit-level recruitment instructions. Being impartial and following regulations are the main virtues; the aim is to ensure that the process is fair, objective, transparent and follows the rules and legislation. Tenure track committees represent new controlling bodies through which universities influence their research fields and the recruitment of academics (Pietilä, 2015). The administrators and HRM experts typically act as gatekeepers, as applying the tenure track model is said to require strong HRM (Kivistö et al., 2019). Bureaucratic power cannot be ignored in any recruitment process as that could cause the recruitment to be illegal. By creating more detailed tenure instructions, universities have become stronger organisational actors, but, at the same time, these bureaucratic powers can "limit the freedom of departments to respond to field-specific needs" (Pietilä, 2015, p. 387).

Managerial power (Parsons, 1991) in higher education belongs to high-level managers, such as deans and department heads. Managerialism, describing the ideology of management, spread to universities and other public sector organisations from business, emphasising competition, marketization of public sector services, monitoring performance and outcome measurements (Deem, 2004; Deem & Brehony, 2005; Klikauer, 2015). As management turned into an 'ism', it had to have a proper ideology, one targeted at the future: "It has become common to see ideology as a set of ideas that constitute goals, expectations and actions"; it is a vision (Klikauer, 2015). Therefore, managerialism and managerial power in recruitment emphasise future potential. The managerial ethos (Kallio et al., 2016) is based on the basic assumption that managers should have freedom to manage because they are accountable for organisational performance (Vedung, 2010). The approach also emphasises the role of management in setting systems and metrics related to organisational goals and for allocating resources (Kallio et al., 2016).

The change towards more managerial practices in personnel policies in Finnish universities has occurred gradually with the introduction of practices related to performance-based management (Deem, 2004; Kallio et al., 2016). The first culmination point was a change of legislation on civil servants in 1992, when universities were granted the right to terminate, establish and alter the positions and vacancies of their own staff members. Further steps were taken until 2010, when universities were given the independent status of an employer (Kuoppala et al., 2015; Pekkola, 2014). New public management and managerial practices were introduced into higher education institutions, also with the assumption that a managerialist approach was more effective in carrying out strategies and organisational change, a view especially embraced by polytechnics and corporate higher education institutions where managerial practices were adopted more eagerly than in more traditional universities (Allen, 2003; Gibbons et al., 1994, p. 82; Santiago & Carvalho, 2012).

Managerial power is primarily applied by university managers and is therefore associated with institutional or disciplinary strategies. Deans and department heads represent the organisational culture and are inclined to define and adhere to the profiling and strategic aims of the university. Deans are identified as key persons using power, particularly in the recruitment of more senior-level positions (Välimaa et al., 2016). However, Pietilä (2015) found that with hierarchical governance structures, deans and department heads faced tensions, especially if the tenure track procedures and instructions were not well defined.

The managerial approach has been contrasted with the traditional academic collegial practices at universities to the extent that some claim that more business-like and managerial approaches fragment higher education institutions and set academics against professional managers (Allen, 2003). If we consider managerial power to be more strategically focused and future oriented in the recruitment process, recruitment decisions would be based on the profiling of disciplines and fields according to the university's mission. Tenure track is said to entail a stronger role of university management and to be committed to the institutional mission (Kivistö et al., 2019, p. 121; Pietilä, 2015) and has been identified as a means to "introduce and strengthen strategic research fields" (Kivistö et al., 2019, p. 133). (See also Arnhold et al., 2018; Välimaa et al., 2016.) Often, this strategic defining of the position is made before opening the call, but it also occurs when justifying the choice of a particular candidate. It should be noted that academic excellence is entwined in this discussion, so the profiling areas of the institution are those where there is (an expectation of) world-class research.

The third power is *professional power* (Goode, 1957; Weber, 1919), utilised by professors and academics and which in the recruitment process is anticipated to focus primarily on academic excellence and evaluating (past) academic merits. The prevalent assumption is that decisions should be made by academics. This view permeates the attitudes and behaviour of many of the senior academics who also might resist the strategic organisational approach through managerial power, partially because it is perceived as an attack on their professional identity and power (Allen, 2003, p. 85). The open vacancy model is said to promote a "stronger role of the academic profession in recruitment" (Kivistö et al., 2019, p. 121). In the recruitment process, professors represent the disciplinary tradition, and they also largely define what the discipline is about when they act in the tenure group. The disciplinary definition of power is often located outside of the organisation in an international community of scholars. These senior academics of a particular discipline also act as gatekeepers for the discipline. Similarly, the external academic peer review evaluation is one form of professional power. Although it empowers academics, it is also used by university managers (Musselin, 2013).

The picture below (Fig. 16.1) illustrates how these three powers can occur simultaneously in the recruitment process but still conflict in practice. *Bureaucratic power* emphasises the rules and regulations of the recruitment process and supports the merit-based evaluations of the candidates. *Managerial power* is applied by the university managers, who emphasise the strategic goals of the organisation in the recruitment processes and thus pay more attention to the potential of the candidate;

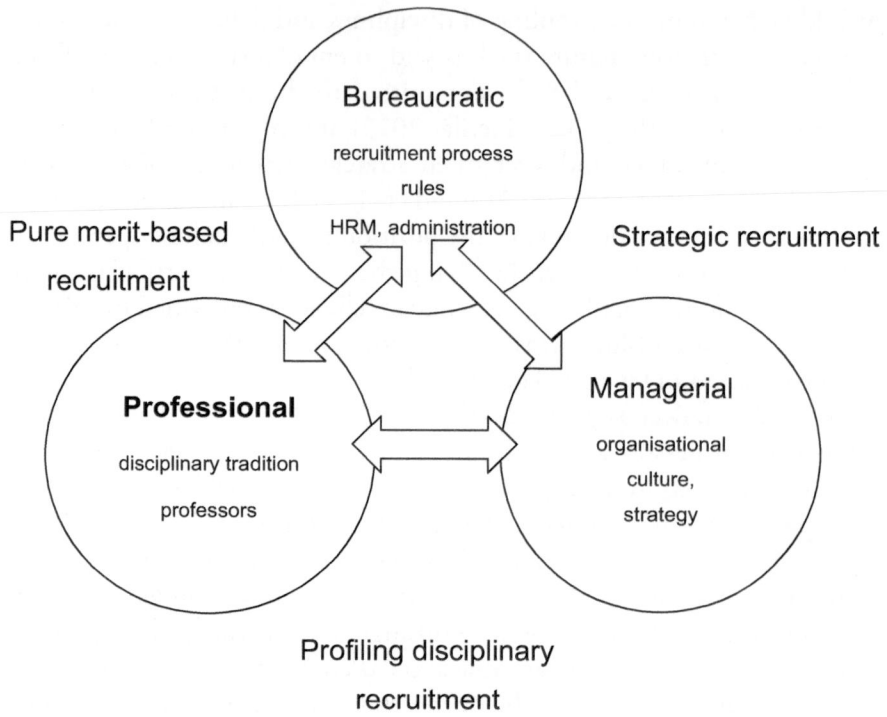

Fig. 16.1 Different powers and persons utilising power in university recruitment

recruitment is strategic and aims to find future talent. *Professional power* counts on scientific evaluation based on the disciplinary tradition made by professors; recruitment is directed towards disciplines aiming to secure continuity in the faculty.

These powers have implications for recruitment. A combination of bureaucratic and managerial power can be associated with strategic HRM, in which the academic managers and HR professionals play an important role in recruitment. In theory, with the combination of bureaucratic and professional power, recruitment can support the academic oligarchy. Merit is defined by professionals and verified by the bureaucratic process (Clark, 1983). Further, this combination of managerial and professional recruitment could lead to merit-based recruiting that supports the disciplinary profiling planned by academic managers.

However, it should be borne in mind that the different powers may have different interests and compete rather than complement each other. Merit-based recruitment supports the selection of senior academics with accumulated merits, talent recruitment supports the selection of "hungry" academics with verified potential (e.g. external funding, networks and top publications) and profiling recruitment supports the selection of candidates within the right field of study.

The Recruitment Process in Tenure Track

In TUT and UTA, different recruitment processes have generally been used, with positions filled through both open vacancy and tenure track recruitment. Our main focus is on the tenure track process at TUT with comparisons on the tenure track used at UTA. In the following, we will look at the process at TUT and UTA, starting with the opening of the position to the justification of the selected applicant.

Defining the Position

The tenure track process starts with defining the need for recruitment and defining the specific (sub) discipline in which the recruitment is targeted. The definition of the academic discipline comes from the unit level (the department, in our cases), but it must be confirmed at the institutional level. At this stage, this is already a strategic allocation of resources and a possibility for negotiating the needs between the department and the organisation. These aims may be parallel, but there may also be conflicting interests as the opening of the position is "a political endeavour, involving negotiations between multiple actors" (van den Brink et al., 2010, p. 1463). Collegial tripartite bodies have lost their importance in the recruitment processes. Who has the power to define the open position depends on the level of the position. Postdoctoral researcher and university lecturer positions are mainly defined by the head of the department, whereas the working group has an important role in higher-level open vacancies and tenure track recruitments (Välimaa et al., 2016). In any case, the definition of the open position is a negotiation between different level managers (e.g. department heads and the dean) and within the working group. Pietilä (2015) found that, with hierarchical governance structures, deans and department heads face tensions, especially if the tenure track procedures and instructions were not well defined. Tenure track committees represent new controlling bodies by which universities influence their research fields and the recruitment of academics. By creating the tenure instructions, universities become stronger organisational actors, but at "the same time they limit the freedom of departments to respond to field-specific needs" (Pietilä, 2015, p. 387). The recruitment process rules need to be followed, and thus each open position needs to be justified. The powers utilised at this point can be studied by looking at the justification of opening a (new) position, whether it is filling a position that has been left vacant due to retirement or job change (as usual with vacancies) or a new tenure position based on institutional profiling. When the position is opened for tenure track, the tenure level is not strictly defined but set on a broader scale as it can be targeted at assistant, associate or full-professors or all levels at the same time. The level of recruitment can be defined more exactly according to the qualifications of the applicants. The level of the tenure recruitment may also vary depending on the reasons why the position has been opened. When tenure track was recently adopted at TUT, there was also tenure recruitment at more senior levels (full professor). A more senior applicant might be desired due to retirement. In this case, the tenure track recruitment can resemble an open vacancy. Recruiting tenure track at the assistant professor level would seem to be a more strategic approach for a long-term staff development plan.

For tenure track recruitment, a tenure working group is set up to conduct the recruitment process, a practice similar in most open vacancy recruitment at similar levels. At TUT the tenure working group included internal representatives, such as department heads, professors from the field, HR experts, external academics and also industry representatives. The industry representative was a peculiarity of the technical university, not used at UTA.

Advertisement and Application

The field of the open position and the applicant criteria should be clearly stated in both the open vacancy and the tenure recruitment call. The criteria are important as they are used for evaluating the applicants, and the recruitment decision needs to be justified. After the decision is made to open a tenure track position, the bureaucratic power of the administration and HRM takes on the process. The opening of the position, advertising the call, receiving applications and so on are part of normal recruitment and HR processes, regulated and managed similarly in most organisations.

Evaluation

After the applications have been received, the primary evaluation and shortlisting of the candidates is first done by the tenure working group to define which candidates are evaluated by the external experts. The external academic evaluation is a common practice in the tenure track, although there might be exceptions when only the working group makes the evaluation.

We looked at the recruitment criteria, how they were evaluated and whether there were differences in the importance placed between the criteria, evaluation and emphasis in the working group or the external evaluation. Generally, the criteria were related to the three tasks of universities: research, teaching and outreach activities. When comparing merit, documented attainments are considered most important, but in tenure track, especially, potential compatibility with the working environment and particular substance skills is also emphasised (Välimaa et al., 2016, p. 46). From a meritocratic perspective, we expect the academic task, research and publications to be the most important. These are also criteria that both the working group professors and the external experts are likely to emphasise.

We can also ask what parts of the evaluation are considered to be an assessment of qualifications (merits, which are public) or of the person (personality and character, which are confidential). The application documents are similarly public in both types of universities, and there is no difference in whether the application documents are seen as employment- or administrative-related preparatory documents. Only the observations of an individual's personal characteristics, such as psychological assessments, are confidential (Section 24 of the Act on the Openness of Government Activities). However, the actual practices of transparency and accountability of these processes in organisations vary; generally, universities have been somewhat reluctant to disclose detailed information on the actual recruitment process. It is also noteworthy that, if the assessment is not open or criteria are obscure, it increases the risk of bias in all the evaluation phases (van den Brink et al., 2010, p. 1459). Even if the assessment criteria are stated, some of them can be disputable, such as potential. Similarly, criteria may vary according to the different tenure levels the candidates are evaluated for. As the editors note in their introduction of this volume (Forsberg et al. 2021), a variety of biases may affect peer review including

> epistemic bias; values and beliefs (O'Meara et al. 2016); gender bias and stereotypical judgement (van den Brink et al., 2010) and reputation of alma mater, habitus and networks, to mention a few.

Decision and Justification

As noted earlier, the criteria set in the tenure call are important because they not only guide candidate evaluation but also should be used to justify the selection. We also attempt to determine whether the decisions were justified concerning the criteria set in the tenure call or in relation to other aspects. In addition, it is necessary to justify the decision from a legal perspective, particularly if in cases of suspicion of discrimination. The possibility of bias in recruitment and evaluation is nowadays better recognised, and gender aspects have been included in recruitment protocols (van den Brink et al., 2010).

In the following, we will analyse the tenure track cases, focusing on the cases of the technical university in particular, and examine the UTA cases in order to compare them to the TUT case.

TUT Findings

According to the TUT tenure instructions, "the aim is to attract and to keep competent, creative and inspiring research and teaching staff at TUT and in that way enforce the status of TUT as a high-level and international research university…." This emphasises managerial power, where the strategy, position and ranking of the university are seen as an integral aim of the tenure process. It may be said that this reflects the introduction of new public management, which has brought corporate culture aspects like managerialism into higher education. According to the documents on setting up the tenure system at TUT, there is a focus on evaluating the potential of the candidates and the main criteria should be the potential of the candidate to advance on the tenure track to more demanding positions. This statement also reinforces managerial power as the temporal target is in the future and potential is emphasised. The definition of tenure already establishes a certain power balance, but, in the following, we will look at how the powers are manifested in the actual process of three different tenure processes at TUT—one in 2012 and two in 2016. In the 2012 and the 2016 calls, external evaluation was used, while the 2016 instance was carried out by the tenure working group only, which left room for managerial power, represented by the dean and the department head, in the process.

Tenure Call

In the TUT instructions, the disciplinary boundaries are not enforced rigidly; this is to allow for many applicants. This might further weaken the professional power of academics (professors), who are prone to guard disciplinary boundaries and define tenure calls in a particular (sub) field. This is perhaps added to the tenure track instructions because the degree of specialisation in engineering is generally high. In the TUT cases, computer science also seems closely connected to another discipline: signal processing. The division between these fields is shifting, and the 2012 tenure track was transferred to the Department of Signal Processing based on the suggestion of the Department of Computer Science. Based on some of the materials during this process, the definition of the field narrowed the potential applicant pool from the perspective of computer science but broadened it towards signal processing. The desire for a broader field might be contradictory, its aim to strategically define the recruitment concerning the department and university profile. In this TUT case, one candidate withdrew their application because the position became so closely defined to signal processing that it did not fit their profile. Even if the field of the open positions need to be defined so that they attract enough applicants, defining the subfield in more detail may be justified by the department's profiling.

Setting up the Working Group

The tenure working group that prepares the tenure track position opening usually has one or two external experts, who may be academics or, in TUT, also industry representatives. In the TUT examples, there was one external academic and one industry representative in the 2016 case, where the evaluation of the candidates was performed by the working group only. There were no external experts in the tenure group in the 2012 case. In the other 2016 case, the external academic representative for the TUT group was chosen by UTA, maybe anticipating the approaching merger or as a reflection of the thematic closeness of the fields in the two institutions. Industry representatives should be high-level experts in the fields but are not required to hold a PhD. In the 2016 TUT cases, there was an industry representative who had a doctoral degree and an expert who did not. Although experts with doctoral degrees may be seen as knowledgeable of academic criteria, industry representatives are not that likely to align with the professional power of academics. They are nominated to the working group to bring a different viewpoint, that of the industry stakeholders. Their role and views seem to be more important in a technical university in industry-related fields than in other kinds of disciplines. The close relationship with industry stakeholders is a particular identity feature of the technical university as an organisation (Vellamo et al., 2019).

Evaluation

According to TUT's instructions, the evaluation should be open and equal, and the evaluators should have the highest possible expertise. Let us assume this refers to both the internal evaluation done by the tenure working group as well as the external evaluation. In addition, fairness and transparency may also be seen as traits of bureaucratic power and the legality of the process. These traits are important in the justification of the evaluation: if the evaluation is fair and transparent, it is possible to examine how and on which explicitly stated grounds the evaluation is based.

External scientific evaluation is used in the tenure track for the short-listed candidates chosen by the tenure working group. External evaluation can be seen as a form of peer review despite the differences in the recruitment models. Peer review has been the primary institution of modern science evaluation, and its use has extended to more evaluation practices. According to Musselin (2013), scientific evaluation empowers academics since university managers are dependent on it in many university processes. The external evaluation peer review process is moderated by the definition of criteria and qualifications set in the call for application and, more generally, by the recruitment process descriptions. The external academic evaluators are selected for their expertise in the field of the open position, but, since they are nominated by the recruitment group, there may internal power struggles in this process.

External evaluators are experts representing the professional power of the international scientific community. When the tenure track system was introduced at TUT, gender and other diversity factors were almost completely missing from the tenure process description (TUT tenure process Academic Board decision 22.11.2010). In TUT, the Human Resources Strategy for Researchers (European Commission. (n.d.). Human resources strategy for researchers HRS4R) evaluations of 2012–2014 paid attention to the gender imbalance of the external experts and a recommendation was given to include more female evaluators and also to pay attention to the gender balance in the tenure working group. In our case from 2016, there was also a female evaluator, an effort to ensure diversity in the evaluation in practice perhaps. It is noteworthy that the legislation does not directly require the representation of both genders in the evaluators but refers to more generally taking into account gender equality in all decision-making. Additionally, because of the merger process and new joint tenure model, TUT's tenure actions on gender balance and addressing gaps in employment due to family leave were postponed. In the TUT case examples, there were few female applicants; in one (2016), a female applicant was sent for external evaluation but was not selected.

Although there are set criteria for the external evaluation, there is still considerable room for interpretation of the criteria and evaluating the merits of each candidate. Moreover, the weight placed on each aspect may differ according to the evaluator, and the recruitment committee may steer the selection decision in a different direction than the external evaluators. In TUT, the tenure working group seemed to place different weighting on the evaluation criteria depending on the tenure level. Emphasis on a certain criterion was justified by the profiled need of the department (2016), the complementarity of expertise within the unit's faculty (2016) or if it was mentioned in the tenure call (2012).

According to the TUT tenure track criteria, research papers are evaluated for the number and level of journal publications, and the candidates' citations and H-index are listed. In both TUT 2016 recruitments, there were also Excel sheets summarising applicants' quantitative data. However, based on these cases, it is hard to tell how much weight this had in the recruitment. Some evaluators seemed to emphasise these aspects in their evaluations to justify the academic research excellence (or lack of it) of the candidate. In the 2016 case, where the external evaluators had noted the number of high-level publications as a weakness of the candidate, the tenure group justified the number as "a consequence of a long period of work in industry."

In TUT, experience as a project leader and attained research funding are seen as important factors for the highest tenure positions. This is also mentioned in the evaluation of the candidates, but somewhat surprisingly also at the assistant professor level (2016 case).

Related to both teaching and education, the supervision of master's and doctoral theses is seen as an important merit for those applying for higher tenure or full professor positions. However, there may be different institutional policies regarding who is allowed to supervise doctoral theses, and, in one of the 2016 external evaluation statements, the candidate's lack of supervision experience was reportedly due to the policy of the university in which the candidate was currently employed.

Industrial experience seems to be important to TUT. In the 2016 external evaluation case, a candidate who had a good academic track record but had spent an entire career in academia was evaluated as weak in terms of industrial experience. The candidate selected by the tenure group had strong industry experience, a moderate academic track record and a "not very thorough teaching record" (TUT 2016 tenure track external evaluator statement) Apparently, a lack in one category could be compensated by achievements in another. Industry experience seemed to be more important than academic merit and teaching skills for the tenure working group in this case. This may be a particular feature related to the discipline and the technical universities, and more acceptable in the engineering field (Whitchurch, 2012, p. 6). However, this emphasis was also justified by the criteria set beforehand: industry experience was specifically mentioned in the tenure call even though the importance of the different criteria listed was not indicated in the call. Emphasising industrial experience is also contradictory to the trend observed in the national tenure track practices as, it is said, it is difficult to move to tenure track from outside of academia (Välimaa et al., 2016 p. 53).

The shortlisted candidates also underwent psychological tests, the results of which were utilised by the working group in their evaluation. While could be perceived as impartial external evaluation, it also support[s] the perception of the working group on the candidates(). Even so, it seems unlikely that psychological evaluation would significantly affect the final recruitment decision. In some cases, it was mentioned as affirming the perceptions of the tenure group regarding the candidates. In the 2012 case, the psychological tests indicated that only the selected candidate was clearly suitable for the position. Psychological tests may be seen as supporting managerial power and enabling the selection of the most appropriate candidate.

Decision and Justification

In one of the 2016 recruitment processes, although the external evaluators pointed out substantial weaknesses in each candidate's profile, the working group chose to value industry experience over research merit. This seems somewhat exceptional based on previous studies and might not be a typical case, even at TUT. The emphasis was justified by the department's strategy and a profiling factor that sets the department apart from all other Finnish software engineering centres. In the tenure call, industry had a less important role: "practical experience in industry software projects is seen as an advantage" (TUT Tenure call 2016). This could be interpreted so that industry experience would not be a core requirement but rather an additional asset. This is also a case where the recruitment criteria, such as industrial experience, exceeds traditional notions of academic merit and the tenure working group exercised its decision power over that of the external evaluators.

UTA Findings

In UTA, the School of Informational Sciences (SIS) operated until 2017, when it was split into two faculties, and Computer Science began operating within the Faculty of Natural Sciences. After the Tampere University merger in 2019, it was included in the Faculty of Information Technology and Communication Sciences (ITC). Both tenure track calls examined took place in SIS, specifically in the field of data analysis, in 2013 and 2015 as assistant professor (2013) and as assistant professor or postdoctoral researcher (2015). The use of a postdoctoral researcher as a tenure track title reflects the variations in the system nationally as even the terms were not uniform.

Tenure Call

In both recruitment calls, the requirements were specified clearly, and the tenure level(s) for the position was set in the call. The calls included the same requirements: "the person appointed associate professor must hold a doctoral degree, high-level academic qualifications and experience in directing scientific research, be able to provide high-quality, research-based instruction as well as to have a track record of international scientific activities ... [and be] fluent in English." (UTA tenure calls 2013 and 2015)

In the recruitment calls, research and teaching were both mentioned. In the other call (2013), teaching was emphasised as playing a "central role in planning this master's degree programme and [the associate professor] will mainly teach courses" (UTA tenure call 2013). In the other call (2015), the focus was on research, and the recruitment targeted one of the strategic focus areas of SIS, emphasising managerial power and the department's strategy. The position where teaching was emphasised does not represent a typical tenure track position, which usually focuses on research, but seems more like an open vacancy type of position with emphasis on teaching. However, in the new UTA, there are plans to create a new teaching career track parallel to the tenure track.

Evaluation

SIS had evaluation guidelines for tenure track recruitment, addressing that evaluations were being made in three areas: (1) research, (2) teaching and (3) activity in the scientific community and academic leadership. How these dimensions were emphasised in each recruitment depended on the case. In their research evaluation, publications in JUFO (Finnish Publication Forum ranking of scientific journals)-recognised journals were valued. In teaching evaluation, producing materials for teaching, pedagogical training, awards and teaching evaluations were valued. In general, those kinds of activities and merits are recognised in evaluation required at the next level of the tenure track, but teaching seems to have more relevance than in the TUT tenure track cases.

The assessment was based on openness, reason and eligibility in an international comparison. The associate professor should have published high-quality scientific research articles, designed curricula and planned study modules. Moreover, the candidate should have had teaching experience, a recognisable personal research field, an acknowledged position in the research field and have supervised several theses. In addition, if they had received (or pursued) external funding, and had started their own research group, these were considered positively. The list was extensive and similar to the expectations at TUT, except for industry or other work experience, which are not mentioned at UTA. Supervising theses was considered an important aspect for both universities.

In the first recruitment case (2013), the tenure working group short-listed as many as 6 of the 20 applicants (including one woman) for external evaluation. The evaluators completed the first round of evaluation and selected three candidates for interview. Two of the three external evaluators selected the same top three candidates with similar justifications, emphasising research productivity and quality, engagement with the research community, solid research plans, involvement in research projects and having received funding, but they also placed great value on PhD supervisory experience. One of the evaluators placed more emphasis on publications than the others did, and clearly reviewed the qualifications of the candidates more generally. In addition, this evaluator's top two was different from that of the other evaluators. Thus, it seemed that, using the same criteria, the external evaluators ranked the applicants differently. It is also noteworthy that teaching was not highlighted in the external evaluation.

One evaluator (with a more general view) gave some suggestions regarding what to ask the candidates in the forthcoming interviews. The evaluator also pointed out the challenges of the evaluation, such as different publishing cultures in this interdisciplinary field. Another (also with a more general view) listed all the evaluation criteria taken into consideration, saying that all six candidates submitted interesting applications, so "the university will have to decide what is most important and how the candidates would fit in with the rest of the school." This accentuates the university's autonomy, which it legally has in the selection process.

In the second recruitment (2015), the tenure track position was opened at two levels: university researcher and associate professor. Commonly, a university researcher is not included in the tenure track, showing an example of the non-standardised tenure track processes (but standardised later). There were 23 applicants for the position: 17 applied for the position of associate professor, the rest for the position of university researcher. The tenure working group, including the faculty manager, a student, a lecturer, two professors and the HR expert, presented two candidates for the external evaluators. The minutes of the faculty board meeting clarified that this internal group could choose the applicants to send to the external evaluators. This group also selected the external evaluators.

The first external evaluator emphasised research, teaching, activity in the scientific community and academic leadership. Although academic leadership was not mentioned in the call, these mostly aligned with the recruitment call. The other external evaluator stated that one applicant would be successful in the US, whereas the other would not. He stressed the publications (H-index and citations), research and conferences in which they had participated but also examined the applications, especially future research plans. The second evaluator mentioned he was not able to evaluate the teaching since he lacked material (e.g. quotes from students). Despite the differences in evaluation, both external evaluators proposed the same applicant for the position. The assessments were in line with the recruitment call, where the emphasis was more on research than teaching.

Decision and Justification

In the 2013 recruitment process, the applicants ranked as the top three by two of the external evaluators were interviewed. The tenure working group justified its selected candidate by indicating that they had "strong potential based on scientific and teaching merits to proceed in tenure track and become a professor." However, the candidate withdrew their application, so the faculty manager asked the rector to choose the second candidate on the list. This justification was related to teaching experience, active research activity and an innovative vision regarding a new master's degree programme. The candidate was said to have "promising preconditions." It seems that the first candidate had stronger research merit, whereas the teaching experience of the second one seemed stronger, and both criteria were sufficient to justify the selection.

In the 2015 recruitment, the working group made its decision based on the proposals of the external evaluators. These evaluators agreed on the strongest applicant, both emphasising research merits, international collaboration and the preparation of funding proposals, as well as the applicant being graded as "good" at teaching.

It seems that in both these UTA recruitment processes, the power of the tenure working group was strong, especially in the pre-screening of candidates. Additionally, the actual recruitment call directed the recruitment from the beginning. The power of the department head and the working group was strong (Siekkinen et al., 2016). In the 2013 recruitment, the working group sent six applications to the external evaluators but in 2015, only two. In the latter recruitment, the power of the faculty is manifested. They also chose the external evaluators, whose evaluations were in line with the recruitment calls, and the working group followed their views. Teaching was emphasised more in the 2013 recruitment than in 2015, which showed in the evaluations. In both cases, the rector agreed with the working group's proposal.

Conclusions

In this article, we were interested in how bureaucratic, managerial and professional powers were manifested in tenure track recruitment process, especially in the technical university. We examined the differences in the tenure track recruitment process in two universities to find out whether the powers were different according to the organisational form and organisational identity of the technical university.

Labour law regulations do not differentiate between public law universities and foundation universities. Yet, despite the legal aspect of publicity, the processes are not that open and transparent in either university. They seemingly have great autonomy in defining their recruitment processes and do not necessarily always follow their own internal regulations in the process.

We discovered that bureaucratic power was present in both universities' tenure recruitment processes. This could be seen as fulfilling the legal minimum, illustrated, for example, in the impartiality of the external experts required by law. However, the impartiality is only related to the personal relationship of the evaluators with the applicants and does not consider disciplinary or scientific partiality.

Even though tenure track evaluations are based on the three tasks of the university—research, teaching and societal impact—different people evaluate these aspects in different ways, with varying importance placed on different criteria. The two universities seemingly place different priorities on the criteria. Based on previous research you might anticipate that research performance and success in acquiring research funding would be the most valued criteria for tenure track, but we discovered other priorities. While we only looked at a few case examples, some differences emerged in the recruitment processes between the two universities. Our case study showed that, although research is important, in the examined cases industry experience (TUT) and teaching (UTA) were also significant.

When opening the tenure call, the (managerial) power of university- and department-level strategy, often represented by the department head, is emphasised, whereas the tenure working group has significant power in shortlisting the candidates and naming the external evaluators. Often, department heads also have a position in the tenure group. In the TUT model and cases, managerial power seemed to play a strong part in all these phases of opening and defining the tenure, and the same strategy-derived aspects were used in the selection justification. There was even one case where the tenure working group carried out the whole process, and managerial power almost completely excluded (the external) professional academic views from the process, even though the tenure group included two professors from the specific field (one from TUT and one from another Finnish university). The dean and department heads are academics themselves, but not necessarily from the discipline of the tenure position, and they also are considered to implement strategy. Why one of the TUT evaluations was carried out only by the tenure working group was not justified in the process, except by a reference that the evaluation was carried out according to the TUT tenure track instruction. The tenure group in TUT used its power, in one case, by carrying out the evaluation itself and, in another case, by deciding against the evaluations of the external experts, whereas in the UTA cases, the external evaluation group's evaluations were followed.

The cases demonstrate that the decision of the tenure working group can differ from external expert views. In such cases, managerial power was used to justify the decision, with the department's strategic profile mentioned as the main reason for not following the evaluator's recommendations. Bureaucratic power, focusing on fairness and transparency, might have been compromised, and thus one might question whether institutional tenure criteria emphasising research and academic achievement were followed.

Based on our findings, the powers interact and overlap with each other in a tense relationship. Bureaucratic power defines the minimum requirements for the process that must be ensured to make the process legitimate. Within this legal frame, managerial power defines the limits of the use of professional power. Professional power appears to have the most limited power in the process, yet it is still the central power enabling tenure recruitment since it defines merit and evaluates potential. Managerial power can influence the way these professional evaluations are weighted in the final decision and justification, and bring organisational strategy to the decision. All the recruitment is the result of a managerial decision in the sense that the management team has to accept the opening of the tenure and the board makes the final decision based on the tenure group's proposal.

It is possible that managerial power is related to the organisational form of the foundation university but even more so, to the identity of the technical university, which emphasises industry relations. The identity of the technical university is more strategically oriented and manifested in its tenure recruitment. The use of managerial power is revealed in the tenure process documents, but it would require further study and other data (e.g. interviews) to determine whether other powers are more strongly present in the non-documented aspects of the process or whether the managerial powers are also dominant.

References

Act on Equality between Women and Men, 609. (1986). https://www.finlex.fi/en/laki/kaannokset/1986/en19860609_20160915.pdf

Act on the Openness of Government Activities, 621. (1999). https://www.finlex.fi/en/laki/kaannokset/1999/en19990621_20150907.pdf

Allen, D. (2003). Organisational climate and strategic change in higher education: Organisational insecurity. *Higher Education, 46*, 61–92. https://doi.org/10.1023/A:1024445024385

Arnhold, N., Pekkola, E., Püttmann, V., & Sursock, A. (2018). *Focus on performance—World Bank support to higher education in Latvia (Vol. 3): Academic careers (English)*. World Bank Group.

Bruun, N., & von Koskull, A. (2012). *Työoikeuden perusteet*. (2nd ed.) (Juridica; No. 4). Talentum.

Castilla, E. J., & Benard, S. (2010). The paradox of meritocracy in organizations. *Administrative Science Quarterly, 55*, 543–576.

Clark, B. (1983). *The higher education system: Academic organization in cross-national perspective*. University of California Press.

Clark, B. R. (1987). *The academic profession: National, disciplinary, and institutional settings*. University of California Press.

Deem, R. (2004). The knowledge worker, the manager-academic and the contemporary UK university: New and old forms of public management? *Financial Accountability & Management, 20*(2), 107–128.

Deem, R., & Brehony, K. J. (2005). Management as ideology: the case of 'new managerialism' in higher education. *Oxford Review of Education, 31*(2), 217–235. https://doi.org/10.1080/03054980500117827

Ehrenberg, R. G. (2012). American higher education in transition. *Journal of Economic Perspectives, 26*(1), 193–216.

European Commission. (n.d.). *Human resources strategy for researchers (HRS4R)*. Retrieved from https://euraxess.ec.europa.eu/jobs/hrs4r

Finkin, M. W. (1996). *The case for tenure*. Cornell University Press.

Finnish Ombudsman for Equality. (2009). *13.6.2011 TAS432/2010, dnro 473/2009. Statement of the Finnish Ombudsman for equality*. https://www.tasa-arvo.fi/documents/10181/34936/Tampereen+yliopisto_432-10_johtaja-valinnat.2.pdf/680a4266-1a21-4c57-aeb1-0900855787c7

Forsberg, E., Geschwind, L., Levander, S., & Wermke, W. (2021). *Peer Review in Academia. In Forsberg, E., Geschwind L., Levander, S., & Wermke, W. (Eds), Peer Review in an Era of Evaluation. Understanding the Practice of Gatekeeping in Academia*. Anthology Springer Nature.

Gibbons, M., Limoges, C., Nowotny, H., Schwarzman, S., Scott, P., & Trow, M. (1994). *The New Production of Knowledge: The Dynamics of Science and Research in Contemporary Societies*. London: Sage Publications.

Goode, W. (1957). Community within a community: The professions. *American Sociological Review, 22*(2), 194–200.

Kallio, K.-M., Kallio, T. J., Tienari, J., & Hyvönen, T. (2016). Ethos at stake: Performance management and academic work in universities. *Human Relations, 69*(3), 685–709.

Kezar, A., & Sam, C. (2011). Understanding non-tenure track faculty: New assumptions and theories for conceptualizing behavior. *American Behavioral Scientist, 55*(11), 1419–1442.

Kivistö, J., Pekkola, E., & Pausits, A. (2019). Academic Careers and Promotions in Finland and Austria: System and Institutional Perspectives. In Mahat, M., & Tabate, J. (Eds.), *Achieving Academic Promotion*. London: Emerald Publishing.

Klikauer, T. (2015). What Is Managerialism? *Critical Sociology, 41*(7–8), 1103–1119. https://doi.org/10.1177/0896920513501351

Kuoppala, K., Pekkola, E., & Ranta, M. (2015). Tilivirastosta yksityiseksi yliopistoiksi: Työnantaja aseman muutos ja uudet palvelussuhteet. In K. Kuoppala, E. Pekkola, J. Kivistö, T. Siekkinen, & S. Hölttä (Eds.), *Tietoyhteiskunnan työläinen-Suomalaisen akateemisen projektitutkijan työ ja toimintaympäristö*. Tampere University Press.

Levander, S., Forsberg, E., & Elmgren, M. (2019). The meaning-making of educational proficiency in academic hiring: A blind spot in the black box. *Teaching in Higher Education, 1*(19), 1–19.

Macfarlane, B. (2005, October). The disengaged academic: The retreat from citizenship. *Higher Education Quarterly, 59*(4), 296–312.

Merton, R. K. (1973). The normative structure of science. In R. Merton (Ed.), *The sociology of science: Theoretical and empirical investigations* (pp. 267–278). University of Chicago Press.

Mohrman, K., Ma, W., & Baker, D. (2008). The research university in transition: The emerging global model. *Higher Education Policy, 21*, 5–27.

Musselin, C. (2007). Are universities specific organisations? In G. Krücken, A. Kosmützky, & M. Torka (Eds.), *Towards a multiversity? Universities between global trends and national traditions* (pp. 63–86). Transcript.

Musselin, C. (2013, June). How peer review empowers the academic profession and university managers: Changes in relationships between the state, universities and the professoriate. *Research Policy, 42*(5), 1165–1173.

Nielsen, M. W. (2016). Limits to meritocracy? Gender in academic recruitment and promotion processes. *Science and Public Policy, 43*(3), 386–399.

Non-discrimination Act, 1325. (2014). https://www.finlex.fi/en/laki/kaannok-set/2014/en20141325.pdf

O'Meara, K., Bennett, J. C., & Neihaus, E. (2016). Left unsaid: The role of work expectations and psychological contracts in faculty careers and departure. *The Review of Higher Education, 39*(2), 269–297.

Parsons, T. (1991). *The social system.* Routledge.

Pekkola, E. (2011). Kollegiaalinen ja manageriaalinen johtaminen suomalaisissa yliopistoissa. *Hallinnon tutkimus, 1,* 37–55.

Pekkola, E. (2014). *Korkeakoulujen professio Suomessa—kehityskulkuja, käsitteitä ja ajankuvia* (Doctoral dissertation). Retrieved from Tampere University Trepo (2014-11-24T08:54:04Z). http://urn.fi/URN:ISBN:978-951-44-9654-7

Pietilä, M. (2015). Tenure track career system as a strategic instrument for academic leaders. *European Journal of Higher Education, 5*(4), 371–387.

Pietilä, M. (2017). Incentivising academics: Experiences and expectations of the tenure track in Finland. *Studies in Higher Education., 44*(6), 932–945.

Regets, M. C. (2007). *Research issues in the international migration of highly skilled workers: A perspective with data from the United States* (Working Paper No. SRS 07-203).

Rhoades, G. (2010). Envisaging invisible workforces: Enhancing intellectual capital. In G. Gordon & C. Whitchurch (Eds.), *Academic and professional identities in higher education: The challenges of a diversifying workforce.* Routledge.

Santiago, R., & Carvalho, T. (2012). Managerialism rhetorics in Portuguese higher education. *Minerva, 50,* 511–532.

Siekkinen, T. (2019). *The changing relationship between the academic profession and universities in Finnish higher education* (Doctoral dissertation). Retrieved from Finnish Institute for Educational Research, University of Jyväskylä. http://urn.fi/URN:ISBN:978-951-39-7931-7

Siekkinen, T., Pekkola, E., & Carvalho, T. (2016). Change and continuity in the academic profession: Finnish universities as living labs. *Higher Education, 79*(3), 533–551.

Universities Act, 558. (2009). https://www.finlex.fi/fi/laki/kaannokset/2009/en20090558.pdf

Välimaa, J., Stenvall, J., Siekkinen, T., Pekkola, E., Kivistö, J., Kuoppala, K., ... Ursin, J. (2016). Neliportaisen tutkijanuramallin arviointihanke: Loppuraportti. *Opetus- ja kulttuuriministeriön julkaisuja, 2016,* 15.

van den Brink, M., Benschop, Y., & Jansen, W. H. M. (2010, November). Transparency in academic recruitment: A problematic tool for gender equality? *Organization Studies, 31.* https://doi.org/10.1177/0170840610380812

Vedung, E. (2010). Four waves of evaluation diffusion. *Evaluation, 16*(3), 263–277. https://doi.org/10.1177/1356389010372452

Vellamo, T., Siekkinen, T., & Pekkola, E. (2019). Technical education in jeopardy?—Assessing the interdisciplinary faculty structure in a university merger. In M. P. Sørensen, L. Geschwind, J. Kekäle, & R. Pinheiro (Eds.), *The responsible university: Exploring the Nordic context and beyond* (pp. 202–233). Palgrave Macmillan. https://doi.org/10.1007/978-3-030-25646-3_8

Weber, M. (1919). Science as a vocation *Wissenschaft als Beruf,* from *Gesammlte Aufsaetze zur Wissenschaftslehre* (Tubingen, 1922), 524–555. Originally delivered as a speech at Munich University, 1918. Published by Duncker & Humbldot, Munich. http://www.wisdom.weizmann.ac.il/~oded/X/WeberScienceVocation.pdf

Weber, M. (1978). *Economy and society: An outline of interpretive sociology.* University of California Press.

Whitchurch, C. (2012). *Reconstructing identities in higher education: The rise of 'third space' professionals.* Routledge.

Correction to: "Disciplining" Educational Research in the Twentieth Century

Raf Vanderstraeten

Correction to:

Chapter 3 in: E. Forsberg et al. (eds.), *Peer review in an Era of Evaluation*,
https://doi.org/10.1007/978-3-030-75263-7_3

Since the publication of this chapter, it has come to our attention that the author of this chapter failed to properly attribute the work of Alex Csiszar from the text titled "*'Disciplining' Educational Research in the Twentieth Century*" on pages 53–4 and 73. The author acknowledges his error and we publish this erratum to correct the record.

The updated original version for this chapter can be found at
https://doi.org/10.1007/978-3-030-75263-7_3

E. Forsberg et al. (eds.), *Peer review in an Era of Evaluation*,
https://doi.org/10.1007/978-3-030-75263-7_17